New Approaches to the Study of Early Upper Paleolithic 'Transitional' Industries in Western Eurasia

Transitions great and small

Edited by

Julien Riel-Salvatore
Geoffrey A. Clark

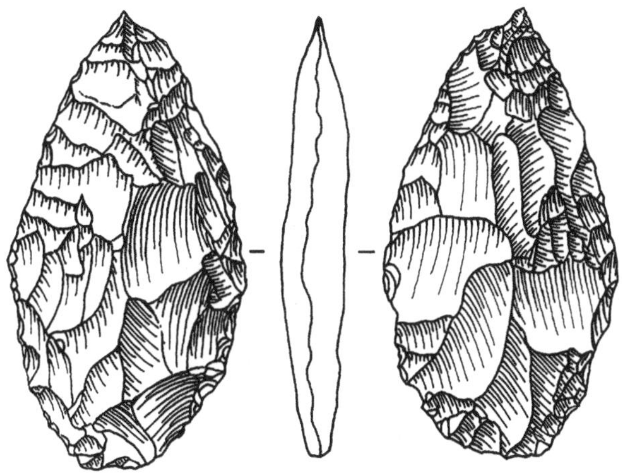

BAR International Series 1620
2007

Published in 2019 by
BAR Publishing, Oxford

BAR International Series 1620

New Approaches to the Study of Early Upper Paleolithic 'Transitional' Industries in Western Eurasia

ISBN 9781407300320 paperback
ISBN 9781407330884 e-book

DOI https://doi.org/10.30861/9781407300320

A catalogue record for this book is available from the British Library

This book is available at www.barpublishing.com

BAR Publishing is the trading name of British Archaeological Reports (Oxford) Ltd. British Archaeological Reports was first incorporated in 1974 to publish the BAR Series, International and British. In 1992 Hadrian Books Ltd became part of the BAR group. This volume was originally published by Archaeopress in conjunction with British Archaeological Reports (Oxford) Ltd / Hadrian Books Ltd, the Series principal publisher, in 2007. This present volume is published by BAR Publishing, 2019.

BAR
PUBLISHING

BAR titles are available from:

BAR Publishing
122 Banbury Rd, Oxford, OX2 7BP, UK
EMAIL info@barpublishing.com
PHONE +44 (0)1865 310431
FAX +44 (0)1865 316916
www.barpublishing.com

Fiorello: *Piano, pianissimo,*
senza parlar,
tutti con me
venite qua.

Coro: *Piano, pianissimo,*
eccoci qua.

Tutti: *Tutto è silenzio;*
nessun qui sta
che i nostri canti
possa turbar.

In Memoriam
Prof. Amilcare Bietti
1937-2006

It is with great sadness that we bid farewell to our friend and colleague Amilcare Bietti, Professor of Prehistoric Ecology in the Department of Animal and Human Biology at the Università di Roma "La Sapienza." In the truest sense of the word, Amilcare was a genial man, combining keen intellect and open-mindedness in a way unique in Italian, indeed international, archaeology. Originally trained as a theoretical physicist, he approached archaeology by integrating a rigorous empiricism with Americanist anthropological theory. One of the leading Italian prehistorians, he excavated over many years such world-renowned sites as Grotta Breuil, Riparo Mochi and the Arene Candide, and regularly contributed to international debates about the Middle–Upper Paleolithic transition and the nature of hunter-gatherer lifeways in the Pleistocene. A man of many interests, including literature, music and history, he was always able to enjoy life to the fullest, and his passion for all of its aspects was a contagious source of inspiration.

Transitions Great and Small

Julien Riel-Salvatore & Geoffrey A. Clark, Editors

Table of Contents

List of Figures . . . . *vi*

List of Tables . . . . *vii*

1. *Asking New Questions and Using New Methods to Understand the Middle–Upper Paleolithic Transition*
Julien Riel-Salvatore and Geoffrey A. Clark (Arizona State University) *1*

2. *Even the Notion of a "Transitional Industry" is a Suspect Typological Construct*
Lawrence Guy Straus (University of New Mexico). *11*

3. *On the Fate of the Neanderthals and the Middle–Upper Paleolithic Transition in Western Europe*
Francis B. Harrold (University of Nebraska, Kearney) *19*

4. *The Châtelperronian of Grotte XVI, Cénac-et-Saint-Julien (Dordogne, France)*
Géraldine Lucas and Jean-Philippe Rigaud (Université de Bordeaux I), Jan F. Simek (University of Tennessee) and Marie Soressi (Max Planck Institute for Evolutionary Anthropology). *33*

5. *"Transitional" Industries from Neandertals to Anatomically Modern Humans in Continental Italy: Present State of Knowledge*
Amilcare Bietti (University of Rome "La Sapienza") and Fabio Negrino (University of Pisa). *41*

6. *New Quantitative Perspectives on the Middle–Upper Paleolithic Transition: The View from the Northern Mediterranean*
Julien Riel-Salvatore and C. Michael Barton (Arizona State University). *61*

7. *The Middle/Upper Paleolithic Interface in Croatia*
Ivor Karavanič (University of Zagreb). *75*

8. *Gulyás Archaeology: The Szeletian and the Middle to Upper Palaeolithic Transition in Hungary and Central Europe*
Brian Adams (University of Illinois) *91*

9. *Hafting and Technological Innovation in the Western Eurasian Middle Paleolithic*
Michael S. Bisson (McGill University) *111*

10. *Issues in the Development of the Early Upper Paleolithic, and a "Transitional" Industry from the Zagros Region*
Deborah I. Olszewski (University of Pennsylvania Museum). *131*

11. *Putting Transition Research in a Broader Context*
Geoffrey A. Clark (Arizona State University) *143*

List of Contributors *179*

List of Figures

4.1. *Location of Grotte XVI.* ... 35

4.2. *Archeostratigraphy of Grotte XVI.* ... 36

4.3. *Bifaces, couche C, Grotte XVI.* ... 37

5.1. *Distribution of the sites mentioned in the text.* ... 42

5.2. *Industry from Riparo Mochi.* ... 44

5.3. *Industry from Via San Francesco, San Remo.* ... 45

5.4. *Industry from Via San Francesco, San Remo (continued).* ... 46

5.5. *Stratigraphy of Grotta della Fabbrica.* ... 47

5.6. *Industry from Grotta Breuil, Layer 4.* ... 48

5.7. *General stratigraphic profile of Grotta di Castelcivita.* ... 49

5.8. *Industry from Grotta della Fabbrica, Layer 2.* ... 51

5.9. *Industry from Grotta della Fabbrica, Layer 2 (continued).* ... 52

5.10 .*Industry from the Uluzzian layers of Grotta di Castelcivita.* ... 53

5.11. *Aurignacian industry from Riparo Mochi.* ... 55

5.12. *Schematic representation of raw material procurement patterns at Late Mousterian, "transitional," and early Aurignacian sites.* ... 57

6.1. *Graphical depiction of patterns for Paleolithic assemblages form Gorham's Cave.* ... 63

6.2. *The correlation between OIS-scale climatic fluctuations and the density of lithic artifacts and incidence of cores and retouched tools in the Gorham's Cave assemblages.* ... 64

6.3. *The relationship between absolute artifact density and frequency of retouched pieces for Paleolithic assemblages from eastern Spain.* ... 65

6.4. *The relationship between absolute artifact density and frequency of retouched pieces for the eight sites from the Salento.* ... 68

6.5. *The relationship between absolute artifact density and frequency of retouched pieces after differentiating between flint and limestone sub-assemblages, for the four sites from the Salento.* ... 69

7.1. *Important late Middle and/or early Upper Paleolithic sites of Croatia.* ... 76

7.2. *Selected artifacts from level G3, Vindija.* ... 77

7.3. *Selected artifacts from level G1, Vindija.* ... 78

7.4. *Selected artifacts from Mujina Pečina.* ... 81

7.5. *Selected artifacts from Šandalja II.* ... 83

8.1. *Location of selected Bükk Mountain Palaeolithic sites.* 92

8.2. *Szeleta Cave stratigraphy.* 93

8.3. *Leaf points from Szeleta Cave.* 94

8.4. *Istállóskő Cave stratigraphy.* 95

8.5. *Istállóskő Cave stratigraphy, 2000 excavations.* 96

8.6. *Szeleta Cave stratigraphy, 1999 excavations.* 98

8.7. *Szeleta Cave, 1999 excavations: hearth..* 99

8.8. *Szeleta Cave, 1999 excavations: obsidian bladelet core.* 99

8.9. *Szeleta Cave, 1999 excavations: level 2/3 boundary.* 100

8.10. *Leaf points from Istállóskő cave.* 100

8.11. *Distribution of felsitic quart porphyry bifaces in the Carpathian Basin.* 102

8.12. *Distribution of Aurignacian and Szeletian sites in the Váh (Vág) River valley near Moravány, west Slovakia.* . . . 104

8.13. *Distribution of Aurignacian and Szeletian sites in Moravia, Czech Republic.* 105

9.1. *Common hafting modes attributed to Middle Paleolithic tools..* 113

9.2. *Hafting modifications at Skhūl* 119

List of Tables

1.1. *Comparison of generalized typological inventories for the Mousterian and the Aurignacian.* 6

4.1. *Grotte XVI. Technological characteristics of lithic assemblages in levels C, B and Aib.* 38

5.1. *Absolute radiocarbon dates for Grotta di Castelcivita* 50

5.2. *Debitage distributions for three lithic samples from Castelcivita* 50

5.3. *Debitage distributions for the industry from Layer 2 of Grotta della Fabbrica..* 54

5.4. *Debitage distributions for the early Aurignacian industry from Layer G at Riparo Mochi* 56

5.5. *Debitage distributions for the industry of Layers 3–4 from Grotta della Fabbrica.* 56

8.1. *Lithic tool types from Szeleta Cave.* 93

8.2. *New Chronometric Dates from Szeleta and Istállóskő Caves.* 97

8.3. *Local and non-local lithic raw materials at selected Hungarian Middle and early Upper Paleolithic sites.* . . . 101

8.4. *Detailed breakdown of proportions of non-local raw materials at selected Hungarian Middle and early Upper Paleolithic sites.* 101

8.5. *Lithic raw materials from Sajóbábony-Méhésztető* 103

9.1. *Inventory of formal flake tools in Skhūl Layer B compared to the sample from that collection in the Redpath Museum, McGill University, and the sample excavated by Jelinek from the dump of the original Skhūl excavations* 116

9.2. *Hafted tool types from Skhūl, McGill Collection* 118

9.3. *Hafted tool types from Skhūl, Dump Collection* 120

9.4. *Secondary Retouch on Tabūn scrapers* 121

10.1. *Final Core Morphology in Levels AA–LL at Warwasi* 135

10.2. *Debitage from Levels AA–LL at Warwasi* 136

10.3. *Tool Blanks from Levels AA–LL at Warwasi* 136

10.4. *Tools from Levels AA–LL at Warwasi* 137

Acknowledgement

The editors wish to thank Cindy Heath-Smith for all aspects of production, including composing, editing, electronic and camera-ready copy. This was a particularly daunting task, given the various electronic formats used in manuscript production, the great number of (relatively obscure) diacritical marks, the occasional sub-standard illustration, and the many languages compiled in the bibliographies. Somehow, she managed to take all this disparate material and transform it into something that looks professional. Transitions Great and Small would not have become a reality without her expertise and dedication.

1

Asking New Questions and Using New Methods to Understand the Middle–Upper Paleolithic Transition

Julien Riel-Salvatore
Arizona State University

Geoffrey A. Clark
Arizona State University

This volume comprises nine original essays first presented in the symposium *Upper Paleolithic 'Transitional' Industries: New Questions, New Methods* on March 21, 2002, at the 67th annual meetings of the Society for American Archaeology. The impetus behind both the symposium and the book is a resurgence of interest in the Middle–Upper Paleolithic transition (hereafter 'the Transition') fueled by recent discoveries and by recent reinterpretations of the nature of change over the Transition interval, generally taken to be the 10 millennia bracketing 40 kya. Although the basic analytical units (indeed, the Transition itself) are artifacts of the historical development of paleolithic archaeology, their persistence into the 21st century can create serious difficulties in understanding a whole range of culture process questions focusing on changing human adaptation over the 120 millennia of the Upper Pleistocene. Questions arising from these reevaluations have highlighted, among other things, the limited explanatory potential of traditional, culture-historical approaches in helping archaeologists to understand the nature and evolutionary implications of those industries (see discussion in Riel-Salvatore & Clark 2001, Clark & Riel-Salvatore 2006). Through the symposium, we sought to bring together an international panel of researchers involved in this debate to address specific questions related to Transition phenomena, and to identify, outline, and assess the credibility of recent discoveries and new methods developed to study them. We enlisted colleagues from the US, Canada, France, Italy, Croatia and Portugal to discuss the following nine questions:

(1) What are the theoretical implications of the notion of 'transitional' industries, and how do we establish criteria for examining transitions?

(2) How do current questions about 'transitional' industries differ from those asked in the past? What preconceptions and assumptions do they reflect?

(3) What discoveries and methodological advances give rise to current questions about such industries?

(4) What hypotheses are these questions designed to assess?

(5) What are the implications of these questions for hominid biological and cultural evolution over the OIS 5a–3 interval?

(6) What new methods are being developed and implemented to tackle these questions?

(7) What are some of the preliminary conclusions generated by the application of these methods, and what are their implications for Transition research?

(8) Are there new questions generated from the implementation of these new methodologies?

(9) How widely applicable are these new methods to geographically, temporally and conceptually distinct 'transitional' industries?

We focused our attention on a very specific aspect of the Transition, namely the lithic industries—the most common kind of data available from these ancient time ranges. This decision was grounded on a number of considerations. First, chipped stone artifact assemblages have been the basis for recognizing and naming paleolithic industries since the inception of the discipline in the last quarter of the 19th century. It is crucial to have a thorough understanding of their defining properties if we are to incorporate them in scientific debate about the later phases of human evolution. Second, unlike archaeozoological research, lithic analysis does not have a commonly understood set of principles to guide it. There are no real counterparts to analytical units like NISPs, MNEs or MNIs (but see Hiscock 2002). The closest we have come to this were type-lists like those of Bordes (1953, 1961), de Sonneville-Bordes and Perrot (1953, 1954, 1955, 1956) or Laplace (1964, 1966), which systematized published information. The study of 'transitional' industries throws into sharp relief the main limitation of these approaches: they were developed to describe the retouched tool components of industries assigned a priori to different periods, and thus do not lend themselves well to comparisons between periods, nor to the study of assemblages at the interface between periods (Olszewski, this volume; Harrold [1988, 1989], and this volume; Grayson & Cole [1998]). The focus on the retouched component (often <5% of the assemblage total) ignores important behavioral information encoded in the frequency distributions of cores and débitage (Riel-Salvatore & Barton [2004], this volume). Typological systematics also cannot address the behavioral connotations of raw material variability within assemblages (see Adams, Bietti & Negrino, Riel-Salvatore & Barton, this volume). In fact, by confounding functional and stylistic aspects of lithic artifact manufacture (Jelinek 1976), typological systematics can mask a discrete behavioral 'signal' in a lithic assemblage, transitional or otherwise (Bisson 2000). The contributors to this volume offer paleoarchaeologists interested in 'transitional' industries new and innovative approaches to these problems—an armamentarium with applications far beyond the limited domain of Transition research.

We were also interested in what lithics as a class of objects could tell us about the Transition. Lithics represent the most abundant and durable aspect of any archaeological record, with much higher time/space resolution than scarce human fossils. Lithics are also the tangible remains of conscious human actions; they can sometimes constitute direct reflections of past hominin behavior that can be reconstructed, given the appropriate combination of methodological sophistication and middle-range theory. It is our opinion that archaeologists need to develop analytical frameworks that permit an understanding of chipped stone artifacts in their own right. That said, however, it is also and equally clear that lithic assemblages can provide even greater interpretive insights when their study is articulated with that of other classes of primary archaeological evidence (e.g., faunal remains, settlement strategies, paleoenvironmental data—see Clark, this volume). Because of their historical importance in the western European paleolithic 'heartland', lithic industries are most often accorded a priori importance and tend to be viewed *en bloc,* as *sui generis* analytical units (e.g., Aurignacian, Châtelperronian, Magdalenian) when in fact they were created (not discovered) in order to solve what were perceived at the time (last quarter of the 19th century, first half of the 20th) to be culture history problems, in particular the temporal ordering of hypothetical, 'history-like' social entities in the years before absolute dating methods first became available. This has had as a consequence a focus on the characteristics of the retouched pieces (often to the near exclusion of other evidence), each object being assigned to its 'appropriate' type or class, often on the basis of a superficial description embodied in a type list. Because of this historical bias in favor of retouched stone artifacts, one of the principal aims of the symposium was to highlight new methods that might expand our analytical horizons beyond those encompassed by the typologies used to describe retouched stone tools.

The contribution by Lawrence Straus (Chapter 2) sets the backdrop for this discussion by challenging the often-implicit assumption that individual 'transitional' techno-complexes (e.g., Châtelperronian, Uluzzian, Szeletian) can be equated with identity-conscious groups of people. Straus argues that the tendency to interpret them in terms of quasi-historical processes is misguided, and notes that we are far from understanding the internal dynamics and the range of variability that characterize individual 'transitional' industries. He focuses on the Châtelperronian—the quintessential transitional industry—to show that we still lack clear definitions of transitional industries, and suggests that their recognition is still largely determined by the presence or absence of allegedly time-sensitive, stylistic marker types (i.e., *fossiles directeurs*) rather than by criteria of known behavioral significance. Having undermined the credibility of the Châtelperronian as a well-defined analytical unit, Straus concludes that the Middle–Upper Paleolithic transition is perhaps best understood as a mosaic process with multiple (and uncorrelated) episodes of technological innovation and transmission following channels cut by alliances of various kinds (e.g., mating, trade networks) (see also Gamble 1986, Clark 1989, Barton et al. 1994).

In Chapter 3, Francis Harrold (one of the original symposium's discussants) argues that constructs are unavoidable in any scientific endeavor and that, in contrast to Straus' contribution, the Châtelperronian, as a set of assemblages technologically and typologically distinct from the Mousterian and the Aurignacian, represents a credible, coherent and analytically useful construct. He reviews lithic, faunal, chronological and symbolic evidence to undermine the empirical basis for the 'mosaic' and 'indigenist' models, leaving—by a process of elimination—some kind of population replacement model (e.g., Davies 2001, Mellars 2005) as the most credible alternative. Harrold is critical of 'acculturation' models, arguing that they encompass a range of scenarios (some more credible than others), but often tend to oversimplify implicit assumptions about authorship (we don't actually know who made most Early Upper Paleolithic industries—see Churchill & Smith [2000], Conard et al. [2004]), and complex patterns in data already in hand. In his view, the current state of paleoanthropological knowledge, especially as bolstered by biomolecular data extracted from a number of Neanderthal specimens (see also Caramelli et al. 2003, but cf. Trinkaus 2005), supports the (near-universal) consensus that Neanderthals and modern humans were separate populations. In the absence of a genetic 'yardstick' against which to measure species differences, however, whether we will ever be able to demonstrate that they were different species is arguable, although Harrold believes that they probably were. Harrold suggests that the Upper Paleolithic 'revolution' (e.g., Mellars & Stringer 1989) might have had a demographic and cultural basis (e.g., Zubrow 1989), but issues of chronological resolution currently impede a more thorough understanding of the Middle–Upper Paleolithic transition.

Chapters 4–8 combine reviews of the current state of knowledge about the transition in France, Italy, Iberia, Croatia, Hungary and the Levant, and methodological innovations that address the some of the issues raised in the context of the original symposium. In Chapter 4, Géraldine Lucas and her colleagues discuss the Châtelperronian assemblage from Grotte XIV to assess that industry's validity as an analytical unit.[1] After voicing some well-founded concerns about the contextual integrity of Châtelperronian assemblages rich in Mousterian tool types, the authors use intrasite spatial analysis, refitting, and technotypological analysis to show that, at least at Grotte XVI, this industry is a unique and indisputable 'transitional' technocomplex that must be understood in its own right. What it means in the larger context of modern human origins research is yet to be determined.

Amilcare Bietti and Fabio Negrino (Chapter 5) review transition-interval lithic evidence from continental Italy as seen through the lens of technological variation. Using in part a *chaîne opératoire* approach (e.g., Boëda 1988), they attempt to reconstruct raw material procurement strategies under the assumption that where and how hominins acquired the stone from which they manufactured their tools should shed light on their relative cognitive capacities . This important re-evaluation of the Italian evidence reclassifies the 'transitional' assemblage from Grotta della Fabbrica as terminal Mousterian, thus relegating the Uluzzian industry to the southernmost third of the peninsula. They marshal evidence for late Mousterian assemblages in the northern two-thirds of the peninsula, document a rather sudden and late appearance of very specialized blade and bladelet technologies (= the Aurignacian?) and, in areas where high-quality raw material is not locally available, a dramatic increase in the frequency of fine-grained exotics. Both patterns are taken by Bietti and Negrino to indicate a rather sudden population replacement in northern Italy coincident with the appearance there of the earliest proto-Aurignacian assemblages, but similar assemblages also show up farther south at Castelcivita (Campania). Their work shows that the combined study of lithic technology and raw material exploitation patterns has much to offer researchers interested in 'transitional' technocomplexes. Whether or not these assemblage types correspond to different kinds of humans is a more arguable proposition (Bietti 1991, 1997).

In Chapter 6, Julien Riel-Salvatore and C. Michael Barton (Chapter 6) also focus on technology, Italy, and the Italian transitional industry, the Uluzzian (after the Bay of Uluzzo, in the 'heel' of the Italian boot) but they expand the research to include comparisons of transitional industries in southeastern Spain and Gibraltar. They propose a new approach to study prehistoric lithic assemblages that incorporates the interpretive potential of débitage (Fish 1979). By plotting the frequency of retouched pieces in a level against its artifact volumetric density, it becomes possible to reconstruct land-use decisions across time, and a pattern of overall behavioral continuity across the traditional Middle–Upper Paleolithic boundary is documented. However, the lithics also display a radical reorganization in the Late Upper Paleolithic (after c. 20 kya), probably linked to different land-use practices in the Early Upper Paleolithic and the Late Upper Paleolithic respectively, and brought on by climatic deterioration during the Last Glacial Maximum. The Italian sites show once again the importance of accounting for pattern in raw material variation in paleolithic behavioral reconstructions. The authors propose that their methodology can tackle issues of behavioral shifts throughout the Late Pleistocene archaeological record of the Old World.

Moving east, Ivor Karavanič's contribution focuses on Croatia, and how a local 'transitional' industry—the Olschewian—fits into the picture (Chapter 7). He looks at the transition at four sites, two in the northwestern interior of Croatia (Vindija, Velika Pečina), and two on the Adriatic Coast (Mujina Pečina, Šandalja II). A comparison of the transition in the two regions shows that, at the Middle–Upper Paleolithic interface, the inland sites contain comparable Olschewian assemblages that combine 'Middle' and 'Upper' Paleolithic tool types along with a bone industry. In contrast, the coastal sites show no evidence of a human presence for the 10 millennia separating the latest Mousterian levels at Mujina Pečina from the 'classic Aurignacian' at Šandalja II (see also Karavanič 2003). While he argues that the data are still too sparse to provide a definite picture of the Transition process, the presence of these different modalities in the Balkans suggests a complex mosaic of distinct adaptations operating at both the cultural and (arguably) biological levels.

In Chapter 8, Brian Adams presents new data on the Szeletian technocomplex derived from recent work at Istállóskö and Szeleta Caves, in the Bükk Mountains of Hungary. New dates suggest that the Szeletian was more recent than previously thought, and likely contemporary with the Aurignacian of the region (see also Adams & Ringer 2004). Adams shows that Szeletian raw material exploitation patterns are comparable to those of Aurignacian sites in Central Europe, and do not resemble those of the local Mousterian. This, and topographical differences in the location of Szeletian and Aurignacian sites, leads him to suggest that the typically 'Aurignacian' split-based bone points and the equally distinctive 'Szeletian' bifacial 'leaf points' might have been functionally equivalent technological alternatives, and that Szeletian assemblages most likely represent Aurignacian special purpose sites. Adams makes the important point that environmental considerations must be fully integrated into reconstruction of behavior at the time of the Middle–Upper Paleolithic transition. Again, a focus on raw material variation helps disentangle previously ignored strands of archaeological evidence that lead to a more refined understanding of 'transitional' industries.

Shifting the geographical and chronological focus to the Mousterian of Mount Carmel, Michael Bisson compares data on secondary retouch in artifact assemblages from Mugharet es-Skhūl, known for its early modern human remains, and Mugharet et-Tabūn, which has so far produced only Neanderthals (Chapter 9). The two caves are only a few hundred meters apart. The intent is to analyze differences in hafting techniques between the two different (or not so different) hominin populations. Bisson identifies rules for the placement of retouch and finds that modern humans are the only ones to systematically violate them, indicating preferential modification for hafting purposes in tools made by those hominins. The adaptive advantages offered by this behavior (if any) are unclear, but modern humans might have been more inclined to experiment with technology than Neanderthals, despite a shared, fundamentally Mousterian behavioral *repertoire*.

Noting the plethora of contemporary lithic industries recognized as 'transitional' across western Eurasia, Deborah Olszewski (Chapter 10) examines the technological and typological parameters of the EUP levels (AA–LL) at the Warwasi rockshelter, in the Zagros Mountains of northwestern Iran. Both technology and typology here show an interesting association of classic 'Middle' and 'Upper' Paleolithic elements: prismatic blade(let) technology, small, radially prepared flake cores, carinated endscrapers and sidescrapers of various types. These industries are evaluated in terms of the criteria necessary and/or sufficient to document a 'transition', and are compared and contrasted with Aurignacian developmental sequences elsewhere. The important issue of typological comparability on either side of the transition is identified as perhaps the most important factor leading to perceptions of relatively abrupt or punctuated change. How typologies (including typologies of technology) are defined and used by workers in different research traditions lies at the heart of why most Middle and Upper Paleolithic industries are seen as so characteristically dissimilar.

The book closes with an essay on epistemological issues by Geoff Clark, originally a discussant in the symposium from which this volume derives (Chapter 11). Clark tries to 'situate' the archaeological Transition within the broader context of the modern human origins debate. Like Straus, he argues that contemporary analysts must recognize that the 'cultures' claimed to be involved in the transition are no more than convenient labels created by prehistorians (in some cases, more than a century ago), and that their meaning has frequently changed over the course of the history of paleolithic research. Because of a near-absence of critical self-consciousness in the more culture-historically inclined research traditions, the meaning assigned to the constellation of basic analytical units used in transition research is not at all clear. His call for an explicit concern with the logic of inference underlying our knowledge claims, and by extension, the meaning of our basic analytical units, neatly sums up the stated intention of the volume.

* * *

Since March 2002, when the symposium took place, edited volumes and monographs concerned at least in part with 'transitional' industries have been published with some frequency (e.g., Svoboda & Bar-Yosef 2003, Goring-Morris & Belfer-Cohen 2003, van Andel & Davies 2003, Zilhão & d'Errico 2003, Brantingham et al. 2004, Finlayson 2004, Straus 2005). This surge of publication might lead one to think that transitional industries, and the Transition itself, had become a threadbare field of inquiry. We respectfully disagree, and would in fact argue that the present volume offers an antidote to the ceaseless pattern searching that dominates so much Paleolithic research, particularly the narrow focus on the chronological dimensions of the debate. Before attempting to understand how they are distributed in time, it is imperative that a consensus be reached (or at least approximated) on how the various transitional industries are to be defined, and what they might 'mean' (if anything) in terms of human adaptation. In the absence of such a consensus, the chronological dimensions of transition research tend to 'take over', and the work can sometimes degenerate into a search for the oldest transitional industries, which are then forced into the ill-fitting conceptual boxes of the culture history approach (Verpoorte 2005). A primary objective of this volume is the effort to unravel implicit biases and preconceptions that bear on the definitions of the basic analytical units used in Transition research. New methodologies are also advocated, but will produce relatively little in the absence of a consensus about what our analytical units might mean. Obviously, culture historians and community ecologists will view an analytical unit like the Uluzzian through very different conceptual filters or lenses.

That said, novel methodologies can sometimes change the perception of pattern, and what pattern might mean, even in the absence of sweeping conceptual change. For example, even the staunchest typologist would be hard-pressed to deny the fundamental importance that scraper reduction models have had on our understanding of Mousterian lithic variability (Dibble 1995; see also Barton 1988, Kuhn 1995). While we don't mean to imply that reduction studies are the key to understanding the Transition (indeed, only one of the essays [Chapter 6] focuses on, and emphasizes reduction analysis), a consideration of how blank size and shape, and the relative intensity of edge modification, can affect artifact form is a good illustration of the kind of rethinking that can result from taking a fresh look at an old collection. Assessing the impact of raw material selection, blank morphology, and edge reduction on interpretations of Mousterian behavior also opens the door for new and interesting questions about 'transitional' and Upper Paleolithic industries.

Artifact reduction is pretty clearly a general aspect of all lithic technologies. A number of studies now document artifact reduction in the Acheulean (McPherron 2000, 2003), Châtelperronian (Holdaway 1991), Aurignacian (Holdaway 1991; Blades 1999, 2001, 2003), and Gravettian (Blades 2003), just to name some of the analytical units that loosely bracket the Middle Paleolithic in western Europe. Lithic reduction has been identified as a significant determinant of stone artifact morphology over time and space, and on all continents (except Antarctica) (see, e.g., Hiscock & Attenbrow [2003]). Reduction studies have also moved far beyond sidescrapers—their original analytical focus—to include notches and denticulates (e.g., Barton 1988, Holdaway et al. 1996), Mousterian points (e.g., Holdaway 1989), endscrapers (e.g., Morrow 1997, Shott 1989, Weedman 2002, Blades 2003, Bietti 2003), burins (e.g., Stafford 1977, Barton et al. 1996, Hiscock 1996), projectile points (e.g., Hoffman 1985, Flenniken & Wilke 1989), and microliths (e.g., Neeley & Barton 1994). Many of these are artifact classes considered to have had 'cultural' significance in regional typologies. The recognition of artifact reduction as an important factor in understanding chipped stone artifact form and function has led to reconsiderations of the preconceptions and assumptions underlying typologies and, in some cases, their cognitive and behavioral underpinnings (e.g., Sackett 1988, Bisson 2000, Marks et al. 2001).

The best-known case study is the typology of the Middle Paleolithic (Bordes 1961). Once artifact reduction was identified as a major determinant of stone tool form (Dibble 1984), the Middle Paleolithic typology was shown to contain far fewer formal types than the 62 originally proposed by Bordes since the 27 sidescraper types (44%) were reduced to a single category, and the variation 'explained away' by differential reduction intensities (e.g., Dibble 1989). Termed 'the finished artifact fallacy' by Noble and Davidson (1993), this was taken by some as evidence that Middle Paleolithic hominids had a more limited typological *repertoire* than those of the Upper Paleolithic, thus underscoring the truly different nature of 'modern' behavior (Noble & Davidson 1996; Bar-Yosef 2002, 2003; Klein 1999, 2003; Mellars 1996, 2005). Despite the numerous studies that unambiguously show artifact reduction in typically 'Upper Paleolithic' types like endscrapers and burins (only some of which are referred to above), no such critical re-evaluation of the Upper Paleolithic typology has ever been proposed. Is the Upper Paleolithic actually 'richer' typologically than the Middle Paleolithic, or is its implied richness partly or wholly attributable to differences in the resolution of the typologies used to describe the two analytical units? We examine this question briefly below by scrutinizing some of the Upper Paleolithic diagnostics, as defined in the 92 types in the typology developed by de Sonneville-Bordes and Perrot (1954, 1955, 1956).

We begin by discarding all carinate types (6, 6.5%), shown by recent studies to have been mainly bladelet cores (e.g., Almeida 2001; Lucas 1997, 2001; Hays & Lucas 2000, 2001; Chazan 2001a, 2001b). This reduces the total number of retouched types recognized by the typology to 86 (n = 86). Next, we consider all remaining endscraper types as more or less reduced versions of a single generalized 'endscraper concept'. This operation reduces 14 types (16.3%) to one. Recognizing that burins (beyond occasionally serving as bladelet cores or as a form of hafting modification) have been shown to have had only graving and cutting functions (Barton et al. 1996; see also Tomaskova 2005) removes an additional 19 types (22.1%) from the list. If the generalized microlith production scheme presented by Neeley and Barton (1994:283) is regarded as credible, the five microlith types (5.8%) collapse into a single, polymorphic 'slotted inset' category. Likewise, under a lithic reduction model, the seven awl types (8.1%) can in all likelihood be reduced to one, and the same is probably true for truncations (10 types, 11.6%). In aggregate, then, the 92 distinct morphological types identified by de Sonneville-Bordes and Perrot are reduced to c. 36. However, to err on the side of conservatism, we should adjust for comparability by eliminating from the comparison all Upper Paleolithic combination tools (e.g., endscraper/burins), which don't exist in the Middle Paleolithic typology despite being present in many Mousterian assemblages (Bordes 1961) Finally, some Upper Paleolithic tools are time-sensitive, stylistic marker types confined to certain 'cultures' (e.g., Solutrean points). It makes little sense to include Gravettian, Solutrean or Magdalenian *fossiles directeurs* in the type list used to describe Aurignacian assemblages, and it would probably be advisable to include only a single set of these period-specific diagnostics when comparing type richness to that of the Mousterian.

The impact of these considerations taken together is shown in the revised Mousterian and Aurignacian type lists presented in table 1. Even allowing for more subtle type-group definitions, it is pretty clear that we are not dealing here with type lists that differ by an order of magnitude. In fact, they are very similar to one another, with only a small subset of types being confined to the Aurignacian. While we acknowledge that the approach used here to assess comparability is 'coarse-grained' and somewhat *ad hoc,* we advocate its adoption because it squares well with the levels of resolution encoded in the commonly used Paleolithic typologies, and believe that both the fundamental point of the exercise, and its basic conclusion, remain broadly credible. Using a single theoretical approach and applying it equally across typologies shows that the differences between the Middle and the Upper Paleolithic are not as great as has been commonly supposed (*contra* esp. Bar-Yosef 2002, Mellars 2005). While we do not suggest that all the contributors to the book would necessarily agree with our assessment of the factors driving Paleolithic technological variation, we think the various contributions show a willingness to revisit accepted views and employ innovative methodologies to analyze the record of typologically defined 'cultures' according to a single set of criteria. This, in our view, will permit direct comparisons of Mousterian, 'transitional,' and Aurignacian assemblages and will throw into sharp relief both behavioral differences and similarities across the Transition. Whether through the analysis of raw material use and transfers, technological patterns, lithic production methods, or reduction intensity, we believe the work presented here offers a sampler of some of the most promising directions in which Transition research should move in the coming years.

Table 1.1. Comparison of generalized typological inventories for the Mousterian and the Aurignacian.

Mousterian	Aurignacian
Levallois product	Endscraper
Mousterian point	Awl
Sidescraper	Burin–graving
Endscraper	Burin–cutting
Burin–graving	Audi knife
Burin–cutting	Font-Yves point
Awl	Backed blade
Backed knife	Truncation
Truncation	Retouched piece (1 or 2 edges)
Notch/denticulate	Aurignacian blade
Bec	Strangulated blade
Misc.retouched piece	Retouched point
Hachoir	Pick
Tanged tool	Notch/denticulate
Chopper	Bec
Chopping tool	Splintered piece
Foliate point	Sidescraper
Other	Retouched microlith/slotted inset
	Backed bladelet
	Dufour bladelet
	Other

Endnote

1. This chapter was originally published as Lucas et al. 2003, and is reprinted here by courtesy of the authors and the Portuguese Institute of Archaeology.

References

Adams, B., & Á. Ringer
2004 New C^{14} dates for the Hungarian Early Upper Palaeolithic. *Current Anthropology* 45:541–551.

Almeida, F.
2001 Cores, tools, or both? Methodological contribution for the study of carinated lithic elements: the Portuguese case. In *Questioning the Answers: Resolving Fundamental Problems of the Early Upper Paleolithic* (M. A. Hays & P. T. Thacker, eds.), pp. 91–97. BAR International Series 1005. Archaeopress, Oxford.

Barton, C. M.
1988 *Lithic Variability and Middle Paleolithic Behavior: New Evidence from the Iberian Peninsula.* BAR International Series 408. British Archaeological Reports, Oxford.

Barton, C. M., G. A. Clark, & A. E. Cohen
1994 Art as Information: Upper Palaeolithic Art in Western Europe. *World Archaeology* 26:185–207.

Barton, C. M., D. I. Olszewski, & N. R. Coinman
1996 Beyond the Graver: Reconsidering Burin Function. *Journal of Field Archaeology* 23:111–125.

Bar-Yosef, O.
2002 The Upper Paleolithic revolution. *Annual Review of Anthropology* 31:363–393.
2003 Reflections on selected issues of the Upper Paleolithic. In *More than Meets the Eye: Studies on Upper Palaeolithic Diversity in the Near East* (A. N. Goring-Morris & A. Belfer-Cohen, eds.), pp. 265–273. Oxbow Books, Oxford.

Bietti, A.
1991 Normal science and paradigmatic biases in Italian hunter-gatherer prehistory. In *Perspectives on the Past: Theoretical Biases in Mediterranean Hunter-Gatherer Research* (G. A. Clark, ed.), pp. 258–281. University of Pennsylvania Press, Philadelphia.
1997 The transition to anatomically modern humans: the case of peninsular Italy. In *Conceptual Issues in Modern Human Origins Research* (G. A. Clark & C. M. Willermet, eds.), pp. 132–147. Aldine de Gruyter, New York.
2003 Caratteristiche tecnico-tipologiche del "Romanelliano" di Grotta Romanelli (Castro Marina; Lecce). In *Grotta Romanelli nel Centenario della sua Scoperta (1900–2000)* (P.F. Fabbri, E. Ingravallo & A. Mangia, eds.), pp. 43–58. Congedo Editore, Galatina, Italy.

Bisson, M. S.
2000 Nineteenth Century tools for Twenty-First Century archaeology? why the typology of François Bordes must be replaced. *Journal of Archaeological Method and Theory* 8:1–48.

Blades, B. S.
1999 Aurignacian lithic economy and early modern human mobility: new perspectives from classic sites in the Vézère valley of France. *Journal of Human Evolution* 37:91–120
2001 *Aurignacian Lithic Economy: Ecological Perspectives from Southwestern France.* Kluwer/ Plenum Press, New York.
2003 End scraper reduction and hunter-gatherer mobility. *American Antiquity* 68:141–156.

Boëda, E.
1988 Le concept Levallois et évaluation de son champ d'application. In *l'Homme de Neandertal 4: La Technique.* (M. Otte, ed.), pp. 13–26. ERAUL, Liège.

Bordes, F.
1953 Essai de classification des industries «moustériennes.» *Bulletin de la Société Préhistorique Française* 50:457–466.
1961 *Typologie du Paléolithique Inférieur et Moyen.* Éditions du CNRS, Paris.

Brantingham, P. J., S. L. Kuhn, & K. W. Kerry (editors)
2004 *The Early Upper Paleolithic beyond Western Europe.* University of California Press, Berkeley.

Caramelli, D., C. Lalueza-Fox, C. Vernesi, M. Lari, A. Casoli, F. Mallegni, B. Chiarelli, I. Dupanloup, J. Bertranpetit, G. Barbujani, & G. Bertorelle
2003 Evidence for a genetic discontinuity between Neandertals and 24,000-year-old anatomically modern Europeans. *Proceedings of the National Academy of Sciences* 100:6593–6597.

Chazan, M
2001a Bladelet production in the Aurignacian of Hayonim Cave, Israel. *Paléorient* 27:81–88.
2001b Bladelet production in the Aurignacian of La

Ferrassie (Dordogne, France). *Lithic Technology* 26:16–28.

Churchill, S. E., & F. H. Smith
2000 Makers of the Early Aurignacian of Europe. *Yearbook of Physical Anthropology* 43:61–115.

Clark, G. A
1989 Romancing the stones: biases, style, and lithics at La Riera. In *Alternative Approaches to Lithic Analysis* (D. O. Henry & G. H. Odell, eds.), pp. 27–50. American Anthropological Association, Washington, DC.

Clark, G. A., & J. Riel-Salvatore
2006 Observations on systematics in Paleolithic archaeology. In *Transitions Before the Transition: Evolution and Stability in the Middle Paleolithic and Middle Stone Age* (E. Hovers & S. Kuhn, eds.), pp. 29–56. Springer, New York.

Conard N. J., P. M. Grootes, & F. H. Smith
2004 Unexpectedly recent dates for human remains from Vogelherd. *Nature* 430:198–201.

Davies, S. W. G.
2001 A very model of a modern human industry: new perspectives on the origins and spread of the Aurignacian in Europe. *Proceedings of the Prehistoric Society* 67:195–217.

Dibble, H. L.
1984 Interpreting typological variation of Middle Paleolithic scrapers: function, style or sequence of reduction? *Journal of Field Archaeology* 11:431–436.
1989 The Implications of Stone Tool Types for the Presence of Language during the Middle Paleolithic. In *The Human Revolution: Behavioral and Biological Perspectives on the Origins of Modern Humans* (P. Mellars & C. Stringer, eds.), pp. 415–432. Edinburgh University Press, Edinburgh.
1995 Middle Paleolithic scraper reduction: background, clarification and review of the evidence to date. *Journal of Archaeological Method and Theory* 2:299–368.

Finlayson, C. J.
2004 *Neanderthals and Modern Humans: An Ecological and Evolutionary Perspective.* Cambridge University Press, Cambridge.

Fish, P. R.
1979 *The Interpretive Potential of Mousterian Debitage.* Arizona State University Anthropological Research Papers No. 16, Tempe.

Flenniken, J. J., & P. J. Wilke
1989 Typology, technology, and chronology of Great Basin Dart Points. *American Anthropologist* 91: 149–158.

Gamble, C. S.
1986 *The Palaeolithic Settlement of Europe.* Cambridge University Press, Cambridge.

Goring-Morris, A. N., & A. Belfer-Cohen (editors)
2003 *More than Meets the Eye: Studies on Upper Palaeolithic Diversity in the Near East.* Oxbow Books, Oxford.

Grayson, D. K., & S. C. Cole
1998 Stone tool assemblage richness during the Middle and Early Upper Palaeolithic in France. *Journal of Archaeological Science* 25:927–938.

Harrold, F. B.
1988 The Châtelperronian and the early Aurignacian in France. In *The Early Upper Paleolithic: Evidence from Europe and the Near East* (J. F. Hoffecker & C. A. Wolf, eds.), pp. 157–191. BAR International Series 437, Oxford.
1989 Mousterian, Châtelperronian and Early Aurignacian in Western Europe: continuity or discontinuity? In *The Human Revolution: Behavioural and Biological Perspectives in the Origins of Modern Humans* (P. Mellars & C. Stinger, eds.) pp. 677–713. Princeton University Press, Princeton.

Hays, M. A., & G. Lucas
2000 A technological and functional analysis of carinates from Le Flageolet I, Dordogne, France. *Journal of Field Archaeology* 27: 455–465.
2001 Experimental investigations of aurignacian Dufour bladelets. In *Questioning the Answers: Resolving Fundamental Problems of the European Upper Paleolithic* (M. A. Hays & P. T. Thacker, eds.), pp. 109–116. BAR International Series 1005. Archaeopress, Oxford.

Hiscock, P.
1996 Transformations of Upper Palaeolithic implements in the Dabba industry from Haua Fteah (Libya). *Antiquity* 70:657–664.
2002 Quantifying the size of artefact assemblages. *Journal of Archaeological Science* 29:251–258.

Hiscock, P., & V. Attenbrow
2003 Early Australian implement variation: a reduc-

tion model. *Journal of Archaeological Science* 30:239–249.

Hoffman, C. M.
1985 Projectile point maintenance and typology: assessment with factor analysis and canonical correlation. In *For Concordance in Archaeological Analysis* (C. Carr, ed.) pp. 566–612. Waveland Press, Prospect Heights, IL.

Holdaway, S. J.
1989 Were there hafted projectile points in the Mousterian? *Journal of Field Archaeology* 16:79–85.
1991 *Resharpening Reduction and Lithic Assemblage Variability Across the Middle–Upper Paleolithic Transition.* PhD Thesis, Department of Anthropology, University of Pennsylvania, Philadelphia.

Holdaway, S. J., S. P. McPherron, & B. Roth
1996 Notched Tool Reuse and Raw Material Variability in French Middle and Lower Paleolithic Sites. *American Antiquity* 61:377–387.

Jelinek, A. J.
1976 Form, function, and style in lithic analysis. In *Cultural Change and Continuity: Essays in Honor of James Bennett Griffith* (C. E. Cleland, ed.) pp. 19–33. Academic Press, New York.

Karavanič, I.
2003 L'industrie aurignacienne de la grotte de Sandalja II (Istrie, Croatie) dans le contexte de la région de l'Est de l'Adriatique. *L'Anthropologie* 107:577–602.

Klein, R. G.
1999 *The Human Career: Human Biological and Cultural Origins, 2nd Edition.* University of Chicago Press, Chicago.
2003 Whither the Neanderthals? *Science* 299:1525–1527.

Kuhn, S. L.
1995 *Mousterian Lithic Technology: An Ecological Perspective.* Princeton University Press, Princeton.

Laplace, G.
1964 *Essai de Typologie Analytique.* Annali dell' Università di Ferrara, n.s., sez. XV, suppl. II. Ferrara.
1966 *Recherches sur l'Origine et l'Évolution des Complexes Leptolithiques,* École Française de Rome, Mélanges d'Archéologie et d'Histoire, Suppléments 4, Paris.

Lucas, G.
1997 Les lamelles Dufour du Flageolet 1 (Bézenac, Dordogne) dans le contexte Aurignacien, *Paléo* 9:191–219.
2001 The origin of Dufour bladelet torsion. In *Questioning the Answers: Resolving Fundamental Problems of the European Upper Paleolithic* (M. A. Hays & P. T. Thacker, eds.), pp. 99–107. BAR International Series 1005. Archaeopress, Oxford.

Lucas, G., J.-Ph. Rigaud, J. Simek, & M. Soressi
2003 The Châtelperronian of Grotte XVI, Cenac-et-Saint-Julien (Dordogne, France). In *The Chronology of the Aurignacian and the Transitional Technocomplexes: Dating, Stratigraphies, Cultural Implications* (J. Zilhão & F. d'Errico, eds.), pp. 289–299. *Trabalhos de Arqueologia* 33. Instituto Portugêse de Arqueologia, Lisbon.

Marks, A., H. J. Hietala, & J. K. Williams
2001 Tool standardization in the Middle and Upper Paleolithic: a closer look. *Cambridge Archaeological Journal* 11:17–44.

McPherron, S. P.
2000 Handaxes as a Measure of the Mental Capabilities of Early Hominids" *Journal of Archaeological Science* 27:655–663
2003 Technological and Typological Variability in the Bifaces from Tabun Cave, Israel. In *Multiple Approaches to the Study of Bifacial Technologies* (M. Soressi & H. Dibble, eds.), pp. 55–76. University of Pennsylvania Museum Press, Philadelphia.

Mellars, P.
1996 *The Neanderthal Legacy.* Princeton University Press, Princeton.
2005 The impossible coincidence: a single-species model for the origins of modern human behavior in Europe. *Evolutionary Anthropology* 12:12–27.

Mellars, P., & C. Stringer (editors)
1989 *The Human Revolution: Behavioural and Biological Perspectives on the Origins of Modern Humans.* Princeton University Press, Princeton.

Morrow, J. E.
1997 End scraper morphology and use-life: an approach for studying Paleoindian lithic technology and mobility. *Lithic Technology* 22:70–85.

Neeley, M. P., & C. M. Barton
1994 A new approach to interpreting Late Pleistocene microlith industries in Southwest Asia.

Antiquity 68:275–288.

Noble, W., & I. Davidson
1993 Tracing the origins of modern human behaviour: methodological pitfalls. *Journal of Anthropological Archaeology* 12:121–149.
1996 *Human Evolution, Language and Mind: A Psychological and Archaeological Inquiry.* Cambridge University Press, Cambridge.

Riel-Salvatore, J., & C. M. Barton
2004 Late Pleistocene technology, economic behavior, and land use dynamics in Southern Italy. *American Antiquity* 69:257–274.

Riel-Salvatore, J., & G. A. Clark
2001 Grave markers: Middle and Early Upper Paleolithic burials and the use of chronotypology in contemporary paleolithic research. *Current Anthropology* 42:449–479.

Sackett, J. R.
1988 The Mousterian and its aftermath: a view from the Upper Paleolithic. In *Upper Pleistocene Prehistory of Western Eurasia* (H. L. Dibble & A. Montet-White, eds.), pp. 413–426. University of Pennsylvania Museum, Philadelphia.

Shott, M. J.
1989 On tool-class use lives and the formation of archaeological assemblages. *American Antiquity* 54:9–30.

de Sonneville-Bordes, D., & J. Perrot
1953 Essai d'adaptation des méthodes statistiques au Paléolithique supérieur. Premiers résultats. *Bulletin de la Société Préhistorique Française* 51:323–333.
1954 Lexique typologique du Paléolithique supérieur, outillage lithique. 1. Grattoirs. 2. Outils solutréens. *Bulletin de la Société Préhistorique Française* 51:327–35.
1955 Lexique typologique du Paléolithique supérieur, outillage lithique. 3. Outils composites, perçoirs. *Bulletin de la Société Préhistorique Française* 52:76–79.
1956 Lexique typologique du Paléolithique supérieur, outillage lithique. 4. Burins. *Bulletin de la Société Préhistorique Française* 53:408–12, 547–59.

Stafford, B.
1977 Burin manufacture and utilization: an experimental study. *Journal of Field Archaeology* 4: 235–246.

Straus, L. G. (editor)
2005 *Armageddon or entente? The demise of the European Neandertals in Isotope Stage 3.* Special issue of Quaternary International 137:1–89.

Svoboda, J., & O. Bar-Yosef (editors)
2003 *Stránska Skála: Origins of the Upper Paleolithic in the Brno Basin, Moravia, Czech Republic.* American School of Prehistoric Research Bulletins, Harvard University Press, Cambridge.

Tomaskova, S.
2005 What is a burin? typology, technology, and interregional comparison. *Journal of Archaeological Method and Theory* 12:79–115.

Trinkaus, E.
2005 Early modern humans. *Annual Review of Anthropology* 34:207–30.

van Andel, T. H., & W. Davies (editors)
2003 *Neanderthals and modern humans in the European landscape during the last glaciation: archaeological results of the Stage 3 Project.* MacDonald Institute for Archaeological Research. Oxbow Books, Oxford.

Verpoorte, A.
2005 The first modern humans in Europe? A closer look at the dating evidence from the Swabian Jura (Germany). *Antiquity* 79:269–279.

Zilhão, J., & F. d'Errico (editors)
2003 *The Chronology of the Aurignacian and the Transitional Technocomplexes: Dating, Stratigraphies, Cultural Implications. Trabalhos de Arqueologia* 33. Instituto Português de Arqueologia, Lisbon.

Zubrow, E.
1989 The demographic modelling of Neanderthal extinction. In *The Human Revolution: Behavioural and Biological Perspectives on the Origins of Modern Humans.* (P. Mellars and C. Stringer, eds.), pp. 212–231. Princeton University Press, Princeton.

2

Even the Notion of a "Transitional Industry" is a Suspect Typological Construct

Lawrence Guy Straus
University of New Mexico

Abstract

Early prehistorians created the notions of "Middle" and "Upper" Paleolithic," and "transitional" categories such as the "Aurignaco-Mousterian" and "Châtelperronian." These labels reduce variability and simplify situations in ways that seem to lend themselves to "ethnic" explanations. Reconstructing ethnicity may be the most elusive aspect of Paleolithic prehistory, and while ethnic identity may have existed between 50–30 kya, studies of artifact typology and technology are unlikely to enable archaeologists to monitor it. Even the unity of such constructs as "the Aurignacian" or "the Châtelperronian" is suspect. It is argued here that, instead, the so-called Middle–Upper Paleolithic "transition" needs to be studied as a mosaic of biocultural adaptive change played out against the dynamic background of environmental change during OIS 3.

One of the latest fashions in the debate over the Middle-to-Upper Paleolithic transition in Europe is to speak in terms of the existence of three over-arching cultural categories for the period between about 40,000–30,000 radiocarbon years ago: Mousterian, Aurignacian and "transitional industries." The latter include such archeological constructs as the Châtelperronian, Uluzzian, Bachokirian, Olschevian, Bohunician, Szeletian, Lincombian and other "leaf-point cultures." They are essentially defined by characteristics of their lithic (and, to a lesser extent, osseous) industries, which are held—in true normative fashion—to be "typical" across a number of sites within a more or less constrained geographical area. These litho-cultures are generally assumed to be "real" entities in some sense equivalent to self-identified ethnic groups as observed in the ethnographic present.

Given the underlying (usually unstated, but widespread) presumption that humans in Europe during the waning millennia of isotope stage 3 were somehow organized into several macro-cultural groupings (including, for relict Bordesians, some of the various Mousterian facies, such as MTA-B and Denticulate), the paleoanthropological discourse of recent years has increasingly involved a hypothetical game of chess or war maneuvers, consisting of retrodicting the putative movements of these supposed groups around the geographic "game board." Thus, we can "plot" the retreat of retrograde Mousterians into southern Iberia and the Caucasus as well as the rise, survival and eventual demise of their culturally more sophisticated or adaptable cousins—proven or possible Neandertals—in western France (under the guise of *the* Châtelperronian), in Croatian Zagorje and Slovenia (*the* Olschevian), and in Italy (*the* Uluzzian). Also identified is the precocious florescence of Upper Paleolithic-like blade technology (and some working of antler, bone or ivory) among anatomically non-descript, but possibly robust natives of Bulgaria (*the* Bachokirian), Moravia (*the* Bohunician), Northwest Europe (*the* Lincombian), and Central Europe (*the* Szeletian). Above all, we detect the westward "spread" of the first "truly" modern, linguistically unchallenged, symbolically sophisticated, procreatively prodigious, and otherwise adaptively superior group (and species), *the* Aurignacian.

In this vast board game, different scenarios have the Aurignacians alternatively, simultaneously or consecutively killing off the locals, out-competing them, excluding them, out-breeding them, interacting with and influencing them, or incorporating them. Some models have been derived from presumed analogies with the Neolithization process in the Mediterranean Basin and from the history of European colonizations in the Americas, Africa and Australia during the last half-millennium. Concepts such as "genocide", "acculturation" and "independent local development" are bandied about just as if we were describing historical events and processes involving Puebloan and Athabaskan Indians, Spaniards and eventually Anglo-Americans which were lived, observed and recorded during the 16th through 19th centuries in my state of New Mexico in the American West. There is something deeply engrained in archeology—perhaps especially in European prehistory—that seems to require the reduction of artifact interassemblage variability into ideal-type entities that are automatically (and often uncritically) ascribed ethnic cultural significance. The *modus operandi* often involves the initial study of a limited set of materials which are considered to be "typical", and the subsequent generalization of their characteristics onto a wider universe of sites, archaeological levels or assemblages, and this despite the considerable variability they often display and which is either ignored or explained away by reference to such factors as poor excavation techniques, sample size, lithic raw material, or—more rarely—functional differences. Even if there are analytically defensible groups of assemblages, it does not automatically follow that similarities in artifact types or percentages *mean* ethnic commonality among human bands. Such a (tentative) conclusion could only be drawn as a result of a logical argument supported by plausible facts and relevant middle range theory.

Don't get me wrong: I am not denying the possible existence of socially or ethnically identified groups (macro-, regional or linguistic bands) in the mid–late Upper Pleistocene of Europe. Rather, I am saying that I am skeptical of claims that we have actually identified such entities in the constructs that we call "the Châtelperronian" or "the Aurignacian" (to name but two)—at least to the extent of then being able to speak of the acculturation of one by another, their movements, their success or their demise, as though they were real cultures akin to such extant (albeit also contestable) entities as the New Mexico Anglos, Hispanos, Chicanos, Pueblos, Diné, Apaches or Sikhs. Why is it that modern ethnologists recognize the absolute fluidity of ethnic group boundaries, membership, prominence, and even existence over the short term—not to mention complexity of definition—while prehistoric archeologists insist on their long-term permanence and self-evident existence, especially on the basis of *a priori* trait lists that may include an artifactual grab-bag of supposed (but undemonstrated) ethnic markers in assemblages that may have been deposited over several thousands of years? Before engaging further in the passionate debate over the relationship of the presumptive "cultures" that inhabited the European landscape during the ten last millennia of Oxygen Isotope Stage 3, shouldn't we think a little bit more about what these archeological constructs might be and about what they might represent in terms of real people, bands, social relations and *milieux?*

The Middle–Upper Paleolithic Transition as "Mosaic"

When I use the term "mosaic" to describe the cultural landscape of Europe during the so-called Middle-to-Upper Paleolithic transition—as I have done frequently in recent years (e.g., 1996, 1997, 2005)—I mean just that: the situation was complex and variegated, with myriad conditions and possibilities across continental space and through millennial time. Europe is a continent which—although small—lies astride the critical middle latitudes of the planet and which includes long, complex coastlines and great relief (with major mountain chains, prominent river valley corridors and vast plains). Late isotope stage 3 witnessed a marked climatic downturn punctuated by three significant interstadials (Hengelo, Les Cottés and Arcy), each of which included internal fluctuations in temperature and humidity, meaning that not only climate, but also vegetation and faunal resources, as well as water courses, coasts and humanly favorable habitats, changed a great deal during the course of "the transition" (see Davies et al. 2000; van Andel and Davies 2003). Human adaptive responses were also highly diverse and variable. Not all people changed in the same ways in all places; some changed parts of their way of life (including technology), but not others; some did so quickly, others more slowly; still others perhaps not at all. Some aspects of the material culture (from stone tools to ornaments to "art") changed in some regions before they did (if ever) in others. In short, it was a "mix-and-match" world. And the same seems to have been true in regard to human anatomy, although—at present at least—the only known diagnostic human remains that we have for the period between >40,000 years ago and about 32,000 years ago in Europe are *Neandertals*. This was also apparently a very lightly populated world. I think that we have to imagine rather small human bands inhabiting favorable areas scattered across Europe and separated by large empty zones that were only occasionally visited or traversed, if ever. Social contacts were maintained (perhaps often only tenuously) especially to find mates, which was critical for demographic survival. In this context, new technological ideas (notably ones useful in the hunting of game) could sometimes spread from band to band and among regions. On the other hand, it is questionable as to whether humans would want (or could tolerate) exclusive band membership and closed band territories against a backdrop of low population density and relatively benign interpleniglacial climatic conditions.

When one looks closely at the distribution of so-called transitional sites, the numbers for most of the "cultural" groups are unimpressive. For example, the Uluzzian is represented by less than two dozen sites in the southern two-thirds of the Italian Peninsula; the Olschevian (with four sites), Bohunician (about ten sites) and Bachokirian (one or two sites) have even fewer each in their respective areas. Perhaps they represent individual band territories in the broadest sense of the term? But even beginning to demonstrate this would require chronometric control far greater than we presently have. The Châtelperronian—in part by dint of having been recognized under one guise or another longer than the others and mainly in the country (France), where more intense Paleolithic research has been done than in any other—is by far the best studied of the so-called "transitional industries." How "real" or "unitary" is it, and, thus, how likely is it to represent an actual self-identifying "culture" or ethnic group?

The Châtelperronian and its Internal Variability

Based on the excellent synthetic work of Frank Harrold (e.g., 1981, 1988, 1989, 2000) and Jacques Pelegrin (1995), there are some three dozen credible sites (in addition to a roughly equal number of other, isolated, less-well documented or unstratified surface finds of Châtelperron points) that, at one time or another and on a variety of bases, have been attributed to the Châtelperronian (a.k.a. Lower Aurignacian *sensu* Breuil, Lower Perigordian *sensu* Peyrony, or Castelperronian *sensu* Garrod and Delporte). Châtelperronian sites are concentrated in north-central and southwestern France, with smaller numbers of sites in north-Atlantic and (more problematically) northeastern Spain. Despite many apparent similarities with the Uluzzian (including its apparent age and presence of curved backed points/knives) (Gioia 1990), there is a major geographical hiatus between the Châtelperronian sites of the central Pyrenees and the Uluzzian ones of Italy.

The putative phylogenetic relationship between the Lower and Upper Perigordian (a.k.a. Gravettian) and indeed the very reality of the Châtelperronian have been challenged before (e.g., by Thomas Lynch [1966]). But these are really matters of opinion among prehistorians who are heavily influenced by the paradigms to which they subscribe, notably in terms of their respective views of the nature of interassemblage differences and similarities (Clark 1999, 2001, 2002; Clark & Riel-Salvatore 2006; Clark & Lindly 1991). The unity of the Châtelperronian—whatever "it" was—is however a question of considerable interest that I think is amenable to study. And, once explored dispassionately, it is a question that might shed some light on conditions during the so-called Middle–Upper Paleolithic transition.

Despite recent (*in part* possibly useful, but also casuistic and selective) critiques of the record by d'Errico, Zilhão and others (d'Errico et al. 1998; Zilhão & d'Errico 1999a,b), it seems likely that the Châtelperronian dates to between about 38,000 (or 36,000) and about 33,000 radiocarbon years ago (see Mellars [1999] for a list of dates, plus Fortea [1996] for the addition of dates from A Valiña in Galicia, F. David et al. [2001] for a series of new Oxford AMS dates from la Grotte du Renne at Arcy, and J.-Ph. Rigaud et al. [2002] for several coherent dates—some back to c. 38 kya—for the Châtelperronian of Grotte XVI). This is well within the feasible range of radiocarbon dating and many of the determinations are of modern-quality, done by accelerator mass spectrometry on carefully chosen samples. The TL and ESR date averages of about 36,000 B.P. for both Saint-Césaire and Combe-Saunière respectively translate into radiocarbon ages of around 33,000 B.P, which would seem to provide additional support for an age range well after 40,000 B.P. With nine Châtelperronian sites having been radiometrically dated (several with multiple and largely coherent determinations), it would seem hard to credibly call into question the entire age range of the Châtelperronian construct, as defined by the consensus of several generations of prehistorians.

Of course, there are specific problems that remain to be explained (e.g., the fairly wide scatter and lack of strict stratigraphic coherence of the many Grotte du Renne dates, the huge standard error of the single Cueva Morín determination, the large standard errors of all TL and ESR dates, and the single "old" outlier determinations at Grotte du Renne and Le Moustier (45,000 and 43,000 B.P respectively). What is interesting (and paradigmatically telling), to me at least, however, is that—despite their fiercely critical spirit—Zilhão and d'Errico fundamentally do accept the Châtelperronian and early Aurignacian as real cultural entities, with clear ethnic (and even racial—i.e., Neandertal versus Cro-Magnon) meaning and implications. I am not so sure that at least one of their original co-authors (Pelegrin—who is an acknowledged expert on Châtelperronian lithic technology) would be as sanguine as they are concerning the unity of the Châtelperronian concept as a *bona fide* cultural unit. A close reading of Pelegrin's (1995:254–260 *et passim*) intersite comparisons shows that, although "leptolithic" technological and typological characteristics are paramount in the Châtelperronian assemblages in the two sites he analyzed (i.e., Roc-de-Combe and La Côte), other sites can apparently have truly more "hybrid" Upper-cum-Middle Paleolithic industries. This variability is what I would like to explore a little further here (see Bon [2002], Teyssandier [2005] as well as Straus [2003] for recent, detailed critiques of the supposed unity of the early Aurignacian).

What is it that really holds the Châtelperronian together definitionally as an archeological construct? I would argue that—for better or worse—it is the curved backed "Châtelperron" point or knife (a large version of classic

Azilian points which would come along some 20–25,000 years later in western Europe and very similar to Howiesons Poort crescents in southern Africa, which can be possibly two to three times older than their Châtelperronian look-alikes).

The technomic function(s) of the Châtelperron type is/are still debated: Pelegrin (1995) seems to believe that they were actual projectile tips, while Plisson and Schmider (1990) argue that at least the ones from Grotte du Renne were used as hafted knives. The extent to which the Châtelperron point also may have functioned at the sociotechnic level as a cultural marker is highly problematic and unproven. Did merely possessing one or two Châtelperron points in Cantabria make one "a Châtelperronian" like possessors of such points in Périgord or—more remote still—in Île-de-France?

Other characteristics of the Châtelperronian are far less ubiquitous, and even the points are found in highly variable absolute and relative frequencies. As documented by Harrold (1981:16, table 7), Châtelperron points can make up between 2 and 46% of the tool assemblages (35% in the careful excavation of Roc-de-Combe [Pelegrin 1995] and as high as 18% at Grotte du Renne [Plisson and Schmider 1990]). Yet, the mere presence of one or a very few such *fossiles directeurs* can *sometimes* clinch an assignment of a site to the Châtelperronian. For example, in north-Atlantic Spain, Level 10 of Cueva Morín has two "typical" and five "atypical" Châtelperrons among 484 tools; Level VIII of El Pendo has one "typical" and two "atypical" Châtelperrons among 79 tools; Level 9 of Labeko Koba has three "typical" and one "atypical" Châtelperrons among only 12 tools; and Level IV of A Valiña has only one problematic Châtelperron tip among a total of 60 tools (González Echegaray and Freeman 1971; González Echegaray 1980; Arrizabalaga and Altuna 2000; Llana and Soto 1991). Were it not for these rare objects (and the radiocarbon dates of 36,000, 34,000 and 32–34,000 B.P at Morín, Labeko Koba and A Valiña respectively), would these assemblages be called Châtelperronian? On the other hand, there are actually a few more Châtelperron points in El Pendo Levels VIIIa ("Archaic Aurignacian") and VII ("Aurignacian I") than in the supposedly Châtelperronian Level VIII itself (!)(González Echegaray 1980). And there is an "atypical" Châtelperron in Level 18 of El Castillo—labeled (with reason) an "Archaic Aurignacian" (Cabrera 1984) and abundantly dated between 37–40,000 radiocarbon years ago (Cabrera et al. 1996). Similarly, there are three Châtelperron points in early Aurignacian Level H at L'Arbreda in northern Catalonia (dated between 38–40,000 radiocarbon years ago) and two Châtelperrons (one "typical" and one "atypical") in early Aurignacian Level A of nearby Reclau Viver, dated to 40,000 radiocarbon years ago (Canal and Carbonell 1989; Maroto et al. 1996). On the other hand, Els Ermitons, also in Gerona, yielded a single Châtelperron in an industry qualified as "Mousterian", but dating between 33–36,500 radiocarbon years ago and containing numerous burins, endscrapers, perforators and backed pieces, as well as many sidescrapers, denticulates and notches (Maroto et al. 1996). When do Châtelperron points make for a Châtelperronian assemblage—and when do they not? Maybe they somehow do when they are not accompanied by lots of keeled or nosed endscrapers or Dufour nibbled bladelets, or maybe they don't when they come with "too many" Mousterian tools and "not enough" Upper Paleolithic ones... Who decides and how and why?

The classic (but evolving) definition of the Châtelperronian—from Breuil to Peyrony, to de Sonneville-Bordes, Bordes and Delporte—included a combination of typically Middle and Upper Paleolithic tools (i.e., sidescrapers, notches and denticulates *as well as* burins, endscrapers and perforators), as well as the curved backed knives or points (which were seen as deriving from backed Abri Audi knives of the Mousterian of Acheulean Tradition B) and even rare osseous points. Levallois flakes were said to be found in some Châtelperronian assemblages. Controversy has raged (see Pelegrin 1995 for details) as to the possibility of mechanical mixture between underlying Mousterian and overlying Châtelperronian levels in at least some sites (e.g., at La Ferrassie and, perhaps to a lesser degree, at Roc-de-Combe). As Pelegrin and Harrold show, Mousterian tools are however abundant also in Châtelperronian assemblages at sites where there is *no* (or no directly underlying) Mousterian level (e.g., La Grande Roche at Quinçay). On the other hand, some sites *with* underlying Mousterian levels have little Mousterian-like material in the Châtelperronian (e.g., at the type site of Châtelperron itself). In addition, some sites (e.g., la Grotte du Renne) show use of *both* Levallois and prismatic blade techniques for blank production in the Châtelperronian, while others are very much dominated by true (crested) blade production (not that this is exclusively Upper Paleolithic, having been found in many, often very early Mousterian sites, especially in northern France and Belgium; see Révillion and Tuffreau 1994). In general, in Harrold's (1981) sample, the three supposedly characteristic Mousterian tool types (sidescrapers, denticulates and notches) range among Châtelperronian collections from 0 or 1% to 14 or 19%, while endscrapers and burins range from 5 or 3% to 26 or 27%. The extent of "Mousterian-ness" of some Châtelperronian assemblages is, however, much greater still: the uppermost Châtelperronian level (8) at the meticulously excavated site of Saint-Césaire—well separated from underlying Mousterian deposits—yielded 305 tools, of which 122 (40%) are sidescrapers, while there are only eight Châtelperron points (Lévêque 1993). The degree of variability among assemblages that are called Châtelperronian is enormous indeed—both typologically and technologically. It is telling that both characteristically Aurignacian-type tools and small bifaces can also be found in Châtelperronian contexts (as Breuil [1937] himself argued; see also Lévêque 1993). For example, in carefully excavated Cueva Morín Level 10 (González Echegaray and Freeman 1971), there are 26 keeled and nosed endscrapers (5.4% of the assemblage) and a Dufour bladelet (whose downward migration might, however, be easier to argue). In Saint-Césaire there are small bifaces.

Also very significant is the variability in the representation of osseous artifacts (e.g., bone/antler points, elaborately worked awls) or ornaments (e.g., perforated teeth or shells, carved ivory) in Châtelperronian assemblages. In fact, most have none or very few (Harrold [1981:22, table 8] could find such items in only eight sites; since the time of his research, meticulous excavation of Labeko Koba in Guipúzcoa has yielded *one* tiny distal *sagaie* fragment in the Châtelperronian level [Arrizabalaga and Altuna 2000].) There are a few items of adornment (perforated fox canines) at Quinçay, but the *only* site with relatively many osseous artifacts and ornaments is la Grotte du Renne, where many of the items bear close similarities to ones in the early Aurignacian both of France and Belgium (Baffier and Julien 1990; Taborin 1990; White 1993). Can bone/tooth-working (an Upper Paleolithic characteristic *par excellence*) be said to be typical of the Châtelperronian? Hardly. Why is Grotte du Renne so peculiar in this and other respects (e.g., the presence of clear huts)? Is it in any way representative of the Châtelperronian as a whole? But what is the Châtelperronian beyond a collection of assemblages in western and southern France and northern Spain that seem to date to the period between around 36–33,000 radiocarbon years ago and that have large curved backed points or knives made on blades, plus some endscrapers and burins? Beyond this, every aspect of the normative definition seems to be subject to many exceptions, especially concerning the abundance of those points and the presence of, on the one hand, Mousterian types and techniques and, on the other hand, osseous artifacts and ornaments.

The Mosaic Conclusion

The Châtelperronian problem represents, from my point of view at least, a microcosm of the mosaic situation that characterizes the Middle-to–Upper Paleolithic as a whole. Rather than thinking in terms of old typological pigeonholes, we must begin to recognize and understand a vast range of variability in human experiments that were being tried out during the period from around 40,000–30,000 years ago, some of which—such as true blade production—had had antecedents tens or even hundreds of thousands of years earlier in the Middle Paleolithic. Along these lines, it makes sense to interpret the meaning of such phenomena as indicating considerable continuity between the Mousterian and early Aurignacian in Cantabria (Cabrera et al. 1997; Bernaldo de Quirós et al. 2001), the combination of Mousterian-like lithic economy and Aurignacian-type and Mousterian-type tools and leaf points in a 40/38,000 year old level at Trou Magrite (Otte and Straus 1995), the co-association of lithic leaf points, Aurignacian-like split-base Mladeč-type *sagaies,* a burin, an endscraper, an engraved bone and Mousterian-like stone tools with Neandertal remains dated to 28/29,000 B.P in the Olschevian of Vindija Cave (Karavanič 1995; Karavanič and Smith 1998; Smith et al. 1999). Similarly, the Uluzzian with its curved backed stone points, and the plethora of northwest, central and east European industries with stone foliates and ivory or bone points (some classified as Middle and others as Upper Paleolithic) all add to the mosaic picture, as do the different industries (some very old) at Bacho Kiro and Temnata in Bulgaria and Korolevo in Ukraine, and the early Aurignacian, notably at the superbly excavated and dated site of Geissenklösterle.

Why not seriously contemplate the possibility that a few useful inventions (e.g., curved backed stone points, cylindical, rhomboidal or split-base osseous points) were made—perhaps more than once—in various regions of Europe and spread from band to band as a result of social contacts (visits, exchanges, intermarriage, rituals, etc.) because they were perceived as advantageous in the killing and/or butchering of animals? Maybe the many climatic fluctuations and ultimate downturn of late isotope stage 3 prompted people to accept such innovations, as they helped to provide a competitive edge in the deadly game of survival on the European frontier? The new projectile tips were alternative solutions to basic adaptive problems, but they did not necessarily require acceptance of broader cultural packages, and blade manufacture per se was nothing new in reality. Nor is there any necessary relationship between hominid anatomy and tool types. Indeed the fact that there were Neandertals at Saint-Césaire and Arcy-sur-Cure does not necessarily mean that all so-called Châtelperronian assemblages were made by Neandertals—although, in terms of credible, diagnostic paleontological discoveries, they seem (at present at least) to have been "the only game in town" until around 34–35,000 radiocarbon years ago, with the recent demotion of the remains from Hahnöfersand to the Mesolithic, Vogelherd to the Neolithic and Velika Pečina to the Chalcolithic via direct AMS ^{14}C dating.

Rather than a Europe of monolithic ethnic groups invading into or retreating from bounded territories during the late Interpleniglacial, I see a Europe dotted with bands continuously adapting to changing local conditions, each with its peculiarities as determined both by chance (due to relative isolation, small membership sizes, the presence or absence of "geniuses", etc.) and by circumstances (latitude and longitude, lithology, relief, vegetation, fauna, etc.). When they did make social contacts, objects and ideas could possibly spread, take hold and be modified to suit local conditions if they were perceived to be of use or better than what was already at hand. Many ideas and inventions were "brewing" during the Interpleniglacial, and, over several millennia and with significant inter-regional differences, some of these were successful to one degree or another. This we dimly perceive in the very partial archeological record of such phenomena as the Bohunician, Szeletian, Lincombian, early Aurignacian, Bachokirian, Olschevian, Châtelperronian, Uluzzian, etc. I do *not* think we are seeing phenomena equivalent to the demographically denser, temporally constrained Linear Bandkeramic or Cardial cultures, even if we were *sure* how *they* spread.

This chapter is dedicated to the memory of Victoria Cabrera Valdés, 1951–2004.

References

Arrizabalaga, A., & J. Altuna (editors)
2000 Labeko Koba. *Munibe* 52:1–395.

Baffier, D., & M. Julien
1990 L'outillage en os des niveaux châtelperroniens d'Arcy-sur-Cure. In *Paléolithique Moyen Récent et Paléolithique Supérieur Ancien en Europe* (C. Farizy, ed.), pp. 329–334. Mémoires du Musée de Préhistoire d'Ile de France 3, Nemours.

Bernaldo de Quirós, F., V. Cabrera, M. Lloret, & A. Pike-Tay
2001 New kids on the block? Some comments on the Middle–Upper Paleolithic transition in Cantabrian Spain. In *Questioning the Answers: Resolving Fundamental Problems of the Early Upper Paleolithic* (M. Hays & P. Thacker, eds.), pp. 27–38. British Archaeological Reports S-1005. Archaeopress, Oxford.

Bon, F.
2002 *L'Aurignacien entre Mer et Océan.* Mémoires de la Société Préhistorique Française, no. 29. Paris.

Breuil, H.
1937 *Les subdivisions du Paléolithique supérieur et leur signification* (2e édition). Emanuel Grevin et Fils, Lagny.

Cabrera, V.
1984 *El Yacimiento de la Cueva de "El Castillo."* Bibliotheca Praehistorica Hispana 22. Madrid.

Cabrera, V., H. Valladas, F. Bernaldo de Quirós, & M. Hoyos.
1996 La transition Paléolithique moyen-Paléolithique supérieur à El Castillo: nouvelles datations par le carbone-14. *Comptes Rendus de l'Académie des Sciences de Paris,* series 2a, 322:1093–1098.

Cabrera, V., M. Hoyos, & F. Bernaldo de Quirós
1997 The transition from Middle to Upper Paleolithic in the cave of El Castillo. In *Conceptual Issues in Modern Human Origins Research* (G. A. Clark & C. M. Willermet, eds.), pp. 177–190. Aldine de Gruyter, New York.

Canal, J., & E. Carbonell
1989 *Catalunya Paleolítica.* Patronat Eiximenis, Girona.

Clark, G. A.
1999 Modern human origins—highly visible, curiously intangible. *Science* 283:2029–2032; 284:917.
2001 Discussion: The logic of inference in transition research. In Questioning the Answers: Re-solving Fundamental Problems of the Early Upper Paleolithic (M. Hayes & P. Thacker, eds.), pp. 39–48. BAR International Series No. 1005, Oxford.
2002 Observations on paradigmatic bias in French and American paleolithic archaeology. In *The Role of American Archeologists in the Study of the European Upper Paleolithic* (L. Straus, ed.), pp. 19–26. BAR International Series No. 1048, Oxford.

Clark, G. A., & J. M. Lindly
1991 Paradigmatic biases and paleolithic research traditions. *Current Anthropology* 32:577–587.

Clark, G. A., & J. Riel-Salvatore
2006 Observations on systematics in paleolithic archaeology. In *Transitions Before the Transition: Evolution and Stability in the Middle Paleolithic*

and Middle Stone Age (E. Hovers & S. Kuhn, eds.), pp. 29–56. Springer, New York.

David, F., N. Connet, M. Girard, V. Lhomme, J.-C. Miskovsky, & A. Roblin-Jouvé
2001 Le Châtelperronien de la Grotte du Renne à Arcy-sur-Cure (Yonne). Données sédimentologiques et chronostratigraphiques. *Bulletin de la Société Préhistorique Française* 98:207–230.

Davies, W., J. Steward, & T. H. van Andel
2000 Neandertal landscapes—a preview. In *Neanderthals on the Edge* (C. Stringer, R. N. E. Barton & C. J.Finlayson, eds.), pp. 1–8. Oxbow Books, Oxford.

d'Errico, F., J. Zilhão, D. Baffier, M. Julien, & J. Pelegrin
1998 Neandertal acculturation in Western Europe? A critical review of the evidence and its interpretation. *Current Anthropology* 39: S1-S44.

Farizy, C. (editor)
1990 *Paléolithique Moyen Récent et Paléolithique Supérieur Ancien en Europe.* Mémoires du Musée de Préhistoire d'Ile de France 3, Nemours.

Fortea, J.
1996 Le Paléolithique supérieur en Espagne: Galice et Asturies. *Le Paléolithique Supérieur Européen: Bilan Quinquennal,* 1991–1996. ERAUL 76, Liège.

Gioia, P.
1990 An aspect of the transition between Middle and Upper Palaeolithic in Italy: the Uluzzian. In *Paléolithique Moyen Récent et Paléolithique Supérieur Ancien en Europe* (C. Farizy, ed.), pp. 241–250. Mémoires du Musée de Préhistoire d'Ile de France 3, Nemours.

González Echegaray, J. (editor)
1980 *El Yacimiento de la Cueva de "El Pendo."* Bibliotheca Praehistorica Hispana 17. Madrid.

González Echegaray, J., & L.G.Freeman (editors)
1971 Cueva Morín: *Excavaciones 1966–1968.* Patronato de las Cuevas Prehistóricas, Santander.

Harrold, F. B.
1981 New perpectives on the Châtelperronian. *Ampurias* 43:1–51.
1988 The Châtelperronian and the early Aurignacian in France. In *The Early Upper Paleolithic* (J. Hoffecker & C.Wolf, eds.), pp. 157–191. British Archaeological Reports S437. Oxford.
1989 Mousterian, Châtelperronian and early Aurignacian in Western Europe: continuity or discontinuity? In *The Human Revolution* (P. Mellars & C. B. Stringer, eds.), pp. 677–713. University of Edinburgh Press, Edinburgh.
2000 The Châtelperronian in historical context. *Journal of Anthropological Research* 56:59–75.

Karavanič, I.
1995 Upper Paleolithic occupation levels and late-occurring Neandertal at Vindija Cave (Croatia) in the context of Central Europe and the Balkans. *Journal of Anthropological Research* 51: 9–35.

Karavanič, I., & F. H. Smith
1998 The Middle/Upper Paleolithic interface and the relationship of Neanderthals and early modern humans in the Hrvatsko Zagorje, Croatia. *Journal of Human Evolution* 34:223–248.

Lévêque, F.
1993 The Castelperronian industry of Saint-Césaire: the upper level. In *Context of a Late Neandertal* (F. Lévêque, A. Backer & M. Guilbaud, eds.), pp. 23–35. Prehistory Press, Madison.

Llana, C., & M. J. Soto (editors)
1991 *Cova da Valiña.* Arqueoloxía/Investigación 5 (Xunta de Galicia). Santiago de Compostela.

Lynch, T.
1966 The "Lower Perigordian" in French archaeology. *Proceedings of the Prehistoric Society* 32: 156–1998.

Maroto, J., N. Soler, & J. M. Fullola
1996 Cultural change between Middle and Upper Palaeolithic in Catalonia. In *The Last Neandertals, The First Anatomically Modern Humans* (E. Carbonell & M. Vaquero, eds.), pp. 219–250. Universitat Rovira i Virgili, Tarragona.

Mellars, P.
1999 The Neanderthal problem continued. *Current Anthropology* 40:341–350.

Otte, M., & L. G. Straus (editors)
1995 *Le Trou Magrite.* ERAUL 69, Liège.

Pelegrin, J.
1995 *Technologie Lithique: Le Châtelperronien de Roc-de-Combe et de La Côte.* Éditions du CNRS, Paris.

Plisson, H., & B. Schmider
1990 Etude préliminaire d'une série de pointes de

Châtelperron de la Grotte du Renne à Arcy-sur-Cure. In *Paléolithique Moyen Récent et Paléolithique Supérieur Ancien en Europe* (C. Farizy, ed.), pp. 313–318. Mémoires du Musée de Préhistoire d'Ile de France 3, Nemours.

Révillion, S., & A. Tuffreau (editors)

1994 *Les Industries Laminaires au Paléolithique Moyen*. Éditions du CNRS, Paris.

Rigaud, J.–Ph.

2000 Late Neandertals in the South West of France and the emergence of the Upper Palaeolithic. In *Neanderthals on the Edge* (C. Stringer, R. N. E. Barton & J. C. Finlayson, eds.), pp. 27–31. Oxbow Books, Oxford.

Rigaud, J.-Ph., J. Simek, M. Soressi, & G. Lucas

2002 The Middle/Upper Paleolithic "Transition" at the Grotte XVI (Dordogne, France). Paper presented at the 67th Annual Meeting of the Society for American Archaeology, Denver, Colorado.

Smith, F. H., E. Trinkaus, P. Pettitt, I. Karavanič, & M. Paunovič

1999 Direct radiocarbon dates for Vindija G1 and Velika Pečina Late Pleistocene hominid remains. *Proceedings of the National Academy of Sciences* 96:12281–12286.

Straus, L. G.

1996 Continuity or rupture; convergence or invasion; adaptation or catastrophe; mosaic or monolith: views on the Middle to Upper Paleolithic transition in Iberia. In *The Last Neandertals, The First Anatomically Modern Humans* (E. Carbonell & M. Vaquero, eds.), pp. 203–218. Universitat Rovira i Virgili, Tarragona.

1997 The Iberian situation between 40,000 and 30,000 B.P. in light of European models of migration and convergence. In *Conceptual Issues in Modern Human Origins Research* (G. A. Clark & C. M. Willermet, eds.), pp. 235–252. Aldine de Gruyter, New York.

2003 "The Aurignacian?" Some thoughts. In *The Chronology of the Aurignacian and of the Transitional Technocomplexes: Dating, Stratigraphies, and Cultural Implications* (J. Zilhão & F. d'Errico, eds., pp. 11–17. *Trabalhos de Arqueologia* 33. Instituto Português de Arqueologia, Lisbon.

2005 A mosaic of change: the Middle–Upper Paleolithic transition as viewed from New Mexico and Iberia. In *Armageddon or Entente? The Demise of the European Neanderthals in Isotope Stage 3. Quaternary International* 137:47–68.

Taborin, Y.

1990 Les prémices de la parure. In *Paléolithique Moyen Récent et Paléolithique Supérieur Ancien en Europe* (C. Farizy, ed.), pp. 335–344. Mémoires du Musée de Préhistoire d'Ile de France 3, Nemours.

Teyssandier, N.

2005 Résumé de thèse: "Les Débuts de l'Aurignacien en Europe." *Bulletin de la Société Préhistorique Française* 102:211–216.

Van Andel, T., & W. Davies (editors)

2003 *Neanderthals and Modern humans in the European Landscape during the Last Glaciation.* McDonald Institute of Archaeology, Cambridge.

White, R.

1993 A social and technological view of Aurignacian and Castelperronian personal ornaments in SW Europe. In *El Origen del Hombre Moderno en el Suroeste de Europa* (V. Cabrera, ed.), pp. 327–357. UNED, Madrid.

Zilhão, J., & F. d'Errico

1999a Reply. *Current Anthropology* 40:355–360.

1999b The chronology and taphonomy of the earliest Aurignacian and its implications for the understanding of Neandertal extinction. *Journal of World Prehistory* 13:1–68.

3

On the Fate of the Neanderthals and the Middle–Upper Paleolithic Transition in Western Europe

Francis B. Harrold
University of Nebraska, Kearney

Abstract

Currently available evidence indicates that the Neanderthals were replaced in Western Europe by anatomically modern humans during Oxygen Isotope Stage 3. Of the three principal models proposed to explain the archaeological and fossil record of this era—called here the population dispersal, mosaic, and indigenist models — the population dispersal model, relating the replacement to the Middle–Upper Paleolithic transition, best fits the data of that record. Modern humans replacing Western Europe's Neanderthals presumably had some selective advantage(s) over them. A review of evidence concerning subsistence, technology, and durable symbols (e.g., bone beads) highlights notable changes over the transition to the Upper Paleolithic in the latter two categories, and a complex situation in regard to subsistence. Neanderthals were clearly able successfully to hunt sizable mammals, but their adaptation apparently involved low population density and unelaborated technology in a way different from the adaptations of any recent hunter-gatherers. These Neanderthal-modern differences might be explained by differences in cognitive capacities (though the Châtelperronian industry indicates that Neanderthals were capable of at least some types of 'modern' behavior) or by other factors, and no definitive test of rival explanations is yet available. A comparison of the Upper Paleolithic and Neolithic "revolutions" highlights how difficult it will be to sort out the explanations for the biological and cultural transitions of over thirty millennia ago.

In the last two decades, a considerable literature has proliferated around the related subjects of the origin of anatomically modern humans (or AMH), the fate of the Neanderthals and other "archaic" humans, and their connection (if any) with changes in the Upper Pleistocene archaeological record. In the confines of this chapter, I will consider a particular corner of this intellectual territory—the middle part of the Last Glacial period in Western Europe. I will restrict consideration primarily to the zone of the Châtelperronian in France and Northern Spain, for this industry has important implications for the topics raised above. Most geographical definitions of "Western Europe" would include Italy, and thus the territory of the Uluzzian, a "transitional" industry often compared to the Châtelperronian (Bietti and Negrino, this volume). However, the Uluzzian is much less well documented than the Châtelperronian, and I will focus here on this best known of the putatively transitional mid-Last Glacial complexes. Specifically, I will deal with two questions: What happened to the Neanderthals? And how is their fate related to the changes in the archaeological record that we term the Middle–Upper Paleolithic transition?

What Happened to the Neanderthals?

The question of the evolutionary fate of the Neanderthals has been posed since shortly after the original Neanderthal discovery in 1856. Over that period, most of the proposed answers have aligned into two opposing positions: either the Neanderthals were absorbed or driven into extinction by the arrival of modern humans in Europe, or they essentially evolved into modern humans. In the first case, Neanderthal ancestry of later European populations was minor or nonexistent; in the second, Neanderthals were the ancestors of these later people. Intermediate positions combining elements of both explanations have been proposed, but historically, the middle of the spectrum of answers to this question has been less populated than the ends.

The question of the Neanderthals' destiny has of course become linked to the larger question of the origins of anatomically modern humans and their relations to all archaic people, with once again the opposing "Out of Africa" and multiregional models tending to polarize views (Smith &

Harrold 1997). Here, however, I will concentrate on the Neanderthal issue, specifically in Western Europe, and note that continuing fossil finds, as well as subjection of existing remains to new techniques of study, have gradually established several points.

First, whether we designate them a separate species of *Homo* or a subspecies of *H. sapiens,* the Neanderthals were a distinctive, continental-scale human population with deep roots in the European Middle Pleistocene. Their range included much of Europe and extended eastward into Central Asia and southeastward—at least during Oxygen Isotope Stage (OIS) 4—into the Levant. They shared a number of distinctive derived skeletal traits which accreted gradually, as a number of later Middle Pleistocene fossils—most spectacularly, those from the Sima de los Huesos at Atapuerca (Arsuaga et al. 1997)—demonstrate. A number of their characteristics, notably their "hyperarctic" body proportions, are understood as adaptations to cold climate, arguably under conditions of less effective cultural buffering than seen among recent cold-climate humans (Holliday 1997; Churchill 1998; but see Finlayson 2004:82–88 for a contrary view).

Second, morphologically based assessments of Neanderthal distinctiveness have been reinforced by recent studies isolating mitochondrial DNA from Neanderthal fossils and comparing it to mtDNA from modern humans (Krings et al. 1997, 1999, 2000; but see Hawks & Wolpoff 2001, on the Neanderthal status of the Mezmaiskaya remains). These studies found that the Neanderthal individuals tended to differ from humans of any sampled contemporary population (including Europeans) considerably more than any of these populations differ from each other. They also indicated a likely period of several hundred thousand years since the divergence of the ancestral mtDNA lineages leading respectively to Neanderthals (as represented by the admittedly small sample) and contemporary modern people. Modern European samples indicate no mtDNA type judged sufficiently divergent to derive plausibly from Neanderthals (Forster 2004). These results must be treated cautiously. They do not prove that Neanderthals were a separate species, or that they could not/did not interbreed with modern humans. However, they do underline the distinctiveness of the evolutionary trajectory of Neanderthals, and, while they do not rule out some Neanderthal ancestry for later Europeans, they certainly provide no evidence in favor of it.

Third, Neanderthals were by all evidence the only human inhabitants of Europe down to about 40 kya. But by 30 kya (or not long after), Neanderthals had disappeared from the human fossil record. In their place we find AMH differing from the Neanderthals in a broad range of cranial and postcranial characteristics.

Most workers today infer from this accumulated patterning that AMH in a broad sense replaced Neanderthals in Europe (e.g., Howell 1999; Churchill & Smith 2000; Klein 2001). It has become difficult to argue that Neanderthal populations simply evolved into AMH, and even the staunchest multiregional advocates do not now contest an important role for gene flow into Europe in the emergence of European AMH (e.g., Wolpoff et al. 2001). Indeed, Hawks & Wolpoff (2001) make the interesting statement that even if Neanderthals were replaced by AMH without gene flow (which they do not concede), the multiregional model would still be viable. The contrast between this position and the straightforward Neanderthal-AMH evolutionary continuity in Europe propounded three decades ago by Brose and Wolpoff (1971) is striking.

The evidence thus points to a very general answer to the question posed above—that the Neanderthals in Western Europe were replaced by AMH. But there is still much to learn (and much controversy) regarding the details. How and when replacement occurred, how the process varied regionally and locally, how it may be reflected in the archaeological record of human behavior, and how much Neanderthals contributed to the gene pool of Upper Paleolithic Western Europe are all open questions. In regard to the last question, for instance, some "Out of Africa" advocates (e.g., Tattersall & Schwartz 1999) are skeptical of any AMH-Neanderthal interbreeding across what they regard as a species boundary. Others point to occurrence of characteristic Neanderthal traits, both cranial (as in the Mladeč Aurignacian AMH fossils [Churchill & Smith 2000]) and postcranial (the distal limb proportions of the Lagar Velho child [Zilhão & Trinkaus, eds. 2002]), among some Upper Paleolithic AMH, as evidence for significant interbreeding. Even in these cases, though, it is usually accepted that Neanderthals were assimilated by AMH, making a "minority" genetic contribution to subsequent populations.

The Replacement of the Neanderthals and the Archaeological Record

How does the archaeological record of the period centering on 40–30 kya in Western Europe relate to the Neanderthals' fate? Like many others, I think that in this area the replacement is related to the Middle–Upper Paleolithic transition. This is a controversial and many-faceted issue, concerning which I will touch on three preliminary points.

First, some workers (e.g., Clark 1992, 1999; Straus 1995) discount the Middle–Upper Paleolithic transition in Europe as marking a major change in human adaptations or behavior. They point in particular to the lack of good evidence for any great change in subsistence practices across this transition, and argue that if there is any strong inflection in the trajectory of such cultural change in the Upper Pleistocene, it occurs rather around the Last Glacial Maximum (Enloe 2001). They note (as have I [Harrold 1989]) that inclusion of cultural developments from the later Upper Paleolithic into block comparisons of Middle and Upper Paleolithic "trait lists" has confused the issue of how the *early* Upper Paleolithic (EUP) is different from the Middle Paleolithic. And they correctly point out that characteristic Upper Paleolithic lithic techniques (e.g., blade technology) and types (e.g., burins) can be found at various times and places in the Middle Paleolithic.

However, the Middle–Upper Paleolithic transition still demands our attention in the context of this chapter, for two good reasons. First, it saw a clear shift in a number of nontrivial aspects of material culture. Blade technology and new forms of stone artifacts became widely, and to a great extent permanently, a part of the archaeological record. Bone, antler, and ivory entered the record as significant raw materials for artifacts. Items of adornment and even art were also unambiguously present, sometimes in considerable numbers, as well as musical instruments. Second, this transition was apparently linked in Western Europe with the replacement of the Neanderthals. So far, only Neanderthals are associated with Middle Paleolithic deposits. With the exception of the Châtelperronian (discussed below), only AMH remains are associated with Upper Paleolithic assemblages. This combination of a significant change in the archaeological record and these fossil-industrial associations is unlikely to be a coincidence.

A second preliminary point concerns the reality and significance of the industrial complexes or "cultural traditions" by means of which prehistorians have long organized the Paleolithic archaeological record. Straus (this volume) argues that these industries are merely the constructs of prehistorians, wrongly treated as actual entities of the past, and wrongly construed as representing the traces of identity-conscious ethnic groups in that past. In this vein, he particularly sees the Châtelperronian industry, the subject of much debate concerning the fate of the Neanderthals, as a dubious analytical entity. He notes the variability in frequency of occurrence in Châtelperronian assemblages of various classes of artifacts, including Châtelperron knives and other lithic types, bone tools, and ornaments.

Archaeological industries are indeed constructs. So are notions like "social class" and "Oxygen Isotope Stage 3." Science is impossible without constructs. Of course, constructs must be based on empirical evidence in order to be helpful to our understanding of it. In the Paleolithic record, the occurrence and frequency of morphologically different lithic (and other) artifacts is neither constant, continuous, nor random. The record instead tends to show clumps or modes, in which assemblages rather similar in artifact composition tend to have restricted occurrences in space and time. We group such assemblages into industries. I readily agree that the concept of the industry has been interpreted in ways (including as ethnic identity markers) which are misleading, but submit that if we discard large-scale groupings of similar assemblages we are losing information at analytical levels above that of the individual site or assemblage.

I operate here from a perspective similar to that of Sackett (1982) in his discussion of isochrestic style: There are different (i.e., culturally variable) ways to skin a cat, and the shared use of a particular practice may indicate interaction between groups. I do not think that Paleolithic industries represent identity-conscious ethnic groups; for one thing, their spatial and temporal scales are too great (Clark 1989. My interpretation is that they result from much larger networks or spheres of interacting local groups that broadly share a heritage of practices for making and using artifacts, and that may serve to diffuse new practices. Other such networks or spheres, in relative isolation, may develop and transmit over time rather different suites of artifactual practices. Because isolation between such networks could seldom be absolute, industries may be expected to have fuzzy boundaries. And because of all the factors, human and natural, conditioning the formation, preservation, and excavation of the minimal stratigraphic units from which we derive our assemblages, we should expect the assemblages assigned to a Paleolithic industry to be variable in the frequency of different artifact classes. Nor should we be surprised when the occasional "fossil director" of one industry turns up in an assemblage attributed to another. As I have detailed elsewhere (Harrold 1988, 1989), Châtelperronian assemblages are generally readily distinguishable from Mousterian or Aurignacian ones on the basis of the presence/absence and frequency ranges of certain artifact types (not simply Châtelperron points).

The third preliminary issue I would like to address is the controversial one of Neanderthal cognitive capacities. Proposals that Neanderthals did not have the same range of cognitive capabilities as humans today—for syntactical language and/or symbolism, innovativeness, abstract thinking, etc.—are sometimes greeted with charges of "Neanderthal-bashing" (as though any Neanderthals could somehow be vexed by scientific debates in the third millennium A.D.!), and "biological determinism" or "reductionism." Yet, as Klein (2001) notes, we accept that today's human cognitive capacities emerged gradually over time as hominid brains evolved. Given this emergence, at least *some* Pleistocene hominid taxa must have lacked the modern range of such capacities, and we do not know *a priori* which ones they were. Of course, it seems intuitively correct that more ancient and small-brained hominids (who left the simplest behavioral residues) are likelier to have been cognitively different from us than later and larger-brained ones. However, given evidence for the Neanderthals' unique evolutionary trajectory and some behavioral patterns different from those of AMH (see discussion below), it does not seem to me impermissible or unseemly to ask whether such a population may have behaved differently because of differing cognitive capacities. I regard this as an open question. Unfortunately, it is one that will be very hard to settle, especially given the problem of distinguishing between what people habitually do and what they are *capable* of. I suspect that charges of "paleoracism," to use Klein and Edgar's (2002) term, are related to anthropologists' well-founded sensitivity to the devastating legacy of racism (and its scientific misuse) in recent centuries. But the point remains that they contribute neither evidence nor logical argument to the issue at hand, and are not constructive.

If the replacement of the Neanderthals was related to the Middle–Upper Paleolithic transition in Western Europe, then how was it related? There are currently three principal models advanced, which may be called the population dispersal model, the mosaic model, and the indigenist models. They have been described elsewhere in some detail (Harrold 2000; Harrold & Otte 2001), and they, and their problematic aspects, will thus be only briefly characterized here.

The population dispersal model (e.g., Mellars 1998a, 1998b, 2000a, 2000b, 2005; Davies 2001) proposes that AMH entered Europe from the east and/or southeast in the mid-Last Glacial, replacing or absorbing Neanderthal populations. Proponents of the model differ on the degree of Neanderthal contribution to the later Western European gene pool. The model relates the beginning of the regional Upper Paleolithic to this dispersal, with the Aurignacian as an allochthonous industry associated with AMH. The penecontemporaneous Châtelperronian industry is associated with Neanderthals, and is explained as reflecting the impact of the newcomers and their cultures on indigenous Neanderthals, through acculturation or stimulus diffusion. The end of the Châtelperronian before 30 kya in France and northern Spain would thus indicate the extinction/absorption of the Neanderthals there. It is often suggested, though on less secure evidence and dating, that the Uluzzian and Szeletian industries elsewhere in Europe may provide parallels to the Châtelperronian. For an alternative explanation of the Szeletian, see Adams (this volume).

This model is consistent with a great deal of fossil, archaeological, dating, and paleoenvironmental evidence. It is not without problems, however. First, although all human remains associated with the Aurignacian in Western Europe are AMH, they date to the later Aurignacian, around 30 kya or later. Earlier Aurignacian occurrences simply do not yield diagnostic human remains (Churchill & Smith 2000; Henry-Gambier, Maureille & White 2004). Thus we cannot yet identify the people responsible for the early Aurignacian—although, given the associations we have, it is plausible to propose that they were AMH. Nonetheless, an important piece of the case for this model is missing.

There is also the fact that the Châtelperronian is distinguishable from the Aurignacian on both technological and typological grounds. If Neanderthals were being acculturated by AMH, it may be asked, would they not be expected to imitate Aurignacian artifacts, rather than to develop a distinctive Upper Paleolithic industry of their own? Clearly, even if the coming of AMH stimulated the development of the Châtelperronian, there is an original element in the Neanderthal response. On the other hand, White (2001) argues, *contra* d'Errico et al. (1998), that in the most novel aspect of EUP behavior, the manufacturing techniques of Châtelperronian ornaments parallel those of Aurignacian ones, and thus could have been borrowed from them.

According to an alternative, "mosaic" model, events in mid-Last Glacial Western Europe were more complex than just portrayed (e.g., Straus 1995, 1996, 1997, this volume; Clark 1997, 1999, 2001). As noted above, the very notion of readily identifiable industrial complexes is regarded as dubious. All regional Last Glacial populations are envisioned as sharing a largely common repertoire of artifact-production skills, so that industrial variation is primarily due to the demands and constraints of such factors as raw material type and transport, subsistence activities, and type and length of site occupation. Similarly, mosaic model proponents are skeptical that fossils of this period can be clearly sorted into Neanderthal and AMH categories, and feel that such designations oversimplify complex patterns of temporal and geographical variation. In this view, gene flow, *in situ* evolutionary change, invention, diffusion, and convergent trajectories of cultural change all had important roles in the gradual transformation of Middle Paleolithic Western Europe toward the realm of Upper Paleolithic cave artists.

Beyond noting that I find problematic the mosaic model's rejection of the archaeological units and designations of human fossils noted above, there is the further problem of a certain vagueness. Unlike the other two models discussed here, this one does not propose a distinct scenario: it is not specific about the nature and timing of the processes and changes it alludes to. We are presented with a sort of black box, humming with complex evolutionary and cultural processes of change and convergence, but we are unable to know just what is going on inside. With little specification of what happened, where, when, and how, or of why this was a time of cultural and biological ferment, the model in its current state offers limited enlightenment or opportunity for testing.

Finally, there is the "indigenist model" (e.g., d'Errico et al. 1998; Zilhão 2000, 2001, 2002; Zilhão & d'Errico, 2003a, 2003b; d'Errico 2003). It accepts the replacement of Neanderthals by AMH, but redates and reinterprets the Châtelperronian. In this view, all Châtelperronian assemblages antedate all Aurignacian ones, and thus AMH had nothing to do with Châtelperronian origins. Rather, Neanderthals independently developed an Upper Paleolithic of their own. AMH eventually did arrive to supplant the Neanderthals, but not until after a hiatus of several millennia along the "Ebro frontier" in Spain.

This model has received considerable criticism for a strained reinterpretation of chronometric dating and site formation processes at a number sites in its attempt to show that the Châtelperronian is always older than the Aurignacian and quite independent of it (e.g., White 1998, 2001; Mellars 2000a, 2000b, 2005; Mellars et al. 1999; Bernaldo de Quiros et al. 2001; Clark 2001; Harrold & Otte 2001; Conard, Dippon & Goldberg 2003). Model proponents point to the general pattern wherein Châtelperronian components always underlie Aurignacian ones (save for claimed interstratification at three sites) in support of their claim. However, if the Châtelperronian indeed resulted from the reaction of Neanderthals to the arrival of expanding AMH, this stratigraphic succession at any given site is only to be expected. Suffice it here to say that while the interstratification of the industries at three sites has been cast in doubt (d'Errico et al. 1998; Rigaud 2001; Bordes 2003), the pattern of chronometric dates—radiocarbon, AMS, ESR, TL—from many sites (including Geissenklösterle and others in southwestern Germany [Richter et al. 2000; Conard, Dippon & Goldberg 2003; Conard & Bolus 2003] and Arcy-sur-Cure [David et al. 2001]), climatostratigraphic frameworks, and recent magnetic susceptibility work (Harrold et al. 2003) all indicate a significant temporal overlap between the two (though coexistence in any particular locality may have been brief [Hublin 2000]), and an origin of the Western European Aurignacian during the Hengelo Interstadial at close to 40 kya.

There are questions raised by the model but not adequately addressed by its proponents. Why, after tens of millennia of apparently well-adapted cultural stasis, did Neanderthals undergo a "revolution" to an Upper Paleolithic that just happened to share many features with the soon-to-arrive Aurignacian (Hublin 2000)? And then, having demonstrated their parity in both cultural capacity and achievement with AMH, why did Neanderthals relinquish Western Europe to them? Zilhão (2001, 2002) suggests briefly that due to their tropical African origin, AMH would have had higher fertility than Neanderthals, and thus would have outbred them. Whether AMH fertility would still have been higher as populations moved from tropical Africa to glacial-age Europe is at best questionable; certainly, such an argument needs to be better developed and defended.

These three models do not exhaust the spectrum of views on the fate of the Neanderthals; for instance, Finlayson (2004) sees the extinction of Neanderthals and spread of AMH as unrelated events caused by the climatic fluctuations of OIS 3. My sense is that the population dispersal model presently best explains the totality of available data, though further data and analysis may suggest important modifications, or even a considerably different model.

What Was the Advantage?

If AMH replaced Neanderthals in Western Europe, there must have been one or more factors acting as a selective advantage. After all, Neanderthal populations had persisted for tens of millennia in a range that arguably included the climatically harshest regions inhabited by populations of *Homo*. They were presumably adapted to a range of the continent's environments. However, Zubrow's (1989) simulation suggests that even a modest demographic advantage of AMH over Neanderthals could eventually have resulted in the extinction of the latter group. Various hypothesized AMH advantages have been suggested, including higher birth rates, higher infant and child survivorship, more efficient long-range locomotion, lower caloric needs due to lower body mass, more effective hunting methods, superior technology, and enhanced cognitive capacities (or some combination thereof). There is also the suggestion that an important or even decisive role may have been played by the volatile climatic conditions of Oxygen Isotope Stage 3 (Mellars 1998a, 2004; Stringer et al. 2003; d'Errico & Sanchez Goñi 2003). Some of these suggestions are better founded or more potentially testable than others, but it is fair to say that none has been systematically tested.

The putative advantages of AMH over Neanderthals are often summed up under the heading of "modern behavior." There is no standard, well-developed definition of this phrase (the most systematic treatment is in McBrearty and Brooks 2000; see also d'Errico et al. 2003). It is typically meant to denote behavioral characteristics found among recent hunter-gatherers and believed to set them off from archaic humans, variously including syntactic language, abstract thought, planning time depth, symbolism and complex ideology, technological, social and ideological innovativeness, and efficient hunting based on sophisticated understanding of prey behavior. In the absence of an accepted theory and definition of "modern-ness" in behavior, however, I believe it preferable to look at discrete characteristics of the archaeological record, those likeliest to monitor any change over the Middle–Upper Paleolithic transition.

Of course, we do not monitor Paleolithic behavior directly, but infer it from various kinds of archaeological evidence. I will explore three classes of data here: those relating to subsistence, technology, and "durable symbols," to use the phrase of Kuhn and Stiner (2001)—beads, pendants, and other nonutilitarian items that can serve to symbolize individual and group identity. I will then discuss some ideas about why we find the patterning that we do in these categories.

Subsistence: Mellars' (1973) list of characteristics differentiating the Upper from the Middle Paleolithic, repeated and amplified many times in the literature since its original articulation, included the specialized hunting of particular species. Such efficient methods might indeed give one population a competitive advantage over another, especially if, as Binford (1989) has argued, Neanderthals were ineffective hunters who relied heavily on scavenging. However, a sizable body of zooarchaeological research indicates that Middle Paleolithic people must have often been effective hunters capable of bagging prime-age individuals of medium and large prey species, including deer, horse, and bovines (e.g., Chase 1989; Farizy & David 1992; Stiner 1994). Furthermore, there is as yet no demonstration that EUP hunters in Western Europe practiced single-species specialization or other practices that would have conferred on them a competitive advantage (e.g., Bernaldo de Quirós et al. 2001; Enloe 2001; Grayson & Delpech 2002). Neanderthal hunting effectiveness is further suggested by stable-isotope analyses of collagen from several Neanderthal fossils indicating a diet heavily dominated by meat (as we see in modern hunters from cold climates) (Richards et al. 2000, 2001; Bocherens et al. 2001). This evidence indicates not only that there were episodes of Neanderthals successfully exploiting sizable game, but that these episodes were frequent enough to dominate their diet.

With the bone points of the early Upper Paleolithic (EUP), we see the beginning of a long evolution of projectile point technology, sometimes organic and sometimes lithic, that carries right through the Upper Paleolithic (Knecht, ed., 1997). Since organic armatures tend to be associated with thrown (rather than thrusting) weapons, and pursuit of small and medium rather than the largest game (Ellis 1997; Villa & d'Errico 2001), this might imply a shift in both game pursued and methods of killing. In a related vein, Stiner et al. (2000) have suggested that resource diversification, at least in the form of some level of bird hunting, seems to characterize a number of EUP faunas. Perhaps linked to such a hypothetical shift is the suggestion by Trinkaus et al. (1999) that lower-limb diaphyseal morphology of the EUP Saint-Césaire Neanderthal is consistent with a higher degree of characteristic mobility than is usual for Neanderthals, and similar to that seen in Upper Paleolithic AMH skeletons.

While there are thus intriguing possibilities worth pursuing, it is nonetheless fair to say that no major shift in patterns of human subsistence is currently documented across the Middle–Upper Paleolithic transition. There is little evidence to support either of two contentions related to such a putative shift: (1) that Neanderthals were so ineffective as hunters that they had to rely heavily on scavenging, or (2) that the early Upper Paleolithic saw the widespread adoption of hunting practices specializing on single species.

Technology: The subject of artifactual change over the Middle–Upper Paleolithic transition is a dense and complex one, and I will limit myself to a few observations concerning it. There is first the contentious issue of whether, compared with Upper Paleolithic stone assemblages, Middle Paleolithic ones are simpler or more "monotonous." Kuhn and Stiner (2001) have usefully clarified this issue. They note that *chaîne opératoire* studies have documented considerable variability in the technological practices used by Middle Paleolithic people to prepare and reduce cores to obtain desired blanks. These practices show awareness of the flaking characteristics of different varieties of available raw material. They also indicate economizing behavior balancing such factors as lithic abundance and source locations with anticipated future needs. However, Kuhn and Stiner note that if attention moves from technology to the morphology of retouched artifacts, the characterization of monotony (compared with the Upper Paleolithic) becomes more valid. Scrapers, points, notches, and denticulates in varying proportions dominate assemblages over great ranges of time and space. In Western Europe, even the assemblages of the very earliest Upper Paleolithic (Châtelperronian and Aurignacian) contrast clearly with the Middle Paleolithic in this regard (Harrold 1989). There is additionally the factor of greater innovation and industrial turnover over time in the EUP.

A further strong and oft-noted contrast between Middle and Upper Paleolithic concerns the proliferation of artifacts of bone, antler and ivory in the latter period. Again, it is well documented that such items occur in the Châtelperronian and the early Aurignacian (Harrold 1988, 1989; White 2001), but a recent survey by Villa and d'Errico (2001) found no confirmed examples of bone or ivory points from the Lower or Middle Paleolithic in Europe. It seems that artifacts of this sort were simply not part of the standard technological repertoire of people in Middle Paleolithic Western Europe.

"Durable symbols:" As with bone artifacts above, it is clear that items such as beads, pendants, decorated tools, and art objects such as figurines are effectively absent from the Middle Paleolithic, while they occur in the Châtelperronian and early Aurignacian (occasionally in significant numbers). We commonly interpret such items as symbols of individual and/or group identity, related to more complex social relations within and between human groups. These items bespeak symbolism, creativity, and ideological systems. Some, at least (e.g., the famous theriomorphic figurine from the Aurignacian of Höhlenstadel), imply the incorporation of a supernatural dimension into the symbolic world. Such items also imply complex, syntactical language such as humans speak today.

A source of considerable dispute is whether the near- or total absence of such artifacts in the Middle Paleolithic implies the lack on the part of Neanderthals of any of the cognitive capacities that the artifacts imply. There is the complicating factor that Neanderthals at least occasionally buried their dead, and used coloring agents like ocher and manganese (d'Errico & Soressi 2002). These are kinds of behavior that have symbolic dimensions among ethnographically known hunter-gatherers, and so it is plausible that they also did in the Middle Paleolithic. On the other hand, utilitarian explanations for such behavior can be suggested (e.g., the need to dispose of corpses, and the use of ocher in treating skins). A definitive test to settle this question is hard to envision. At the very least, though, one can say that the Middle–Upper Paleolithic transition sees a dramatic increase in the durable evidence for symbolic behavior.

The foregoing brief overview of three categories of evidence indicates no strong case yet for a big change in what people ate or how they obtained it over the Middle–Upper Paleolithic transition, but distinct shifts in toolmaking behavior and production of durable symbols. How might these patterns relate to the replacement of the Neanderthals? I believe that suggestions by Kuhn and Stiner (1998, 2001) may help illuminate this connection. They observe that the aforementioned morphological "monotony" of Neanderthal stone tools contrasts strongly with a well-established pattern found in both the ethnographic record of recent hunter-gatherers and the archaeological record of the late Upper Paleolithic. As latitude increases, the frequency and elaboration of plant-processing equipment tends to decrease, while the elaboration of technology for procuring and processing animal food increases (Torrence 1983; Binford 2001:387–389). This pattern reflects the decreased availability of edible plants as latitude increases, and the consequent necessity to use technology to procure higher proportions of animal protein in the diet. Some of this technology is used to procure small, often hard-to-catch creatures.

As Kuhn and Stiner note, Middle Paleolithic technology and subsistence do not seem to show this pattern. Neanderthals (in the Near East and Europe) left behind precious little evidence of plant processing, and used relatively simple artifacts (probably wooden spears with stone points) to procure animals, regardless of latitude. The elaboration of bone and antler technology related to animal procurement among later people was absent in higher latitudes during the Middle Paleolithic. Furthermore, archaeofaunas indicate that hunting concentrated on prime-age adult animals of medium to large ungulate species. Neanderthals were successful hunters, but "...a different kind of hunter-gatherer from any presently known" (Kuhn & Stiner 2001:129).

Kuhn and Stiner contend that the likeliest explanation for this pattern is that Neanderthals lived at much lower population densities than either modern or late Upper Paleolithic peoples. Low Neanderthal population density is consistent with a considerable range of evidence, such as the relatively low numbers and extent of Middle Paleolithic sites per unit time (e.g., Mellars 1973), and faunal evidence of low Middle Paleolithic predation pressure on easily collected, sessile animals such as tortoises (Stiner et al. 2000). It is also consistent with the frequent presence of cave bear, wolf, and hyena remains in Middle Paleolithic sites (in contrast to the usual Upper Paleolithic pattern), presumably due to small human groups and infrequent visits (Straus 1982; Stiner 1994). In a related vein, Sorensen and Leonard (2001) have argued that, with their relatively massive and energetically expensive bodies, Neanderthals could have been either highly active or poor foragers, but not both. Perhaps they could have been effective hunters only when exerting light pressure on populations of their high-ranked prey—medium and large ungulates. Because high-ranked foods tend to be the most susceptible to over-hunting, Neanderthals might have needed to maintain low population densities to sustain their characteristic adaptations.

This picture would fit the patterning in three categories of evidence that we isolated above. Thus we would expect Middle Paleolithic faunas to contain evidence of successful hunting of a range of medium and large ungulates. In a landscape far emptier of people than in later times, as long as Neanderthals could procure enough larger animals to feed their small groups with their existing technology, they would not need to elaborate that technology, either to catch more of the larger animals, or to begin to exploit smaller, hard-to-catch ones. At the same time, very small co-resident groups and the rarity of encounter with other groups would not encourage the durable symbolic marking of individual or group identity. Furthermore, small groups and limited interactions between adjacent groups would provide only a limited social payoff for innovation, encouraging technological stasis (Kuhn & Stiner 2001), and discouraging the use of artifacts in other contexts (i.e., symbolic ones) than the purely utilitarian (Kuhn & Stiner 1998).

Kuhn and Stiner (2001) do not feel that dating and other evidence from the early Upper Paleolithic allow determination of whether the record of the early Upper Paleolithic parallels that of the LUP and contrasts with that of the Middle Paleolithic, though they suspect that it does. I would argue that the appearance of durable symbols, utilitarian organic artifacts (including points), and more complex lithic assemblages point strongly in that direction. If so, then the Middle–Upper Paleolithic transition marks a significant inflection point in the archaeological record. As argued above, this inflection would coincide with the process of the replacement or absorption of the Neanderthals. Neanderthals, very thin on the ground, would have been replaced by presumably more numerous AMH. The AMH advantage is usually explained in one of two general ways. The first, most explicitly championed by Klein (2001), posits an enhanced cognitive capability on the part of AMH—specifically, the capacity for phonemic, syntactical language, which underlies the "extraordinary modern human ability to innovate" (Klein 2001:121). Klein sees this ability as due to mutation(s) affecting neural functioning occurring among early AMH in Africa before the spread of AMH around the world and the replacement of archaic populations in other regions. This proposal has attracted criticism because it is, as Klein acknowledges, untestable by any methods currently known. A similar view (Wynn & Coolidge 2003; Coolidge & Wynn 2004) posits subtle but important neuropsychological differences from AMH resulting in Neanderthals having less working memory and more routinized, less symbolically charged behavior. Mithen's (1996) suggestion of compartmentalized Neanderthal minds covers similar ground.

The second commonly offered explanation posits a "social revolution" rather than a cognitive one (e.g., White 1982; Hayden 1993). In the context of replacement, this would entail more complex social organization among AMH than among Neanderthals, of the sort that permitted and selected for such Upper Paleolithic traits as durable symbols of individual and group identity. Klein (2001) replies that this approach has its own problems: why was AMH social organization more complex? If it is a result of increased population density, then why did that happen?

The Châtelperronian, as an EUP industry attributed to Neanderthals, is of obvious relevance to the question of cognitive differences. The indigenist model sees in the Châtelperronian a definitive refutation of cognitive difference, since it posits this industry's development by Neanderthals well before, and thus independent of, the arrival of the Aurignacian. As noted above, however, I find this early dating of the industry untenable. If the Châtelperronian and Aurignacian are penecontemporaneous, as I contend, then the Châtelperronian's "Upper Paleolithic" characteristics could be due to contact with AMH, followed by diffusion and imitation. In this connection, White (2001), *contra* d'Errico et al. (1998), notes that all of the techniques of manufacture and decoration of items of adornment recovered from the Châtelperronian can be found in the Aurignacian (though sometimes in different frequencies), and argues that all could result from contact rather than independent invention. There is also the possibility that at least some of the novel traits of the Upper Paleolithic (especially the durable symbols) could have been triggered among both AMH and Neanderthals as members of both populations encountered each other and began to deal with a more crowded and complex human landscape (d'Errico et al. 1998:S36; d'Errico 2003; Otte 2001). In this vein, Bordes (2002) has made the interesting suggestion that technologically and morphologically unusual Dufour bladelets from the Aurignacian at Le Piage could reflect Châtelperronian influence.

But in any event, the Châtelperronian implies that Neanderthals were capable of at least some kinds of Upper Paleolithic technological and social behavior. To Klein (2001:123), the Châtelperronian for this reason presents the most serious problem besetting his version of the "Out of Africa 2" replacement model. Granted, the vast bulk of Châtelperronian items of adornment, bone and antler tools, the mammoth-tusk "huts" and mass quantities of ocher, derive from only one or two sites (and arguably, from relatively late contexts). But early Aurignacian sites also often fail to yield such artifacts and features. It remains that, after the arrival of AMH in Western Europe, Neanderthals living in what was perhaps the most densely populated area of their range were leaving behind archaeological traces that in important respects resembled those of AMH more than those of their Middle Paleolithic forebears.

Yet, the Châtelperronian and the Neanderthals in this area eventually disappeared as surely as elsewhere in Western Europe—indeed, apparently several millennia before those in southern Iberia (Zilhão 2000; but see Klein 2001:123–124). We are thus back to the old question: the selective advantage(s) of AMH over Neanderthals. Unfortunately, almost none of the alternatives can be logically excluded. We still cannot rule out cognitive differences between AMH and Neanderthals: on the one hand, the Châtelperronian–Neanderthal association makes a vast cognitive gulf between the two populations appear unlikely. On the other hand, the Châtelperronian and the Neanderthals disappeared in due course, and ironically, it was Neanderthals in farther Iberia who seem to have persisted the longest, despite no apparent change in their Middle Paleolithic ways, and perhaps only because of their relative isolation (Hublin 2000). There is the curious lack of elaborated technology in the Middle Paleolithic of higher latitudes noted by Kuhn and Stiner, which may be related to low population density—but why the low density? Physiological factors may be important here, some of which, like more massive (and energetically more expensive) Neanderthal bodies, are known, and others, like differential fertility, are currently speculative. There is also the climatic volatility of OIS 3 (Mellars 1998a), which may have been a necessary—if not sufficient—factor in Neanderthal extinction. Neanderthals had been through periods of widespread climatic volatility before, but not in the face of AMH competition. Thus currently, it can be variably argued that AMH replaced Neanderthals because they were, so to speak, smarter; or because they were culturally equipped with more successful adaptive behavior; or because they had lower body mass or other physiological advantage—or perhaps because of some combination of these factors, possibly including an assist from climatic stresses. A first step toward clarifying this issue would be a fuller and better-dated picture of the day-to-day and year-to-year adaptations and behavior of AMH and Neanderthals in the mid-Last Glacial.

A comparison between the "Upper Paleolithic Revolution" and another momentous shift, the "Neolithic Revolution," is instructive here (Bar-Yosef 1998, 2002). The rise of the Neolithic is linked by Bar-Yosef to adaptive responses to a final Pleistocene climatic crisis. He regards the rise of the Upper Paleolithic, like that of the Neolithic, as an adaptive revolution. Because of improved behavioral and technological adaptations, populations grew, and groups and practices spread from an area of origin in Southwest Asia. Bar-Yosef argues that, because it originated among some (but not all) AMH populations in this region, this was a cultural, not a cognitive revolution. However, as AMH spread to encounter European Neanderthals, the possibility of contact between cognitively different human groups logically appears. The crystallization and subsequent spread of the Neolithic can be traced in some detail. Had the Neolithic begun in OIS 3, I suggest that we would have had much more difficulty in pinpointing its time, place, associated lifeways, spread, and climatic correlates, than we have had, due to the limitations—contamination, calibration, error ranges, and site-specific assumptions—of available dating methods and the associated archaeological record. Our understanding of the process would be a good deal more rudimentary than it is, and I daresay that there would be considerable controversy over points that actually are regarded as settled.

This example underlines the great difficulties we face in seeking to wring from an intractable archaeological record, dating back over 30 kya, highly specific information about the factors conditioning the fate of the Neanderthals. Paleoanthropology is unavoidably a discovery-driven science, and sometimes we cannot lay our hands on the proper data to test a hypothesis. Sometimes, as Klein and Edgar (2002) have observed, we must, as in a legal proceeding, put forth a judgment on available data, as limited as they may be. However, unlike a jury, we can emphasize that our judgments are tentative, modify them in the light of new data, and work to maximize the chances of collecting further data of the right kind to test them. I think that the large-scale archaeological record points to the extinction/absorption of the Neanderthals over several millennia in Western Europe during the period of roughly 40–30 kya. That incoming AMH populations had some selective advantage(s) on their side in their confrontation with Neanderthals is an implication of this conclusion. Just what those advantages were is unclear, and solving this problem will challenge our abilities to reliably infer behavioral patterns from archaeological remains thirty to forty millennia old.

References

Arsuaga, J. L., I. Martinez, A. Gracia & C. Lorenzano
1997 The Sima de los Huesos crania (Sierra de Atapuerca, Spain): A comparative study. *Journal of Human Evolution* 33:219–281.

Bar-Yosef, O.
1998 On the nature of transitions: the middle to upper Palaeolithic and the Neolithic revolution. *Cambridge Archaeological Journal* 8:141–63.
2002 The Upper Paleolithic Revolution. *Annual Review of Anthropology* 31:363–393.

Bernaldo de Quirós, F., V Cabrera, M. Lloret, & A. Pike-Tay
2001 New kids on the block? Some comments on the Middle–Upper Paleolithic transition in Cantabrian Spain. In *Questioning the Answers: Resolving Fundamental Problems of the Early Upper Paleolithic* (M. Hays & P. Thacker, eds.), pp. 27–38. BAR International Series 1005. Archaeopress, Oxford.

Binford, L.
1989 Isolating the transition to cultural adaptations: An organizational approach. In *The Emergence of Modern Humans: Biocultural adaptations in the later Pleistocene* (E. Trinkaus, ed.), pp. 18–41. Cambridge University Press, Cambridge.
2001 *Constructing Frames of Reference: An Analytical Method for Archaeological Theory Building Using Hunter-Gatherer and Environmental Data Sets.* University of California Press, Berkeley.

Bocherens, H., D. Billiou, A. Mariotti, M. Toussaint, M. Patou-Mathis, D. Monjean, & M. Otte
2001 New isotopic evidence for the dietary habits of Neandertals from Belgium. *Journal of Human Evolution* 40:497–505.

Bordes, J.-G.
2002 Châtelperronian/Aurignacian interstratification at Roc de Combe and Le Piage: Lithic Taphonomy and archaeological implications. *Journal of Human Evolution* 42:A7–A8.
2003 Lithic taphonomy of the Châtelperronian/Aurignacian interstratifications in Roc de Combe and Le Piage (Lot, France). In *The Chronology of the Aurignacian and of the Transitional Technocomplexes: Dating, Stratigraphies, and Cultural Implications* (J. Zilhão & F. d'Errico, eds.), pp. 223–244. *Trabalhos de Arqueologia* 33. Instituto Português de Arqueologia, Lisbon.

Brose, D. S., & M. H. Wolpoff
1971 Early upper paleolithic man and late middle paleolithic tools. *American Anthropologist* 73:1156–94.

Chase, P. G.
1989 How different was Middle Palaeolithic subsistence? A zooarchaeological perspective on the Middle to Upper Palaeolithic transition. In *The Human Revolution: Behavioural and Biological Perspectives in the Origins of Modern Humans* (P. Mellars & C. Stringer, eds.), pp. 321–37. Edinburgh University Press, Edinburgh.

Churchill, S. E.
1998 Cold adaptation, heterochrony, and Neandertals. *Evolutionary Anthropology* 7:46–61.

Churchill, S. E., & F. H. Smith
2000 Makers of the Early Aurignacian of Europe. *Yearbook of Physical Anthropology* 43:61–115.

Clark, G. A.
1989 Romancing the stones: biases, style and lithics at La Riera. In *Alternative Approaches to Lithic Analysis* (D. Henry & G. Odell, eds.), pp. 27–50. Archeological Papers of the American Anthropological Association No. 1, Washington, D. C.
1992 Continuity or replacement? Putting modern human origins into an evolutionary context. In *The Middle Paleolithic: Adaptation, Behavior, and Variability* (H. Dibble & P. Mellars, eds.), pp. 183–206. Philadelphia: University Museum of the University of Pennsylvania.
1997 The Middle–Upper Paleolithic transition in Europe: An American perspective. *Norwegian Archaeological Review* 30:25–53.
1999 Highly visible, curiously intangible. *Science* 283:2029–32.
2001 Discussion: The logic of inference in transition research. In *Questioning the Answers: Re-solving Fundamental Problems of the Early Upper Paleolithic* (M. Hays & P. Thacker, eds.), pp. 39–47. BAR. International Series, 1005. Archaeopress, Oxford.

Conard, N. J., & M. Bolus
2003 Radiocarbon dating the appearance of modern humans and timing of cultural innovations in

Europe: new results and new challenges. *Journal of Human Evolution* 44:331–371.

Conard, N. J., G. Dippon, & P. Goldberg
2003 Chronostratigraphy and archaeological context of the Aurignacian deposits at Geissenklösterle. In *The Chronology of the Aurignacian and of the Transitional Technocomplexes: Dating, Stratigraphies, and Cultural Implications* (J. Zilhão & F. d'Errico, eds.), pp. 165–176. *Trabalhos de Arqueologia* 33. Instituto Português de Arqueologia, Lisbon.

Coolidge, F., & T. Wynn
2004 A cognitive and neuropsychological perspective on the Chatelperronian. *Journal of Anthropological Research* 60:55–73.

David, F., N. Connet, M. Girard, V. l'Homme, J.-C. Miskovsky, & A. Roblin-Jouvé
2001 Le Châtelperronien de la grotte du Renne à Arcy-sur-Cure (Yonne). *Bulletin de la Société Préhistorique Française* 98:207–230.

Davies, W.
2001 A very model of a modern human industry: new perspectives on the origins and spread of the Aurignacian in Europe. *Proceedings of the Prehistoric Society* 67:195–217.

d'Errico, F.
2003 The invisible frontier: a multiple species model for the origin of behavioral modernity. *Evolutionary Anthropology* 12:188–202.

d'Errico, F., C. Henshilwood, G. Lawson, M. Verhaeren, A.-M. Tillier, M. Soressi, F. Bresson, B. Maureille, A. Nowell, J. Lakarra, L. Blackwell, & M. Julien
2003 Archaeological evidence for the emergence of language, symbolism, and music—an alternative multidisciplinary perspective. *Journal of World Prehistory* 17:1–70.

d'Errico, F., & M. F. Sanchez Goñi
2003 Neandertal extinction and the millenial scale climatic variability of OIS 3. *Quaternary Science Reviews* 22:769–788.

d'Errico, F., & M. Soressi
2002 Systematic use of manganese pigment by Pech-de l'Azé Neandertals: Implications for the origin of behavioral modernity. *Journal of Human Evolution* 42:A13.

d'Errico, F., J. Zilhão, M. Julien, D. Baffier, & J. Pelegrin
1998 Neandertal acculturation in Western Europe? A critical review of the evidence and its interpretation. *Current Anthropology* 39:S1–S44.

Ellis, C. J.
1997 Factors influencing the use of stone projectile tips: An ethnographic perspective. In *Projectile Technology* (H. Knecht, ed.), pp. 37–78. Plenum Press, New York.

Enloe, J. G.
2001 Investigating human adaptations in the Early Upper Paleolithic. In *Questioning the Answers: Re-solving Fundamental Problems of the Early Upper Paleolithic* (M. Hays & P. Thacker, eds.), pp. 21–26. BAR International Series, 1005. Archaeopress, Oxford.

Farizy, C., & F. David
1992 Subsistence and behavior patterns of some Middle Paleolithic local groups. In *The Middle Paleolithic: Adaptation, Behavior, and Variability* (H. Dibble & P. Mellars, eds.), pp. 87–96. University Museum of the University of Pennsylvania, Philadelphia.

Finlayson, C.
2004 *Neanderthals and Modern Humans: An Ecological and Evolutionary Perspective.* Cambridge University Press, Cambridge.

Forster, P.
2004 Ice Ages and the mitochondrial DNA chronology of human dispersals: a review. *Philosophical Transactions of the Royal Society of London,* B, 359:255–264.

Grayson, D. K., & F. Delpech
2002 Specialized Early Upper Paleolithic hunters in Southwestern France? *Journal of Archaeological Science* 29:1439–1449.

Harrold, F. B.
1988 The Châtelperronian and the Early Aurignacian in France. In *The Early Upper Paleolithic: Evidence from Europe and the Near East* (J. Hoffecker & C. Wolf, eds.), pp. 157–191. BAR International Series, No. 437, Oxford.
1989 Mousterian, Châtelperronian and early Aurignacian in western Europe: Continuity or discontinuity? *In The Human Revolution: Behavioural and Biological Perspectives on the Origins of Modern Humans* (P. Mellars & C. Stringer, eds.), pp. 677–713. Edinburgh University Press, Edinburgh.
2000 The Châtelperronian in historical context. *Journal of Anthropological Research* 56:59–75.

Harrold, F. B., B. B. Ellwood, P. T. Thacker, & S. L. Benoist
2003 Magnetic susceptibility analysis of sediments at the Middle–Upper Paleolithic transition for two cave sites in northern Spain. In *The Chronology of the Aurignacian and of the Transitional Technocomplexes: Dating, Stratigraphies, and Cultural Implications* (J. Zilhão & F. d'Errico, eds.), pp. 301–310. *Trabalhos de Arqueologia* 33. Instituto Português de Arqueologia, Lisbon.

Harrold, F. B., & M. Otte
2001 Time, space, and cultural process in the European Middle–Upper Paleolithic transition. In *Questioning the Answers: Re-solving Fundamental Problems of the Early Upper Paleolithic* (M. Hays & P. Thacker, eds.), pp. 3–11. BAR International Series, 1005. Archaeopress, Oxford.

Hawks, J., & M. H. Wolpoff
2001 Paleoanthropology and the population genetics of ancient genes. *American Journal of Physical Anthropology* 114:269–72.

Hayden, B.
1993 The cultural capacities of Neandertals: A review and re-evaluation. *Journal of Human Evolution* 24:113–46.

Henry-Gambier, D., B. Maureille, & R. White
2004 Vestiges humains des niveaux de l'Aurignacien ancient du site de Brassempouy (Landes). *Bulletins et Mémoires de la Société d'Anthropologie de Paris,* n.s., 161:49–87.

Holliday, T. W.
1997 Postcranial evidence of cold adaptation in European Neandertals. *American Journal of Physical Anthropology* 104:245–58.

Howell, F. C.
1999 Paleo-demes, species clades, and extinctions in the Pleistocene hominin record. *Journal of Anthropological Research* 55:191–243.

Hublin, J.-J.
2000 Modern-Neanderthal hominid interactions: A Mediterranean perspective. In *The Geography of Neanderthals and Modern Humans in the Greater Mediterranean* (O. Bar-Yosef & D,\. Pilbeam, eds.), pp. 157–182. Cambridge: Peabody Museum of Archaeology and Ethnography. Peabody Museum Bulletin 8.

Klein, R. G.
2001 Fully modern humans. In *Archaeology at the Millennium: A Sourcebook* (G. Feinman & T. Price, eds.), pp. 109–135. Kluwer Academic/ Plenum Publishers, New York.

Klein, R. G., & B. Edgar
2002 *The Dawn of Human Culture.* New York: Wiley.

Knecht, H. (editor)
1997 *Projectile Technology.* Plenum Press, New York.

Krings, M., A. Stone, R. W. Schmitz, H. Krainitzki, M. Stoneking, & S. Pääbo
1997 Neandertal DNA sequences and the origin of modern humans. *Cell* 90:19–30.

Krings, M., H. Geisert, R. W. Schmitz, H. Krainitzki, & S. Pääbo
1999 DNA sequence of the mitochondrial hypervariable region II from the Neandertal type specimen. *Proceedings of the National Academy of Sciences* 96:5581–85.

Krings, M., C. Capelli, F. Tschentscher, H. Geisert, S. Meyer, A. von Haeseler, K. Grossschmidt, G. Possnert, M. Paunovič, & S. Pääbo
2000 A view of Neandertal genetic diversity. *Nature Genetics* 26:144–46.

Kuhn, S. L., & M. C. Stiner
1998 Middle Palaeolithic 'creativity:' Reflections on an oxymoron? In *Creativity in Human Evolution and Prehistory* (S. Mithen, ed.), pp. 143–164. Routledge, London.
2001 The antiquity of hunter-gatherers. In *Hunter-Gatherers: An Interdisciplinary Perspective* (C. Panter-Brick, R. Layton & P. Rowley-Conwy, eds.), pp. 99–142. Cambridge University Press, Cambridge.

McBrearty, S., & A. S. Brooks
2000 The revolution that wasn't: A new interpretation of the origin of modern human behavior. *Journal of Human Evolution* 39:453–563.

Mellars, P.
1973 The character of the Middle–Upper Palaeolithic transition in south-west France. In *The Explanation of Culture Change* (C. Renfrew, ed.), pp. 255–276. Duckworth, London.
1998a. The impact of climatic changes on the demography of late Neanderthal and early anatomically modern populations in Europe. In *Neandertals and Modern Humans in Western Asia* (T. Akazawa, K. Aoki & O. Bar-Yosef , eds.), pp. 493–508. Plenum, New York.
1998b Modern humans, language, and the 'Symbolic Explosion'. *Cambridge Archaeological Journal*

8:88–94.

2000a The archaeological records of the Neandertal–modern human transition in France. In *The Geography of Neandertals and Modern Humans in Europe and the Greater Mediterranean* (O. Bar-Yosef & D. Pilbeam, eds.), pp. 35–48. Cambridge, MA: Peabody Museum of Archaeology and Ethnography, Peabody Museum Bulletin 8.

2000b Châtelperronian chronology and the case for Neanderthal/modern human 'acculturation' in Western Europe. In *Neanderthals on the Edge* (C. B. Stringer, R. N. E. Barton & J. C. Finlayson, eds.), pp. 33–39. Oxbow Books, Oxford.

2004 Stage 3 climate and the Upper Palaeolithic revolution in Europe: evolutionary perspectives. In *Explaining Social Change: Studies in Honor of Colin Renfrew* (J. Cherry, C. Scarre & S. Shennan, eds.), pp. 27–43. Cambridge: McDonald Institute for Archaeological Research.

2005 The impossible coincidence. A single-species model for the origins of modern human behavior in Europe. *Evolutionary Anthropology* 14: 12–27.

Mellars, P., M., Otte, L. G. Straus, J. Zilhão, & F. d'Errico

1999 CA Forum on Theory in Anthropology: The Neanderthal problem, continued. *Current Anthropology* 40:341–364.

Mithen, S.

1996 *The Prehistory of the Mind: The Cognitive Origins of Art and Science.* London: Thames and Hudson.

Otte, M.

2001 Contribution moustérienne au Paléolithique supérieur. In *Les Premiers Hommes Modernes de la Péninsule Ibérique* (J. Zilhão, T. Aubry & A. Carvalho, eds.), pp. 9–24. *Trabalhos de Arqueologia* 17. Instituto Português de Arqueologia, Lisbon.

Richards, M. P., P. B. Pettitt, E. Trinkaus, F. H. Smith, M. Paunovič, I. Karavanič

2000 Neanderthal diet at Vindija and Neanderthal predation: The evidence from stable isotopes. *Proceedings of the National Academy of Sciences* 98:7663–7666.

Richards, M. P., P. B. Pettitt, M. C. Stiner, & E. Trinkaus

2001 Stable isotope evidence for increasing dietary breadth in the European mid-Upper Paleolithic. *Proceedings of the National Academy of Sciences* 98:6528–6532.

Richter, D., J. Waiblinger, W. J. Rink, & G. A.Wagner

2000 Thermoluminescence, electron spin resonance and ^{14}C-Dating of the Later Middle and Early Upper Palaeolithic site of Geissenklösterle Cave in Southern Germany. *Journal of Archaeological Science* 27:71–89.

Rigaud, J.-Ph.

2001 A propos de la contemporanéité du Castelperronien et de l'Aurignacien ancien dans le nord-est de l'Aquitaine: Une révision des données et ses implications. In *Les Premiers Hommes Modernes de la Péninsule Ibérique* (J. Zilhão, T. Aubry & A. Carvalho, eds.), pp. 61–68. *Trabalhos de Arqueologia* 17. Instituto Português de Arqueologia, Lisbon.

Sackett, J. R.

1982 Approaches to style in lithic archaeology. *Journal of Anthropological Archaeology* 1:59–112.

Smith, S. L., & F. B. Harrold

1997 A paradigm's worth of difference? Understanding the impasse over modern human origins. *Yearbook of Physical Anthropology* 40:113–38.

Sorensen, M. V., & W. R. Leonard

2001 Neandertal energetics and foraging efficiency. *Journal of Human Evolution* 40:483–95.

Stiner, M. C.

1994 *Honor Among Thieves: A Zooarchaeological Study of Neandertal Ecology.* Princeton University Press, Princeton, NJ.

Stiner, M. C., N. D. Munro, & T. A. Surovell

2000 The tortoise and the hare: Small-game use, the broad-spectrum revolution, and Paleolithic demography. *Current Anthropology* 41:39–73.

Straus, L. G.

1982 Carnivores and cave sites in Cantabrian Spain. *Journal of Anthropological Research* 38:75–96.

1995 The Upper Paleolithic of Europe: An overview. *Evolutionary Anthropology* 4:4–16.

1996 Continuity or rupture; convergence or invasion; adaptation or catastrophe; mosaic or monolith: Views on the Middle to Upper Paleolithic transition in Iberia. In *The Last Neandertals, The First Anatomically Modern Humans: A Tale About the Human Diversity. Cultural Change and Human Evolution: The Crisis at 40 KA BP* (E. Carbonell & M. Vaquero, eds.), pp. 203–18. Universitat Rovira i Virgili, Tarragona.

1997 The Iberian situation between 40,000 and 30,000 BP in light of European models of migration and convergence. In *Conceptual Issues in Modern Human Origins Research* (G. A. Clark & C. M. Willermet, eds.), pp. 235–252. Aldine de Gruyter, Hawthorne, NY.

Stringer, C., H. Pälike, T. van Andel, B. Huntley, P. Valdes, & J. Allen
2003 Climatic stress and the extinction of the Neanderthals. In *Neanderthals and Modern Humans in the European Landscape during the Last Glaciation: Archaeological Results of the Stage 3 Project* (T. van Andel & W. Davies, eds.), pp. 233–240. Cambridge: McDonald Institute for Archaeological Research.

Tattersall, I., & J. H. Schwartz
1999 Hominids and hybrids: The place of Neanderthals in human evolution. *Proceedings of the National Academy of Sciences.* 96:7117–7119.

Trinkaus, E., S. E. Churchill, C. B. Ruff, & B. Vandermeersch, B.
1999 Long bone shaft robusticity and body proportions of the Saint-Césaire 1 Châtelperronian Neanderthal. *Journal of Archaeological Science* 26:753–773.

Villa, P., & F. d'Errico
2001 Bone and ivory points in the Lower and Middle Paleolithic of Europe. *Journal of Human Evolution* 41:69–112.

White, R.
1982 Rethinking the Middle/Upper Paleolithic transition. *Current Anthropology* 23:169–192.
1998 Comment in: d'Errico, F., J. Zilhão, M. Julien, D. Baffier & J. Pelegrin (1998): Neandertal acculturation in Western Europe? A critical review of the evidence and its interpretation. *Current Anthropology* 39:S1–S44.
2001 Personal ornaments from the Grotte du Renne at Arcy-sur-Cure. *Athena Review* 2:41–46.

Wolpoff, M. H., J. Hawks, D. W. Frayer, & K. Hunley
2001 Modern human ancestry at the peripheries: A test of the Replacement Theory. *Science* 291: 293–297.

Wynn, T., & F. Coolidge
2003 The role of working memory in the evolution of managed foraging. *Before Farming* 2:1–16.

Zilhão, J.
2000 The Ebro Frontier: A model for the late extinction of Iberian Neanderthals. In *Neanderthals on the Edge* (C. B. Stringer, R. N. E. Barton, & J. C. Finlayson, eds.), pp. 111–121. Oxford: Oxbow Books.
2001 *Anatomically Archaic, Behaviorally Modern: The Last Neanderthals and their Destiny.* Amsterdam: Stichting Nederlands Museum Voor Anthropologie En Praehistorie.
2002 Comment in: Eswaran, V. (2002): A different wave out of Africa: The mechanism of the modern human revolution? *Current Anthropology* 43:749–774.

Zilhão, J., & F. d'Errico
2003a An Aurignacian "Garden of Eden" in Southern Germany? An alternative interpretation of the Geissenklösterle and a critique of the *Kulturpumpe* model. *Paléo* 15:69–86.
2003b The chronology of the Aurignacian and transitional technocomplexes: where do we stand? In *The Chronology and of the Transitional Technocomplexes: Dating, Stratigraphy, Cultural Implications* (J. Zilhão & F. d'Errico, eds.), pp. 313–349. *Trabalhos de Arqueologia* 33. Instituto Português de Arqueologia, Lisbon.

Zilhão, J., & E. Trinkaus, eds.
2002 *Portrait of the Artist as a Child: The Gravettian Human Skeleton from the Abrigo do Lagar Velho and its Archaeological Context. Trabalhos de Arqueologia* 22. Instituto Português de Arqueologia, Lisbon.

Zubrow, E.
1989 The demographic modelling of Neanderthal extinction. In *The Human Revolution: Behavioural and Biological Perspectives on the Origins of Modern Humans* (P. Mellars & C. B. Stringer, Eds), pp. 212–231. Edinburgh University Press, Edinburgh.

4

The Châtelperronian of Grotte XVI, Cénac-et-Saint-Julien (Dordogne, France)

Géraldine Lucas
Université de Bordeaux I, Institut de Géologie et Préhistoire du Quaternaire

Jean-Philippe Rigaud
Université de Bordeaux I, Institut de Géologie et Préhistoire du Quaternaire

Jan F. Simek
University of Tennessee

Marie Soressi
Max Planck Institute for Evolutionary Anthropology

*Abstract**

Considered from its apparition as an industry belonging to the Upper Paleolithic, the Châtelperronian corresponds with the beginning of this period in southwestern Europe. The presence of Mousterian elements in Châtelperronian assemblages has been interpreted as showing an affiliation between the Mousterian of Acheulian Tradition and this industry. Moreover, it seems that Châtelperronian assemblages found on top of Mousterian occupations are richer in Mousterian elements than what is observed in sites where Mousterian occupations are lacking. Because this may indicate mixing between the archaeological levels, it appeared necessary to check the integrity of the Châtelperronian level identified within the sequence of Grotte XVI, where several Mousterian levels have been documented. In this perspective, we used information about the site formation processes as well as chronological, stratigraphic (three-dimensional plots, refittings) and technotypological data. The results of these analyses show that this level is homogeneous and that contaminations with the levels beneath and above are relatively uncommon.

Introduction

Called *Lower Perigordian* by Denis Peyrony, *Lower Aurignacian* by Henri Breuil, the Châtelperronian (or Câstelperronian) has been considered by all who have studied it as a completely Upper Paleolithic industry ever since it was first identified. Appearing above the regional Mousterian sequence in southwest Europe, the Châtelperronian was first interpreted as either an initial stage of the Perigordian or a distinct and original Upper Paleolithic industry. Regardless of its industrial origin, its maker was presumed to be anatomically modern *Homo sapiens sapiens*. For F. Bordes, the taxonomic lineage "Lower Perigordian (i.e., Châtelperronian) -Upper Perigordian" had its origins in the terminal stages of the Mousterian of Acheulian Tradition, and he even went so far as to suggest, in the absence of any compelling fossil evidence, that anatomically modern humans were responsible for the entire sequence (Bordes, 1969). The later discoveries at Saint-Césaire, and the vital debate they engendered, falsified Peyrony's and Bordes' interpretations and empirically confirmed an unanticipated association between a Neandertal and a Châtelperronian assemblage. Thus, understanding the origin and development of the Châtelperronian has always been important to understanding the origin of the Upper Paleolithic and its relation to the appearance of modern humans in western Europe.

The scientific community accepted the techno-typological assignment of the Châtelperronian to the Upper Paleolithic early on with little dispute. When the author of the Châtelperronian was shown to be Neandertal, however, many prehistorians, lacking confidence in the Neandertals' ability to produce blade industries, quickly tried to shift the definitional beginning of the Upper Paleolithic to the earliest Aurignacian. The maker of the Aurignacian, they believed, was surely anatomically modern, our direct

* Reprinted from Lucas et al. (2003: 289–298), with the authors' permission.

ancestor, the only one capable of really modern technological and symbolic behavior (Mellars, 1989; White, 1982; Demars & Hublin, 1989).

This essentialist view, human biological evolution inherently tied directly to technological innovation, has been seriously contradicted by the facts of the archeological record on the one hand and by numerous examples of technological convergence unrelated to biological change that have occurred throughout prehistory and that are of the order seen at the beginning of the Upper Paleolithic. Nevertheless, this concept survives and is still invoked by some in current debates on the emergence of the Upper Paleolithic. In this paper, we would like to present new data from the Grotte XVI (Dordogne, France) that we believe are quite germane to this discussion and consider their implications for the origins of the Upper Paleolithic in southern France.

Techno-typological Characteristics of Châtelperronian industries

Properly considered as an Upper Paleolithic industry, the Châtelperronian is not without similarities, both typological and technological, to the preceding Mousterian technocomplexes from which many believe it issued. This rooting in the Middle Paleolithic, whether it be in the Denticulate Mousterian (as Leroi-Gourhan had it) or in the Mousterian of Acheulian Tradition (the view of Peyrony and Bordes), was supported by several major techno-typological traits common to the Châtelperronian and these Mousterian variants: blade and flake-blade production, the fashioning of backed knives, and the presence of Upper Paleolithic tool-types (endscrapers, burins, becs, etc.). At the same time, frequent stratigraphic superposition of the Châtelperronian over the Mousterian encouraged the idea of techno-typological (if not cultural) continuity, conforming to the new biological continuity suggested by the Saint-Césaire discovery. The simultaneous presence of Mousterian tool forms and newer techno-typological elements like blade production and backed points confirmed for some the transitional character of this industry. However, the presence of these same forms, sometimes numerically quite important, was also seen as evidence for assemblage mixing, either due to less than precise excavation techniques or to geological processes like cryoturbation or solifuction. A rigorous taphonomic critique of Châtelperronian assemblages was, therefore, warranted.

In an earlier paper, one of us (Rigaud, 1996) was able to show the following:

(1) Châtelperronian assemblages *rich* in lithics of Mousterian manufacture were systematically superimposed stratigraphically on a Mousterian layer.

(2) Châtelperronian levels *poor* in Mousterian artifact forms were either lacking Mousterian levels below or isolated from lower Mousterian layers by other Châtelperronian units rich in Mousterian types.

This coincidence is quite suspicious, and calls into question the stratigraphic integrity of many Châtelperronian assemblages, especially given that they were excavated with less than modern technical control on site formation. In that same publication, some reservations were developed concerning the homogeneity of the Châtelperronian levels from Saint-Césaire itself, based on information presented in a recent report on the site (Guilbaud et al., 1994). According to that report, Level *Ejop-superieur,* which contained the Neandertal remains and the Châtelperronian assemblage from the site, comprised two distinct spatial zones: one under the rockshelter overhang, the other on the open-air platform outside the shelter. The great majority of blade products and most of the Upper Paleolithic tool-types (becs, burins, Châtelperronian points), were found in the interior area, associated with the human remains and with a faunal assemblage dominated by reindeer (*Rangifer* sp.). Middle Paleolithic tool-types (sidescrapers, bifaces) were concentrated outside the shelter on the platform, associated with a fauna dominated by bovids. This spatial segregation of objects led the original authors to argue as follows:

> Although temporal association between the two zones cannot be demonstrated, research has not, until now, distinguished any stratigraphic differentiation that might reflect change through time ... This view, however, cannot rule out the possibility that the materials were deposited sequentially in time. For example, material found in front of the rockshelter reflects a basically archaic technology with few Upper Paleolithic attributes, and may have been deposited earlier than those covered by the rockshelter overhang, with significant proportions of technological elements of both upper and middle Paleolithic. (Guilbaud et al., 1994).

In our view, the Châtelperronian assemblage from Saint-Césaire, rich in both Middle and Upper Paleolithic artifact types when considered as a whole, may be a mixture produced by geological processes given the spatial variation noted above. Such contamination may not have been perceptible during excavation in the varied contexts involved. Only fine scale research to establish stratigraphic and chronological relations across the various site areas will resolve this inherent ambiguity (Bordes, 1998).

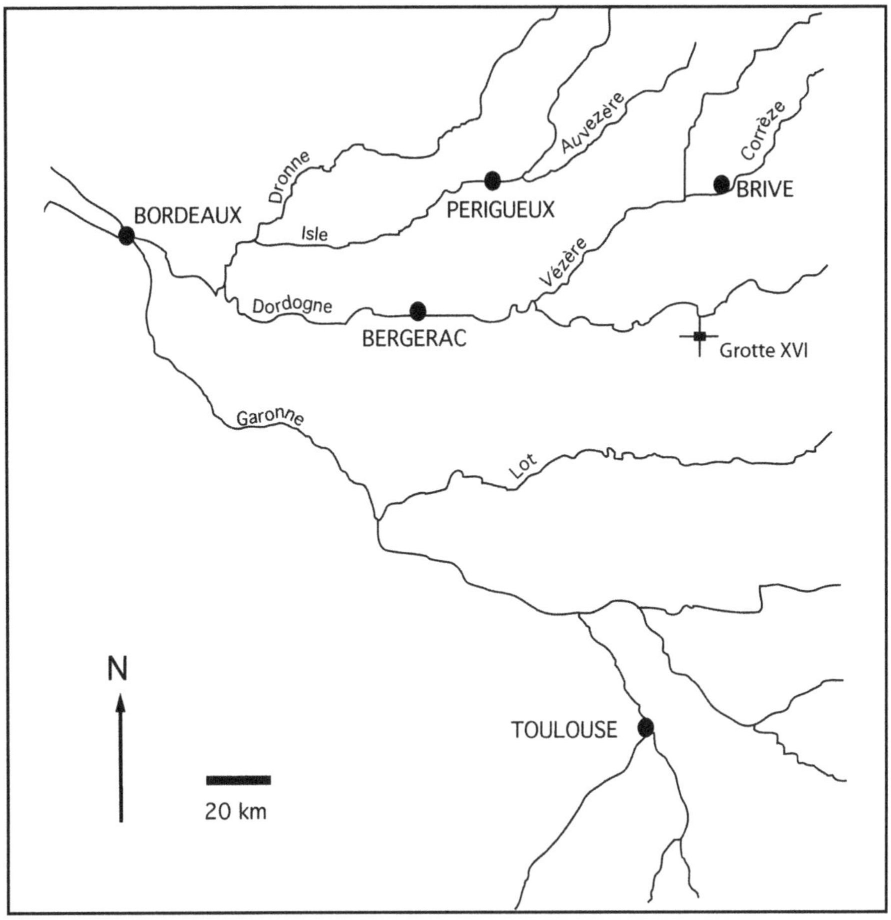

Fig. 4.1. Location of Grotte XVI.

the maximum likelihood age is 64,600 ± 3100 B.P. (Guilbert et al., 1999).

Overlying this Mousterian level is Couche **B** divided, when possible, into **Bc** and **Bf**, based on a slight change in sediment color only sometimes visible during excavation. Couche **B** has been identified over nearly the entire excavated area in the main chamber of Grotte XVI, i.e., some 50 m^2; on average, the stratum is around 10 cm thick. Three ^{14}C age determinations have been obtained for Couche **B**: >39,800 B.P. (AA 2,674); 38,100 ± 1670 (AA 2,997); 35,000 ± 1200 (GifA 95,581)

This layer is overlain by two Aurignacian levels, **Aib** and **Abb**, which themselves have been dated to 29,740 ± 510 B.P. (GifA 94,201), 29,285 ± 420 B.P. (AA 6841) and 28,140 ± 405 B.P. (AA 6840).

Overlying the Aurignacian levels is a single Gravettian layer, Couche **Abc**, which has a ^{14}C age of 26,340 ± 470 B.P. (AA 2670) and then a Solutrean level, **As**, which has a series of ages: 20,460 ± 260 B.P. (AA 2993); 20,410 ± 380 B.P. (AA 2991); 20,280 ± 220 B.P. (AA 2992); 20,230 ± 270 B.P. (AA 2669); 20,070 ± 330 B.P. (AA 2668).

Finally, a Magdalenian level, Couche **Oa**, with two ^{14}C ages—12,530 ± 105 B.P. (AA 6842) and 12,285 ± 100 B.P. (AA 6843), completes the Grotte XVI sequence.

In short, we do not advocate rejecting all Châtelperronian assemblages rich in Mousterian artifact types out of hand because of potential problems with interstratigraphic mixture. We do believe that the archeological integrity of such assemblages must be seriously assessed before they are interpreted in behavioral or evolutionary terms. As an illustration of such assessment, we turn now to the Grotte XVI, where a Châtelperronian level has been found that displays many of the "Transitional" characteristics discussed above.

The Chronological Data from Grotte XVI

Grotte XVI is situated in the Le Conte cliffs at the confluence of the Ceou and Dordogne rivers south of Sarlat (figure 1). The cave has yielded an archeological sequence beginning with a number of Mousterian levels, the last of which, Couche **C**, is richest in artifacts; **C** is also defined by an organic rich zone at the top of the stratum, evidence for a multi-phase and cyclical production of spatially extensive fires over the course of occupation (Rigaud et al., 1995).The assemblage associated with these fires is a Mousterian of Acheulian Tradition that has been dated by TL to an age between 58,000 ± 5200 and 69,900 ± 4600 B.P.;

The Lithic Assemblage from Couche B

The stone tool industry found in Couche **B** has strong Mousterian components, with sidescrapers (most simple with convex edges) composing more than 40% of the total tool count. Sidescrapers *sur face plan* are common, and Quina elements are present but rare. Notches and denticulates make up 20% of the assemblage. Endscrapers and burins are present in small numbers, as are Châtelperronian points (N = 4). We have proposed a Châtelperronian attribution for this collection, but, given the reserves we have already outlined, we too must develop arguments that will allow us to confirm or deny the homogeneity of this assemblage and show that it is clearly independent of the archeological materials contained in the levels above and below Couche **B**.

The Stratigraphic Integrity of the Artifact Layer in Couche B

Spatial Integrity

A number of vertical projection diagrams have been produced over the years for many locations within the Grotte XVI, many of them including Couche **B** artifacts in the projection. These diagrams, generated using a special DATADESK protocol programmed by F. Lacrampe (1996), allow the examination of fine scale stratigraphic distributions of artifacts in relation to geological layers. None of these projections has allowed us to identify definitively continuous sterile layers separating **B** from Couche **Aib** above and Couche **C** below. Nevertheless, vertical variation in the density of artifacts shows that a distinctive archeological level is surely present in Couche **B** (figure 2).

Refitting of Lithic Artifacts

At the present point in our studies of the Grotte XVI materials, we have applied only the method developed by J-G. Bordes for refitting lithic artifacts (Bordes, 1998). This method specifically concerns blades and blade fragments, and it thus has been applied only to the Châtelperronian and Aurignacian levels from Grotte XVI, there being far too few blade products in Couche **C** for appropriate application there. Of the 14 conjoins so far made, nine are within the Aurignacian level **(Aib)**, three are within Couche **B**, and two refits link the very base of **Aib** and the very top of **B**. We cannot, therefore, exclude the possibility of localized mixing at the boundary between layers. For the most part, however, mixing concerns only a limited part of Couche **B**, and overall archeological assemblage integrity seems good when measured this way. We acknowledge that these results are preliminary and based on rather crude assessment criteria; in the future, we will expand our refitting program to examine finer scales of possible interstratigraphic mixture.

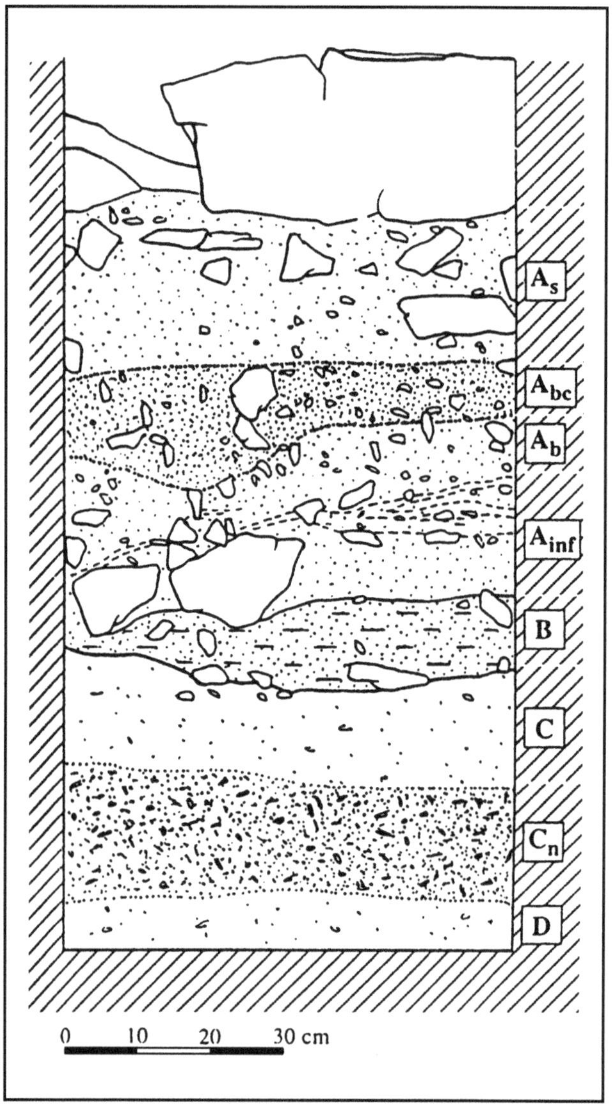

Fig. 4.2. Archeostratigraphy of Grotte XVI (after Guibert et al., 1999).

Techno-typological Integrity of the Couche B Assemblage

Supports for Tools

Tool blanks in the Couche **B** assemblage are principally flakes, some Levalloisian, along with blades (the blade index is 7%, much weaker than for level **Aib** above where the index surpasses 22%). Mousterian Couche **C** contains a number of elongated non-Levalloisian flakes but few true, regular blades that, according to Pelegrin (1995: 64 *et passim*) have negative scars on the dorsal face strictly parallel to the long axis of the piece resulting in a rectilinear shape oriented with the dorsal scar pattern.

Techniques of Percussion

In this analysis, we have classified a piece of debitage more than 3 cm in maximum dimension as produced by hard hammer percussion when the point of percussion is extruded or when the cone of percussion is clearly set back from the striking platform. In contrast, we have attributed pieces to soft hammer production that show lipped platforms and diffuse bulbs (and that do not exhibit extruded striking points) (Ohnuma & Bergman, 1982; Pelegrin, 1995, 2000; Wenban-Smith, 1989, 2000). Among Couche

B flakes larger than 3 cm in length, 79% were produced with hard hammer percussion; the remaining 21% show characteristics of soft hammer percussion removal. This is in marked contrast with the Aurignacian level immediately overlying **B**, where soft hammer debitage predominates. In Couche **C** below, there is little evidence for soft hammer percussion, which apparently was used only for the production of bifaces (an interesting fact beyond the scope of this paper), such as those illustrated in figure 3.

Individualization of Levels Based on Artifact Surface Alteration

In Couche **C**, more than 50% of the lithics exhibit alterations produced by exposure to fire (fissures, potlids, changes in color and structure). This is obviously due to the extensive fire features that cover the Couche **C** surface. The Couche **C** fire traces extend under almost the entire Couche **B** excavated area. Yet only some 7% of the

Fig. 4.3. Bifaces, couche C, Grotte XVI (drawings by J.-G. Marcillaud).

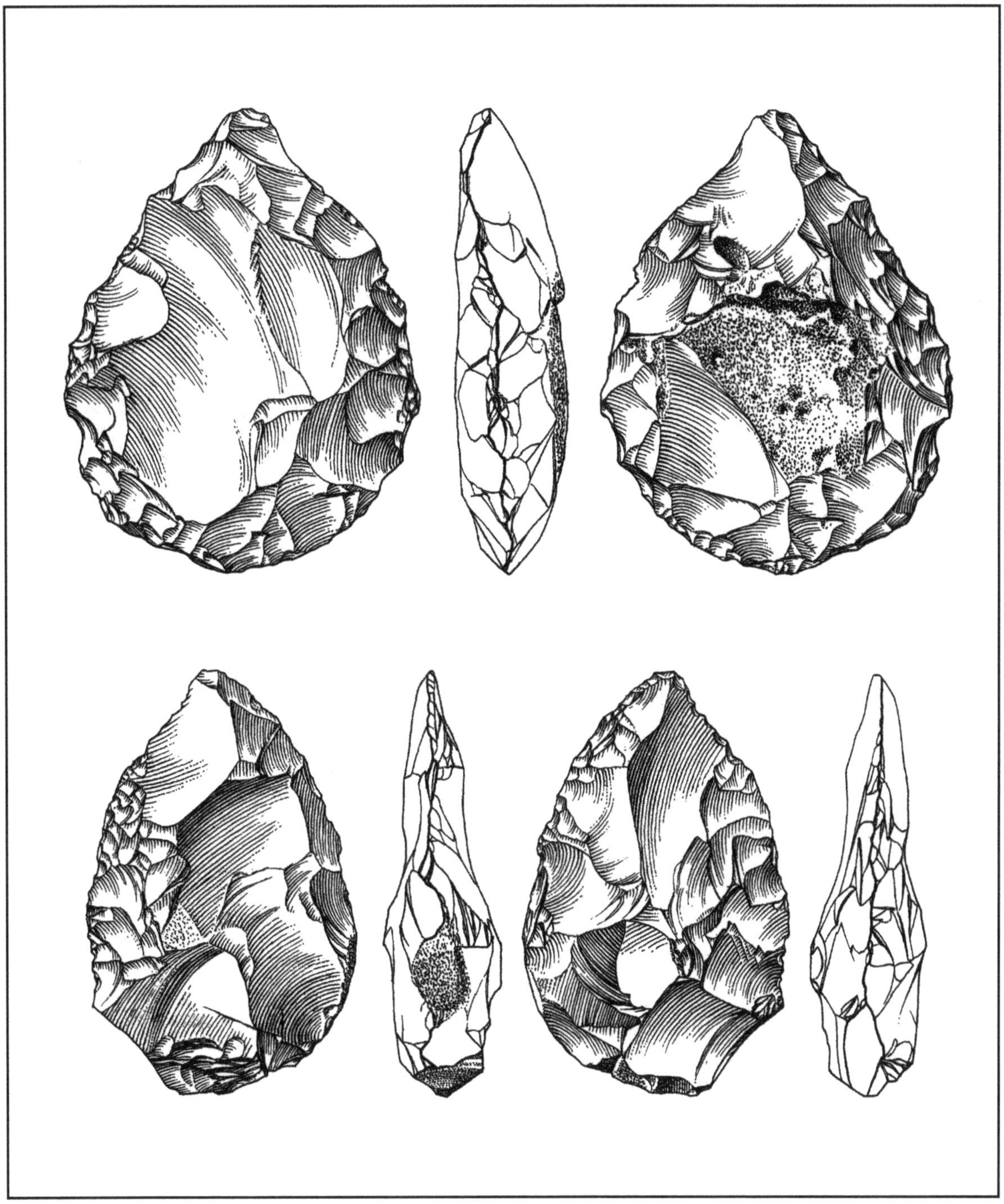

lithics from Couche **B** show traces of fire alteration. Thus, it seems that Couche **C** materials have contributed in limited measure to the Couche **B** assemblage, but for the most part, these are separate assemblages.

Discussion

The vertical distribution of artifacts from levels **C**, **B**, and **Aib** shows that there are no sterile layers clearly separating individual archeological "occupations," which is probably the norm in most cave and rockshelter sequences and in many open air sites as well. On the other hand, vertical variation in the density of artifacts argues for the presence of distinct sheets of artifacts, each one having very little relation to the one above and the one below as indicated by the limited fitting we have documented at the boundaries between layers.

The distinctive technological and typological character of Couche **B**, the use of soft hammer percussion, and the different state of artifact surfaces between **B** and **C**, are also criteria that allow us to define Couche **B** as an independent archeological assemblage from **Aib** above and **C** below (table 1). Importantly, the presence of Châtelperronian points only in this level, along with the not insignificant number of Upper Paleolithic tool-types, allows us to suggest as the best working hypothesis we have in the present state of our analyses that the archeological assemblage from Couche **B** is attributable to the Châtelperronian. Finally, ^{14}C age determinations by AMS indicate an age between 35,000 ± 1200 and 38,100 ± 1670 B.P. for Couche **B**, clearly more ancient than the overlying Aurignacian layers, dated to between 30,250 and 27,735 B.P., and much younger than Couche **C** below, dated by TL to minimally 58,800 ± 5200 B.P. (Guilbert et al., 1999). We note that all ^{14}C ages are punctual dates, each made on a single piece of compact large mammal bone; there is very slight probability of "aging" of the determinations resulting from these samples, either through contamination from organic materials in Couche **C** below or otherwise. Even considering the prudence with which we must approach radiometric ages produced with different techniques (^{14}C and TL) and any ^{14}C data older than 35,000 years, we believe that the Grotte XVI age series permits us to determine that the Couche **B** assemblage cannot be a mixture of **Aib** and **C** artifacts. There are no overlapping dates among the adjacent strata, as there should be if mixture is an agent in producing the **B** assemblage, and in fact, the ages for Couche **B** are internally consistent and place the layer unequivocally between the levels above and below in time. We take this as strong evidence for the stratigraphic integrity of the Couche **B** archeological assemblage.

A comparison of the Couche **B** lithics from Grotte XVI with the industry from Level 8 of Roc de Combe (Pelegrin, 1995), only 15 km away, shows that Levalloisian flakes are present in both these Châtelperronian assemblages. At Grotte XVI, the Levallois technique was practiced in place, as Levalloisian flakes are found in association with Levalloisian cores and core preparation products. The situation is different at Roc de Combe, where Levalloisian flakes found in the Châtelperronian were probably introduced to the site after being produced elsewhere; Pelegrin (1995:86) has even suggested that they were possibly scavenged from earlier materials exposed by erosion or other effects in front of the site. At Grotte XVI, blade blanks for tools are less numerous than at Roc de Combe, but at the latter site, every category of Upper Paleolithic tool-type has roughly equal proportions of blade and flake supports. Only Châtelperronian points, far more frequent at Roc de Combe than at Grotte XVI, are made exclusively on blades; in other words, Châtelperronian blade production at Roc de Combe may have been oriented in large part toward the manufacture of Châtelperronian points. At Roc de Combe, 94% of the tool blanks were removed from their cores using soft hammer percussion (Pelegrin 1995:193). At Grotte XVI, no more than 15% of the tool blanks were extracted in this fashion. The relative paucity of Châtelperronian points found at Grotte XVI may be linked, at least in part, to the reduced effort at blade production exhibited by the Couche **B** assemblage.

Based on his analyses of lithic industries from the site of Grande Roche de la Plemartie at Quinçay, Lévêque (1987) proposes four "evolutionary phases" in the Châtelperronian sequence. The first, an Archaic Châtelperronian or Proto-Châtelperronian, is characterized by numerous sidescrapers, denticulates, and a few bifaces associated with burins, endscrapers on blades, and backed pieces (Châtelperronian points). An early Châtelperronian, rich in Châtelperronian points, follows this archaic phase; an evolved Châtelperronian in which Châtelperronian points are "finer" with more rectilinear backs overlies the archaic level. Finally at Quinçay, a Châtelperronian with regressive characteristics completes the sequence. On the face of it, the typological composition of the Archaic phase at Quinçay appears quite similar to the Châtelperronian of Grotte XVI Couche **B**; in the absence of more precise taphonomic data from Quinçay and a more accurate

Table 4.1. Grotte XVI. Technological characteristics of lithic assemblages in levels C, B and Aib.

Level	C	B	Aib
Technique	Hard Hammer	Hard Hammer > Soft	Soft hammer > Hard
Method of Debitage	Production of elongated non-Levallois flakes	Levallois + Blades	Blades + Bladelets
Blade Index	13%	7%	22%

chronological context all around, however, similarities between these assemblages must remain evocative rather than demonstrated.

Conclusions

Our taphonomic, techno-typological, and chronometric analyses all indicate that the assemblage from Grotte XVI Couche **B** represents a homogeneous collection with little contamination from the assemblages above and below it in the site stratigraphy. Characterized by typological and technological traits that are clearly Mousterian, the assemblage still shows some development of blade production technology associated with the presence of Upper Paleolithic tool-types. ^{14}C determinations suggest an age between 33,800 and 39,770 for the assemblage. Given the nature of this industry and its chronology, we propose to consider it as an early, even initial phase of the Châtelperronian.

In coming analyses, we hope to compare the Couche **B** data with other Châtelperronian levels that are well dated and taphonomically comprehended. We will also look systematically at chronologically late Mousterian levels (mostly MTA) that have techno-typological characteristics "antecedent" to the Châtelperronian. These comparisons, we believe, will yield important information on the technological and typological modalities that were associated with the emergence of the first Upper Paleolithic cultures in Europe.

References

Bordes, F.
1969 Colloque sur l'origine de l'homme moderne. Mouvements scientifiques. *Quaternaria* 11:281–282.

Bordes, J.-G.
1998 *L'Aurignacien 0 en Périgord: analyse des données, un exemple d'application d'une méthode de quantification des remontages d'intérêt stratigraphique (Caminade est, couche G).* Mémoire de Diplôme d'Études Approfondies en Anthropologie et Préhistoire, Université Bordeaux 1.

Demars, P.-Y, & J.-J. Hublin
1989 La transition néanderthaliens/hommes de type moderne en Europe occidentale: aspects paléontologiques et culturels. In l'*Homme de Néanderthal 7: l'Extinction,* (J. Kozlowski, ed.) pp. 23–37 ERAUL No. 34, Liège.

Grayson, D., F. Delpech, J.-Ph. Rigaud, & J. Simek
2001 Explaining the development of dietary dominance by a single taxon at Grotte XVI, Dordogne, France. *Journal of Archaeological Science* 28:115–125.

Guibert, P., F. Betchel, M. Schvoerer, J.-Ph. Rigaud, & J. F. Simek
1999 Datation par thermoluminescence de sédiments chauffés provenant d'une aire de combustion moustérienne (Grotte XVI, Cénac-et-Saint-Julien, Dordogne, France). *Revue d'Archéométrie* 23:163–175.

Guilbaud, M., A. Backer, & F. Lévêque
1994 Technological differentiation associated with the Saint-Césaire Neandertal. *Préhistoire Européenne* 6:187–196.

Lacrampe-Cuyaubère, F.
1996 *Profils et plans de répartition de vestiges archéologiques: techniques de construction et d'analyse avec Data Desk (version 5.0 pour Macintosh).* Mémoire de Diplôme d'Études Approfondies en Anthropologie option Préhistoire. Université de Bordeaux 1.

Lévêque, F.
1987 Les gisements castelperroniens de Quinçay et de Saint-Césaire. Quelques comparaisons préliminaires. Stratigraphies et industries. Préhistoire de Poitou-Charentes, Problèmes actuels. *CTHS, 1987, 111e Congrès National des Sociétés Savantes, Pré-et Protohistoire,* pp. 91–98. Poitiers: UISPP.

Lucas, G., J.-Ph. Rigaud, J. Simek, & M. Soressi
2003 The Châtelperronian of Grotte XVI, Cenac-et-Saint-Julien (Dordogne, France). In *The Chronology of the Aurignacian and the Transitional Technocomplexes: Dating, Stratigraphies, Cultural Implications* (J. Zilhão & F. d'Errico, eds.), pp. 289–299. *Trabalhos de Arqueologia* 33. Instituto Portugêse de Arqueologia, Lisbon.

Mellars, P. A.
1989 Major issues in the emergence of modern humans. *Current Anthropology* 30:349–385.

Ohnuma, K., & C. Bergman
1982 Experimental Studies in the Determination of Flaking Mode. *Bulletin of the Institute of Archaeology* 19:161–170.

Pelegrin, J.
1995 *Technologie lithique: le Châtelperronien de Roc de Combe (Lot) et La Côte (Dordogne)*. Cahiers du Quaternaire. Paris: CNRS.
2000 Les techniques de débitage laminaire au Tardiglaciare: critères de diagnose et quelques réflexions. In *L'Europe Centrale et Septentrionale au Tardiglaciare. Confrontation des Modèles Régionaux* (B. Valentin, P. Bodu & M. Christensen, eds.), pp. 73–86. Nemours: APRAIF (Mémoires du Musée de Préhistoire d'Ile-de-France 7).

Rigaud, J.-Ph.
1996 L'émergence du Paléolithique supérieur en Europe occidentale. Le rôle du Castelperronien. In *The Origin of modern man.* XIIIème Congrès International des Sciences Préhistoriques et Protohistoriques, Forlì, 8–14 sept. 1996. Forlì: ABACO, pp. 219–223.

Rigaud, J.-Ph.; J. Simek, & T. Gé
1995 Mousterian fires from Grotte XVI (Dordogne, France). *Antiquity* 69:902–912.

Wenban-Smith, F. F.
1989 The use of canonical variates for determination of biface manufacturing technology at Boxgrove Lower Palaeolithic site and the behavioural implications of this technology. *Journal of Archaeological Science* 16:17–26.
1999 Knapping technology. In *Boxgrove. A Middle Pleistocene Hominid site at Eartham Quarry, Boxgrove, West Sussex* (M. B. Roberts & S. A. Parfitt, eds.), pp. 384–395. London: English Heritage (Archaeological Report 17).

White, R.
1982 Rethinking the Middle/Upper Palaeolithic transition. *Current Anthropology* 23:169–192.

5

"Transitional" Industries from Neandertals to Anatomically Modern Humans in Continental Italy: Present State of Knowledge

Amilcare Bietti
University of Rome "La Sapienza"
Istituto Italiano di Paleontologia Umana, Roma

Fabio Negrino
University of Pisa
Istituto Italiano di Paleontologia Umana, Roma

Abstract

There are three kinds of "transitional" industries between Neandertals and anatomically modern humans in continental Italy: (1) particular late Mousterian assemblages found mainly in northern and central Italy; (2) the Uluzzian of southern Apulia (in the Salento region); and (3) assemblages combining some (or few) Uluzzian features with original technological characteristics, found in central and southern Italy (Tuscany and Campania). The principal features of these industries, which are all very distinct from the subsequent early Aurignacian ones, have so far been described almost exclusively on morpho-stylistic typological grounds. The present paper examines these Italian "transitional" assemblages according to a more technological perspective that incorporates dimensions of variability in raw material procurement.

The "transition" from Neandertals to anatomically modern humans (AMH) in Europe stimulates an ever increasing interest, in part because a relatively large set of absolute dates indicates the contemporaneousness of the two hominin populations for several thousand years. The process of this "transition" has been the subject of a number of detailed analyses conducted from different perspectives. The existing human fossil record has led to the well-known debate between supporters of a sudden replacement of the Neandertals by AMH of African origin (see e.g., Stringer 1994) and, on the opposite side, researchers espousing the Multiregional Evolution model (e.g., Wolpoff & Caspari 1996), with supporters of the "Hybridization and Replacement" model falling somewhere in between (see e.g., Aiello 1993).

Due to the scarcity of human remains in Europe for the interval between 40–30kya, however, it perhaps unsurprising that most studies concerned with the "transition" have focused on the analysis of the behavioral differences between the two hominin groups, especially as they concern lithic industries and the appearance of ornaments and "art" items; a few contributions have also examined settlement and subsistence strategies (i.e., hunting practices, site function, etc.—for a late Mousterian example in Italy, see Stiner 1994). The multiple perspectives employed in such studies have led to fierce debates such as, for instance, the ongoing discussion about the bone and ivory tools and the personal ornaments from the French Châtelperronian site of Arcy-sur-Cure. These have been attributed to Neandertals on the basis of the human skull discovered at Saint-Césaire, and claimed by some (e.g., Mellars 1996, 2000) to be the result of contacts between contemporary groups of Neandertals and Aurignacian-bearing AMH. This view has recently been opposed by the hypothesis of Châtelperronian Neandertals having developed Upper Paleolithic technology through their own autonomous evolutionary process (or better "trajectory," according to its proponents, who adopt more "à la mode" postmodern or postprocessual definitions [d'Errico et al. 1998]).

Focusing specifically on Italy, we find that the bone industry of "transitional" assemblages (defined as Uluzzian

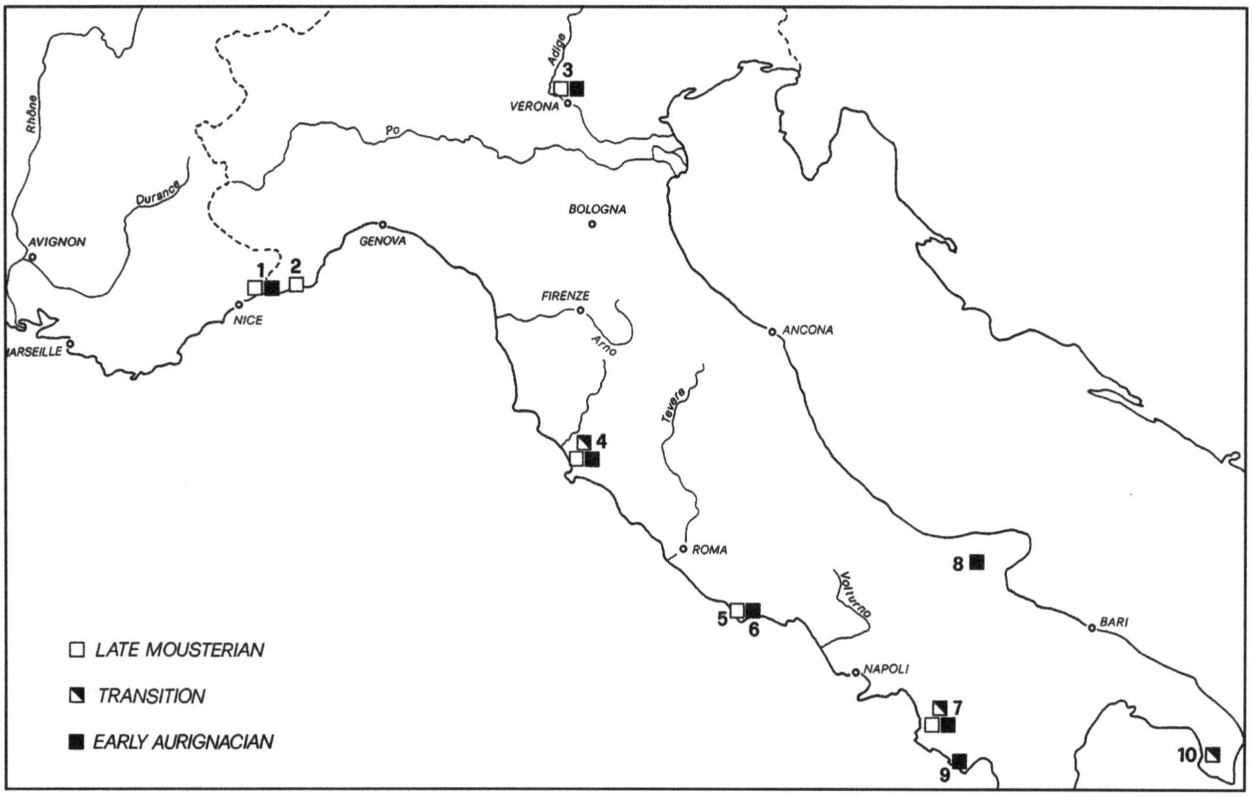

Fig. 5.1. Distribution of the sites mentioned in the text. *1.* Riparo Mochi; *2.* Via San Francesco; *3.* Grotta di Fumane; *4.* Grotta della Fabbrica; *5.* Grotta Breuil; *6.* Grotta del Fossellone; *7.* Grotta di Castelcivita; *8.* Grotta Paglicci; *9.* Grotta della Cala; *10.* Grotta del Cavallo.

sensu lato) is proportionally sparser, which has resulted in lithic assemblages—and more precisely the morphological and stylistic characteristics of their formal tools—being the main subject of discussions about the "transition." Besides a number of journal articles describing individual sites (e.g., Palma di Cesnola 1966, 1967, for Grotta del Cavallo; Pitti et al. 1976, for Grotta della Fabbrica) and an in-depth monograph on Grotta di Castelcivita by P. Gambassini (1997), the topic has been discussed in several reviews in textbooks (e.g., Palma di Cesnola 1993), in meeting proceedings (e.g., Gioia 1990, Mussi 1990) and in monographs of selected papers (e.g., Bietti 1997, Kuhn & Bietti 2000).

In this paper, we propose to revise some concepts and to critically challenge some accepted opinions about the status of some Italian "transitional" assemblages. This effort is based on new observations on and analyses of previously published materials, especially as concerns their technological features and patterns of raw material procurement. The assemblages included in this study all come from stratified sites; we will discuss in more detail the surface finds from central Italy attributed to the Uluzzian on typological grounds later in this paper. Due to the absence of Uluzzian or related assemblages in the north and most of the central part of the peninsula, we also consider as "transitional" a few Late Mousterian sites from those areas. Figure 1 shows the distribution of the sites relevant to our analysis. It is worth stressing that there is no interstratification in any of the sites that contain more than one industry. In other words, late Mousterian strata *always* underlie "transitional" ones, which always underlie Aurignacian ones (Bietti 1997, Kuhn & Bietti 2000), as is the case elsewhere, for instance in France where J.-G. Bordes' (2003, see especially figure 17) extensive interstrata refittings at Le Piage have largely disproved claimed interstratifications at Roc de Combe and Le Piage.

In the next section, we describe the late Mousterian assemblages used in our study, while our sample of "transitional" ones will be discussed in the subsequent section. While Aurignacian assemblages are not strictly relevant in the context of this study, we nonetheless discuss some of them in our concluding section, in order to emphasize the marked differences that exist between them and "transitional" assemblages.

The Late Mousterian in Italy

Starting in northern Italy, we only mention in passing the Mousterian (layer I) of Riparo Mochi, since it is not dated and is separated from the overlying Early Aurignacian of layer G by a semi-sterile level some 40–60 cm thick. It has been correctly assumed (Kuhn & Stiner 1992) that the few artifacts found in layer H represent a mixture from layers I and G. In fact, during the 1998 excavation campaign at Riparo Mochi, we identified a structured Aurignacian hearth *inside* layer H (Alhaique et al. 2000), so that claims for such mixture appear fully justified. The lithic industry of layer I can be divided roughly into two components: a lower one, where Levallois technology is dominant, and an upper one which shows a noticeable decrease in the frequency of this reduction technique (Kuhn & Stiner 1992:83–86). The raw material used in both components is almost totally local, namely flint coming from the Ciotti conglomerates (figure 2, nos. 2 & 4) located at the top of the hill just above the Balzi Rosi cliffs and siliceous limestone an quartzite from San Remo, located about 20 km away (figure 2, nos. 1 & 3). There is a single artifact on rhyolite from Esterel, near Cannes, and another on flint from Trevan, a source in southeast France about 100 km distant from Riparo Mochi.

The site of Via San Francesco (figure 1, no. 2) was discovered during rescue excavations in the center of the town of San Remo (Isetti 1961; see also Tavoso 1988). As detailed elsewhere (e.g., Bietti 1997, Kuhn & Bietti 2000), stone tools and faunal remains were found practically only in a single layer, and the lithic industry shows a very peculiar combination of "Upper Paleolithic" features together with standard Levallois reduction technique (figures 3 & 4), and this in spite of its typological attribution to a generic "Denticulate Mousterian" (see discussion in Kuhn & Bietti 2000:51). The Via San Francesco industry also displays a *chaîne opératoire* of Upper Paleolithic type, with "true" prismatic cores with prepared striking platforms (figure 3, nos. 1 & 2; made on siliceous limestone), as well as several blades (figure 4, nos. 1–6; all on siliceous limestone) and even crested blades (figure 3, no. 4 on quartzite) and some distinctive obliquely truncated blades (figure 4, nos. 5 & 8). The raw material is almost exclusively local, coming mainly from the San Remo formations and, to a lesser degree, from the Ciotti conglomerates. Two exceptions are a burin on purple-brown flint coming from Vaucluse, near Avignon (figure 4, no. 9), and another burin on greenish-grey flint, most probably of local origin (most probably a variety of Ciotti flint) (figure 4, no. 7). The Upper Paleolithic core exploitation technique coexists with the Levallois method, as shown by the presence of centripetal cores (e.g., figure 3, no.3; made on siliceous limestone) and by the facetted butt of some of the blades in the assemblage (e.g., figure 4, nos. 1 & 5).

Always in northern Italy, the Mousterian layers of Grotta di Fumane (figure 1, no. 3) are separated by layer A4I from the early Aurignacian layers A3–A1, which are AMS ^{14}C-dated to between 40 +4/-3 and 31.6 ± 0.4 kya (Bartolomei et al. 1992). Layer A4I contains a few Mousterian artifacts, two Dufour bladelets and a backed tool that, in our opinion, is likely a typical Mousterian backed knife (Bartomlomei et al. 1992: figure 14, no.1). It thus appears that the assemblage from layer A4I is probably the result of mixing, as is the case for layer H of Riparo Mochi (see above). In any case, the underlying Mousterian levels display both Levallois and "discoid" reduction processes (Peresani & Sartorelli 1998) and can reasonably be considered Late Mousterian on the basis of some recent AMS ^{14}C dates that overlap, within the error, with the Aurignacian ones (M. Peresani, personal communication). The raw material is exclusively local flint of excellent quality.

Further south, in central Italy, Grotta della Fabbrica, located near the town of Grosseto in Tuscany (figure 1, no. 4), shows a very complicated stratigraphic sequence due to severe postdepositional disturbances that have eroded away part of the deposit and mixed material from the upper strata with that from the lower ones (Pitti et al. 1976). Besides some intrusive (but clearly recognizable) disturbed material, however, the innermost part of the cave preserves a rather intact depositional sequence. This serves as the basis of the section depicted in figure 5, which presents the stratigraphic profile along a W–E section, modified from a drawing by C. Tozzi (but see also the profile published in Pitti et al. 1976). We can see that a Mousterian industry is present in layers 1a–1c, while the "transitional" industry comes from layer 2 and early Aurignacian assemblages come from layers 3 and 4. Additionally, there also are, at the top of the sequence, some remnants of an Epigravettian deposit (i.e., layer 5). The layer 1a–1c industry is rather scarce in the undisturbed part of the deposit, and since there is up to now no published account of this assemblage (apart from a simple NISP list of the mammalian fauna presented by C. Sorrentino [in Pitti et al. 1976:184]), we do not discuss it further in this paper.

In Latium, the key late Mousterian site is Grotta Breuil (figure 1, no. 5), where the few human fossils for the transition interval in Italy have been found. Detailed descriptions and discussion of this site can be found in several other papers (e.g., Alhaique et al. 1998, Bietti 1997, Kuhn & Bietti 2000; and references therein). As concerns the site's age, we have average LU ESR dates of 36.6 ± 2.7 kya for layers 3–4 (presumed, since it is a gamma-ray spectrometer date on a sample from the uppermost part of the deposits, collected before regular excavation) and of 33 ± 5 kya for layers 6–7. In both cases, the EU determinations coincide within the errors. There are extensive studies of the site's zooarchaeological record (Stiner 1994) and of its lithic industry's techno-typological aspects (Bietti & Grimaldi 1996, Kuhn 1995, Rossetti & Zanzi 1990–91). Overall, we can discern in the upper layers (layers 3–5) a trend toward more elongated, blade-like blanks (figure 6), made on pseudo-prismatic cores (see Kuhn & Bietti 2000: figure 4, nos.8–9) although in those

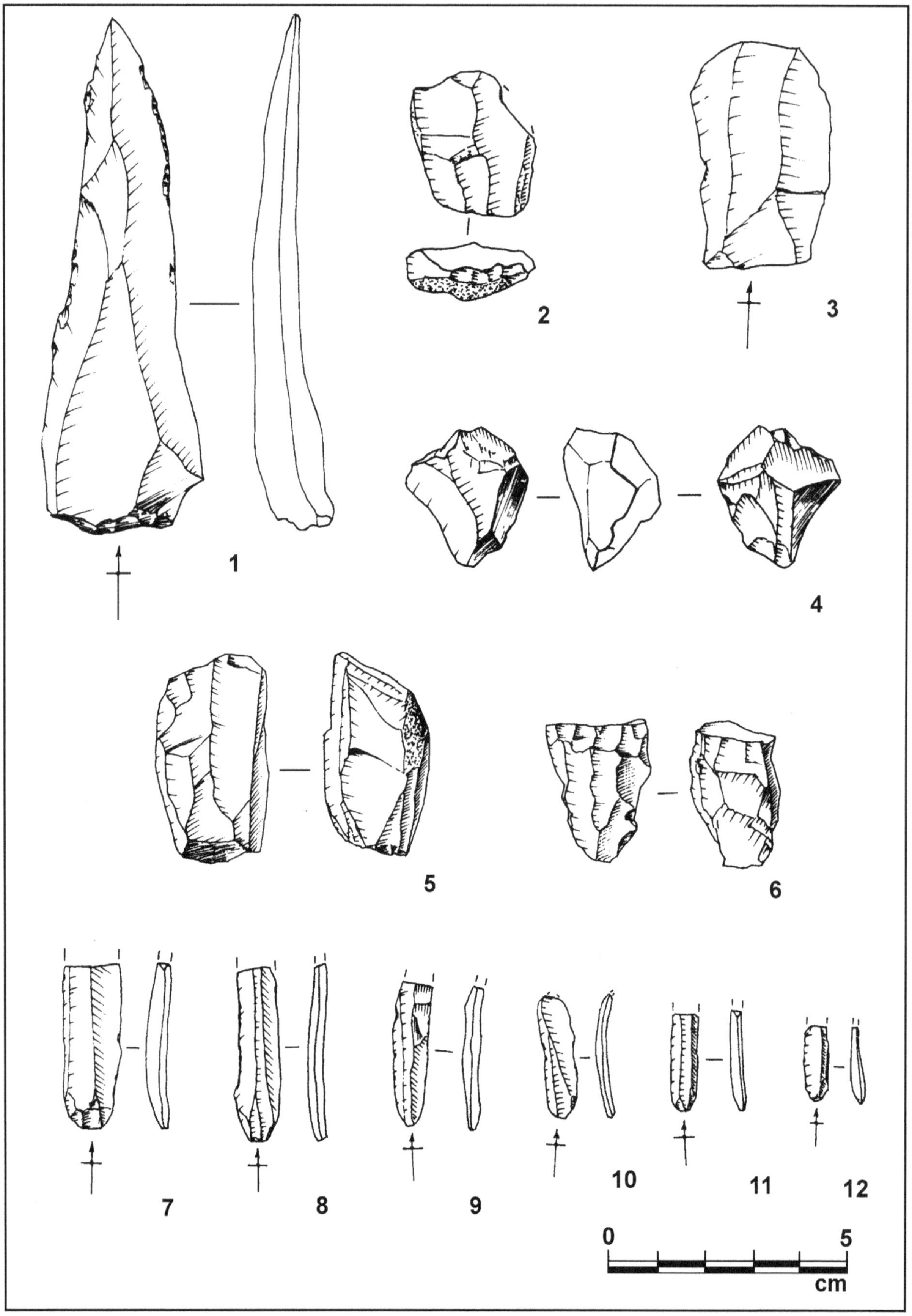

Fig. 5.2. Industry from Riparo Mochi. *1–4,* from Layer I (Mousterian); *5–12* from Layer G (early Aurignacian: no. *5* on Vaucluse flint, nos. *6–11* on local Ciotti flint, no. *12* on "Scaglia" flint).

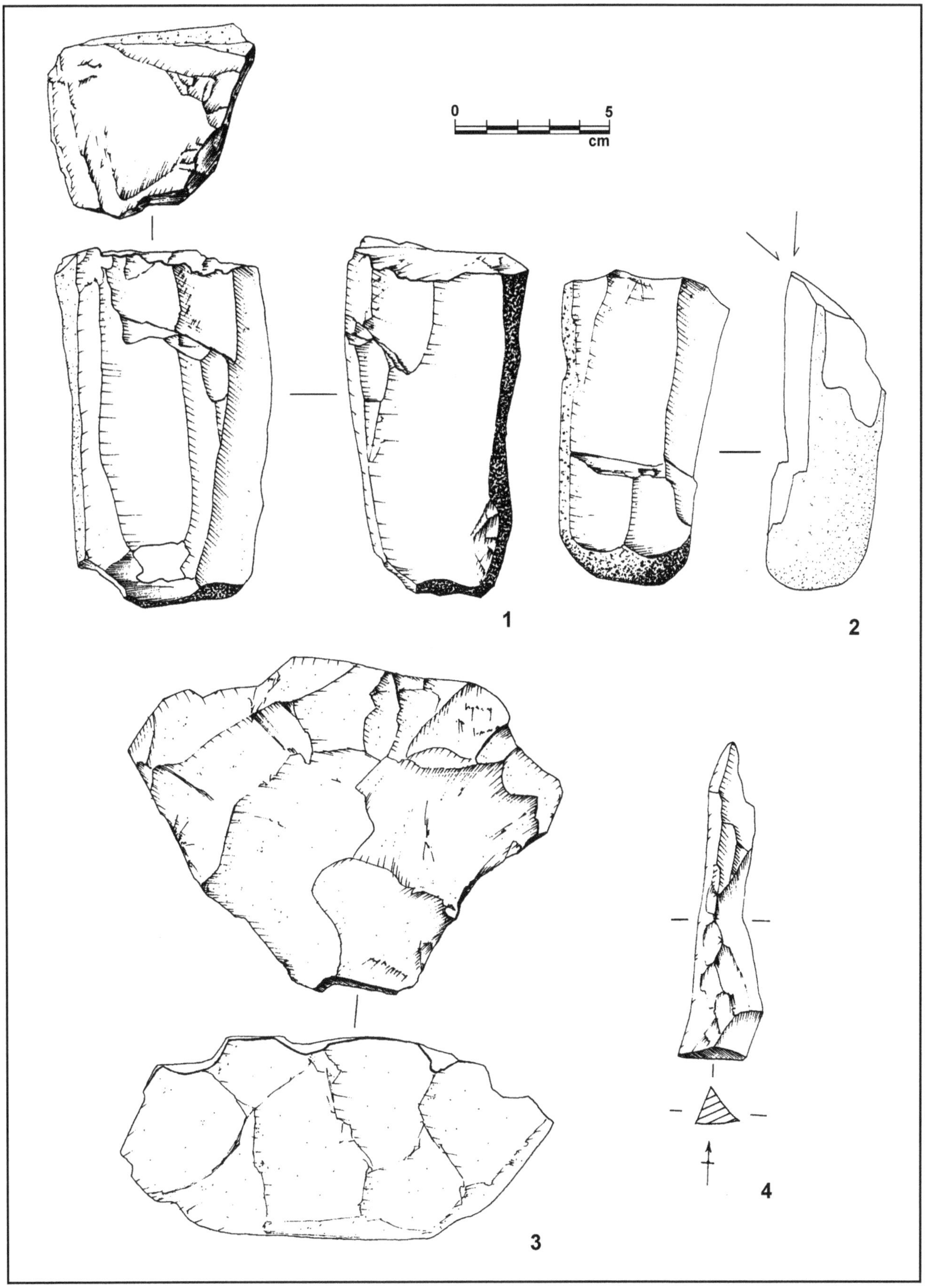

Fig. 5.3. Industry from Via San Francesco, San Remo.

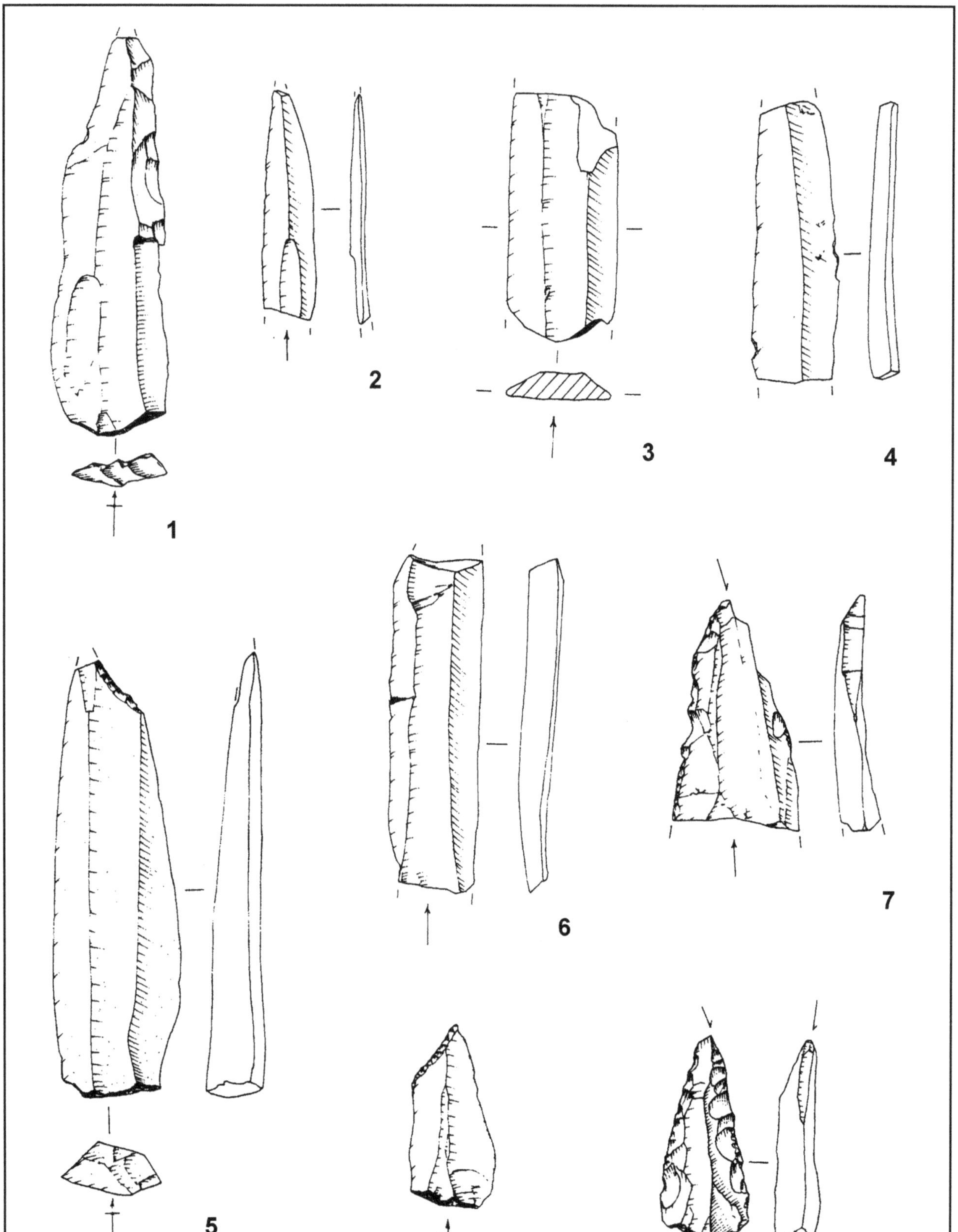

Fig. 5.4. Industry from Via San Francesco, San Remo *(continued)*.

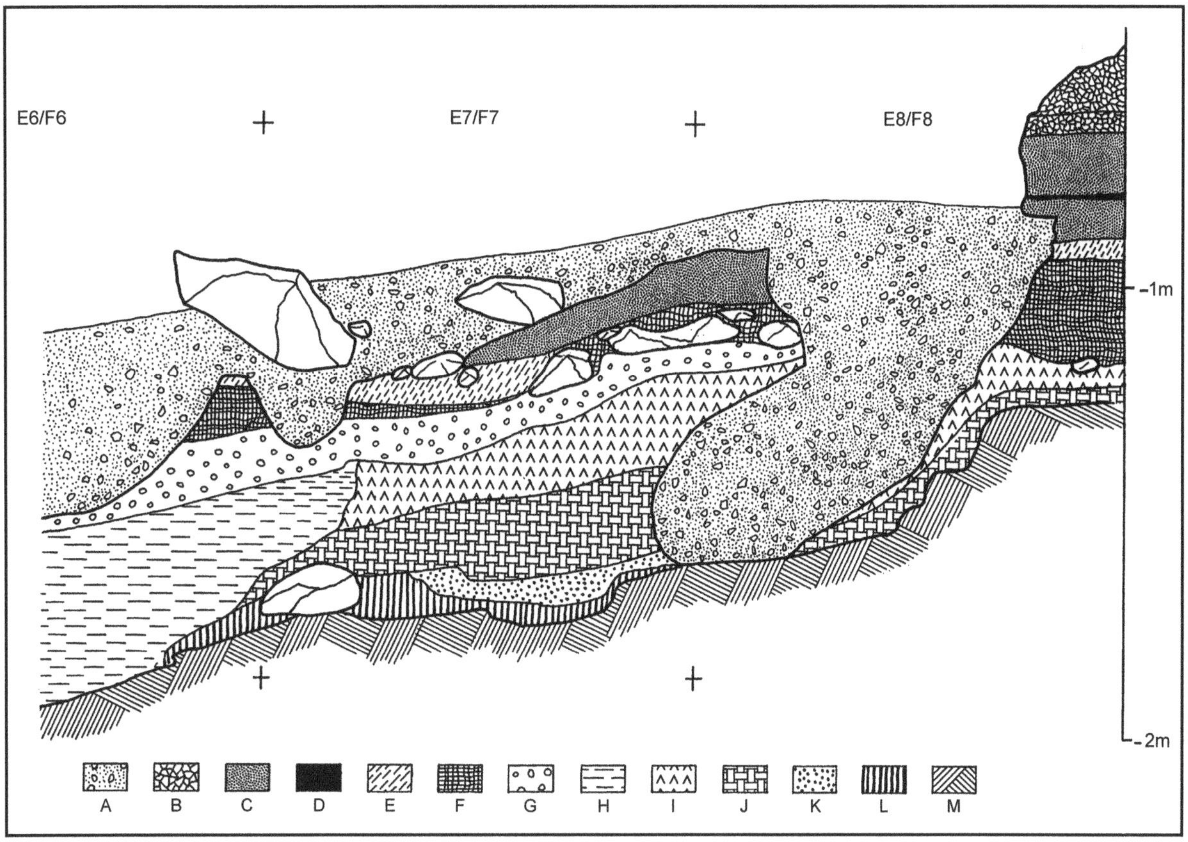

Fig. 5.5. Stratigraphy of Grotta della Fabbrica.
A: postdepositional filling of an eroded deposit; *B:* layer 5 (Epigravettian); *C:* Upper and Middle layer 4 (Aurignacian) separated by *D:* a thin stalagmite; *E:* Lower layer 4 (Aurignacian); *F:* layer 3 (Aurignacian); *G:* layer 2 ("Uluzzian"); *H–K:* layers 1a,1b,1b',1c (Mousterian); *L:* loose and fragmented stalagmitic layer; *M:* Bed-rock. (Modified after Pitti et al. 1976).

layers too, this technique coexisted with the centripetal-Levallois technique (see for instance figure 6, nos. 16, 23–26). The blade-like blanks from the upper Mousterian layers are less striking and "spectacular" than the ones from Via San Francesco (figure 4) for the simple reason that, at Grotta Breuil, the locally available raw material consists of small flint pebbles derived from ancient conglomerates, with little evidence of any exotic type of flint that would come in large nodule size. Preliminary analysis of the lithic industry from layers 7 and 8 at Grotta Breuil show a less pronounced blade-like production technique, with a higher incidence of the "percussion on anvil" (i.e., bipolar) reduction technique, reminiscent of that also reported in the upper Mousterian levels (e.g., layer 37) from nearby Grotta del Fossellone (figure 1, no. 6; see Bietti & Grimaldi 1996).

In southern Italy, we find the very important and well-dated sequence of Grotta di Castelcivita (figure 1, no. 7). Figure 7 shows the site's stratigraphic profile and the results of the sedimentary analysis conducted by M. P. Fumanal (1997). The list of absolute dates is incomplete, however, and table 1 presents the complete list (Gambassini 1997:146). As can be seen based on these dates, the denomination of late Mousterian is entirely justified at this site, at least starting from cut 30, and we also note that the dates from cuts 30–29 and 28–27 almost coincide within two standard deviations. The Mousterian industry shows a striking Levallois character (Gambassini 1997:102–105) and is made exclusively on local raw materials (flint, quartzite and some jasper), all of excellent quality. For the purposes of this study, we examined the debitage of two Mousterian samples (cuts 28–27 and cuts 21–19), and our results are shown in the first three columns of table 2. The predetermined flakes (*sensu* Boëda 1994) are almost all Levallois and the cores in both samples are almost all of recurrent centripetal Levallois type, with a few unidirectional specimens.

The "Transitional" Assemblages

As we have pointed out in the previous sections and in other publications (e.g., Bietti 1997, Kuhn & Bietti 2000), there are no Uluzzian assemblages in northern Italy, not even in layer A4I at Fumane (*contra* the statement of d'Errico et al. 1998: S16). In fact, the Uluzzian has always been defined on purely typological grounds, namely the

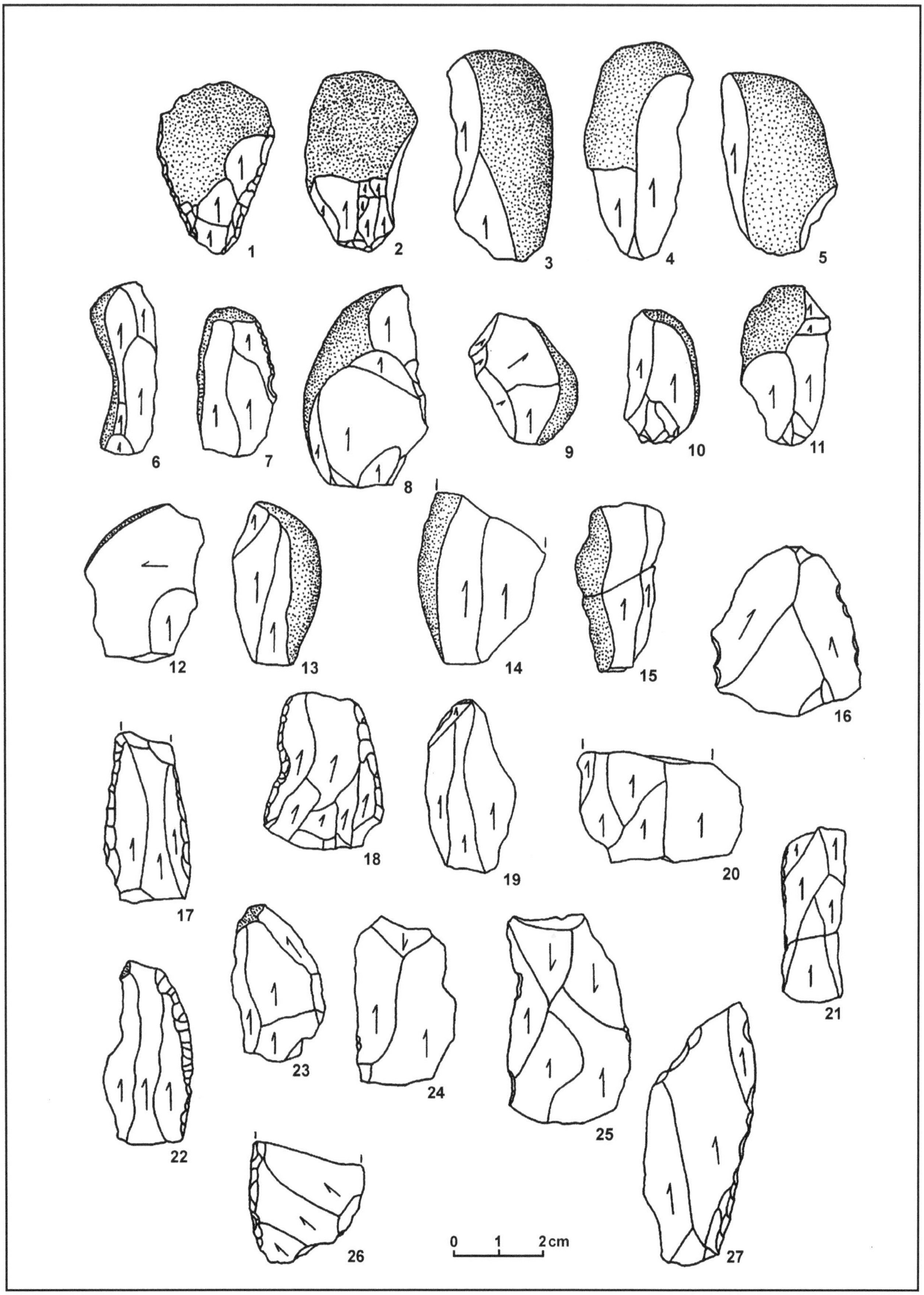

Fig. 5.6. Industry from Grotta Breuil, Layer 4 (from Bietti and Grimaldi 1996).

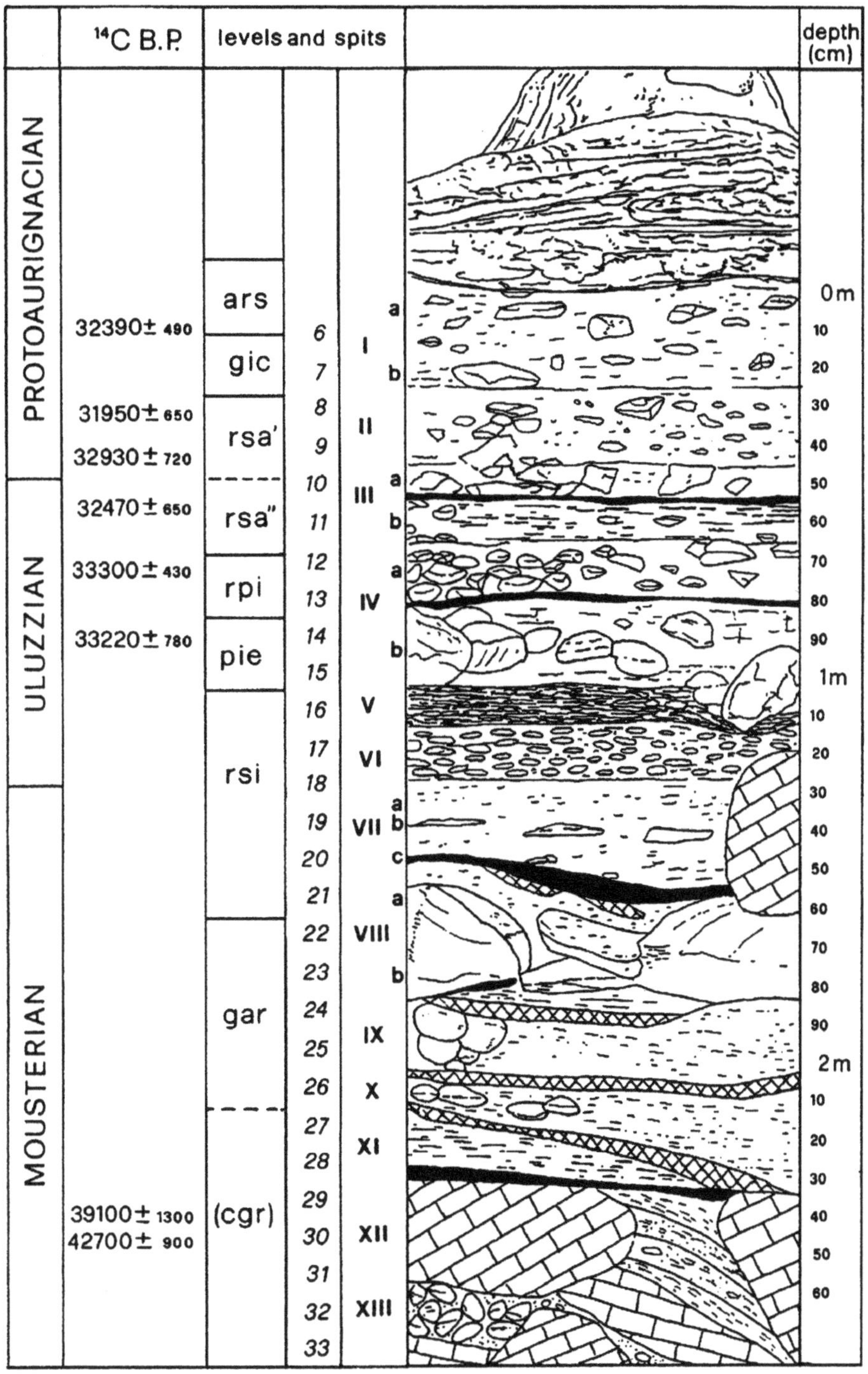

Fig. 5.7. General stratigraphic profile of Grotta di Castelcivita (from Fumanal 1997).

presence of backed tools ("Uluzzian crescents" or *semilune*) and of abundant splintered pieces (a.k.a. *pièces "esquillées"* or *"écaillées"*). In our opinion, in the absence of more detailed technological analyses, this definition should be restricted *only* to the type site of Grotta del Cavallo (figure 1 , no. 10) and, probably to the nearby sites of Grotta di Uluzzo, Riparo di Uluzzo C and Grotta Mario Bernardini in the Salento region of Apulia (Palma di Cesnola 1993, and references therein; see also discussion in Bietti (1997:139–140). In the Bay of Uluzzo, the industry is characterized by typical curved backed tools which are in many instances real crescents, and Palma di Cesnola (1993) argues that there is a typological evolution of the industry (i.e., "early," "middle" and "upper" phases) that is most evident at Grotta del Cavallo. The raw material is exclusively local and, in the lower part of

Table 5.1. Absolute radiocarbon dates for Grotta di Castelcivita (from Gambassini 1997).

Reference number	Provenience	Date (industry)
Beta 58184	Castelcivita, cut 6	32,390 ± 490 BP
CAMS 4622		(Proto-Aurignacian with micropoints)
F 105	Castelcivita, cut 8	31,950 ± 650 BP
		(Proto-Aurignacian with Dufour bladelets)
F 72	Castelcivita, cut 9	32,930 ± 720 BP
		(Proto-Aurignacian with Dufour bladelets)
F 71	Castelcivita, cuts 10-11	32,470 ± 650 BP (Uluzzian)
F 106	Castelcivita, cut 12	> 34,000 BP (Uluzzian)
GrN 13985	Castelcivita, cut 12	33,300 ± 430 BP (Uluzzian)
F 107	Castelcivita, cut 14	33,220 ± 780 BP (Uluzzian)
GrN 13983	Castelcivita, cuts 27-28	33,800 ± 1,300 BP (Mousterian)
GrN 13982	Castelcivita, cuts 29-30	39,100 ± 1,300 BP (Mousterian)
GrN 13984	Castelcivita, cuts 29-30	42,700 ± 900 BP (Mousterian)

Table 5.2. Debitage distributions for three lithic samples from Castelcivita; the numbers in parentheses indicate the quantity of artifacts made on limestone.

	Mousterian			Uluzzian	Aurignacian
Cuts	28–27	21–19	Total	11–10	9–8
Predetermined flakes	107	42	149	43	-
Blades and bladelets	27	2	29	7	175
Naturally backed knoves	2	-	2	-	-
Rejuvenation flakes	1	-	1	-	-
Flakes	141	17	158	284 (110)	412
Core-like *esquillées*	-	-	-	40 (4)	20
Esquillées on flake, etc.	-	-	-	72 (9)	24
Dièdres	-	-	-	7	1

the sequence (i.e., layer EIII), the assemblages are mainly made on locally available thin slabs of siliceous limestone, the same raw material used in the underlying Mousterian levels, which are separated from layer EIII by a layer of sterile volcanic deposits (Palma di Cesnola 1966). S. Milliken, in a very good review of the Italian Mousterian (Milliken 1999–2000) proposes an exotic origin (from Basilicata) of the flint (or "chert" according to her definition) of the Mousterian and transitional assemblages from the Salento. We think, however, that the provenance of the flint in Southern Apulia is still a matter of debate. While Milliken (1999–2000:53) suggests that even the small pebble industry of Mousterian sites on the Adriatic coast (e.g., Polignano, Monopoli) could be imported from Basilicata, it is in our opinion very unlikely that small pebbles would have been imported over long distances. Rather, pebbles are usually collected in nearby riverbeds, well exposed by erosion during glacial times. Moreover, some recent studies of the geology of Apulia show that formations similar to the ones of the Gargano are present in the Murge Baresi formation, and maybe also in the Murge Salentine formation (Ricchetti & Pieri 1999: figures 10 & 11). This question will only be resolved definitely by intensive and systematic geological surveys in the region.

We will not extensively discuss these Uluzzian assemblages here, since they have been the subject of a large number of detailed and well-illustrated papers. We will only mention the existence of two deciduous human molars from layers EIII and EII-I at Grotta del Cavallo, and an infinite conventional ^{14}C date of > 31 kya for layer EII-I.

In southern Tuscany, layer 2 of Grotta della Fabbrica (see stratigraphy in figure 5) has been defined as Uluzzian by Pitti et al. (1976) essentially on the basis of its large number of splintered pieces (figure 8, nos. 5–8; figure 9, nos. 11 & 13) and of *only two* curved backed tools. After a careful reexamination of the tools that came from the likely *in situ* portion of the site, on e of us (A.B.) determined that only one of these curved backed tools (see figure 11, no. 8; made on local jasper) comes from this undisturbed part of layer 2, while the other (see figure 11, no. 7; made on very good quality flint, probably of "Scaglia" type) does not bear any indication as to its stratigraphic provenience.

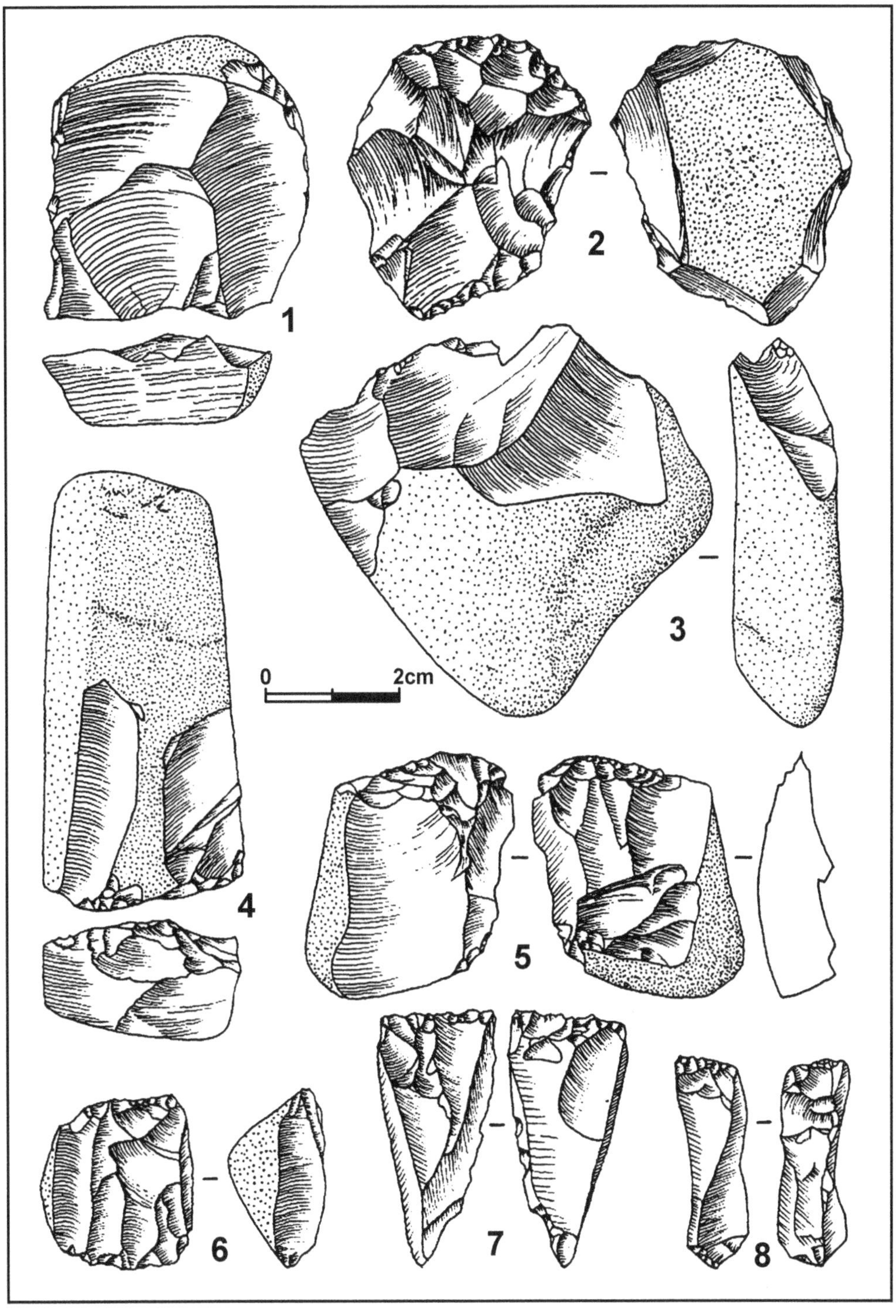

Fig. 5.8. Industry from Grotta della Fabbrica, Layer 2 (from Pitti et al. 1976).

The technological and raw material classification of the *in situ* artifacts from layer 2 of La Fabbrica is shown in table 3. The local raw material comes mainly from the alluvial deposits of the Ombrone River (just to the north of the cave). The jasper is of rather poor quality with quartz and siliceous limestone inclusions; it has been discovered in outcrops south of the city of Siena, not very far from the source of the Ombrone River. The quartz is of extremely bad quality (note the large number of technologically unidentifiable fragments reported in table 3), but this material comes from deposits located near the cave, which opens in a limestone projection of the Uccellina Mountains. Exotic raw material is very rare and is represented by good quality jasper which presumably comes from outcrops near modern-day Livorno and by "Scaglia"-type flint, which comes from outcrops located mainly in

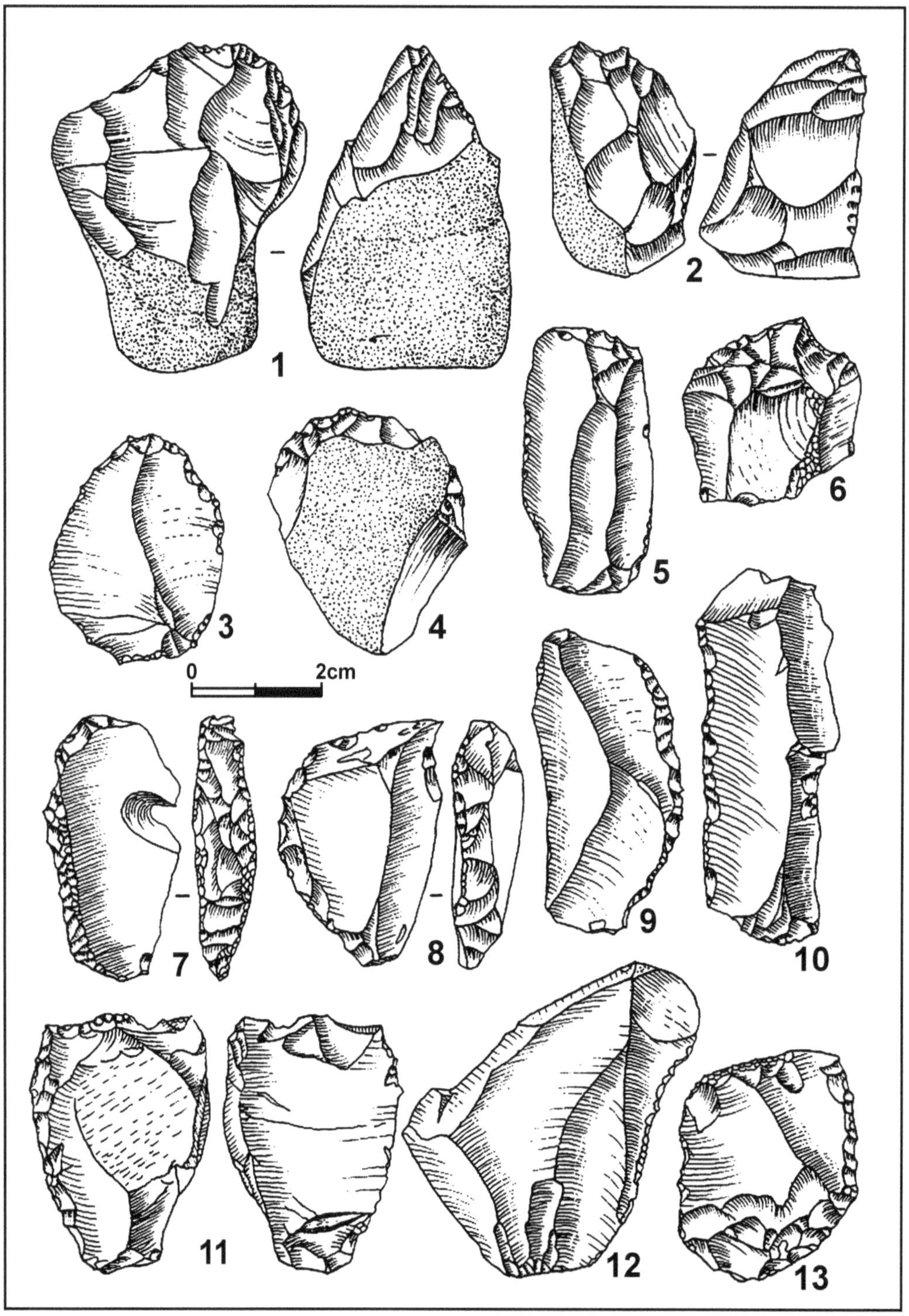

Fig. 5.9. Industry from Grotta della Fabbrica, Layer 2 (continued; from Pitti et al. 1976).

the Apennine Mountains, and in particular in the Tiber valley, at least 100 km east of the site. From a technological point of view, besides some rough pseudo-prismatic cores (e.g., figure 10, no. 1), the assemblages also contains classic centripetal Levallois cores (e.g., figure 10, no. 2), while the remaining cores are mostly shapeless or fragmentary (see table 3). As expected based on the presence of centripetal cores, the lithic industry of layer 2 also contains many predetermined flakes (e.g., figure 11, nos. 3 & 12), as well as some blades and bladelets (see figure 11, no. 10, a Levallois blade). Based on these data, it is our opinion that the assemblage from layer 2 at La Fabbrica shows very little similarity to the "classic" Uluzzian of southern Apulia and that it should therefore instead be considered as one of the various aspects of the late Mousterian in central Italy.

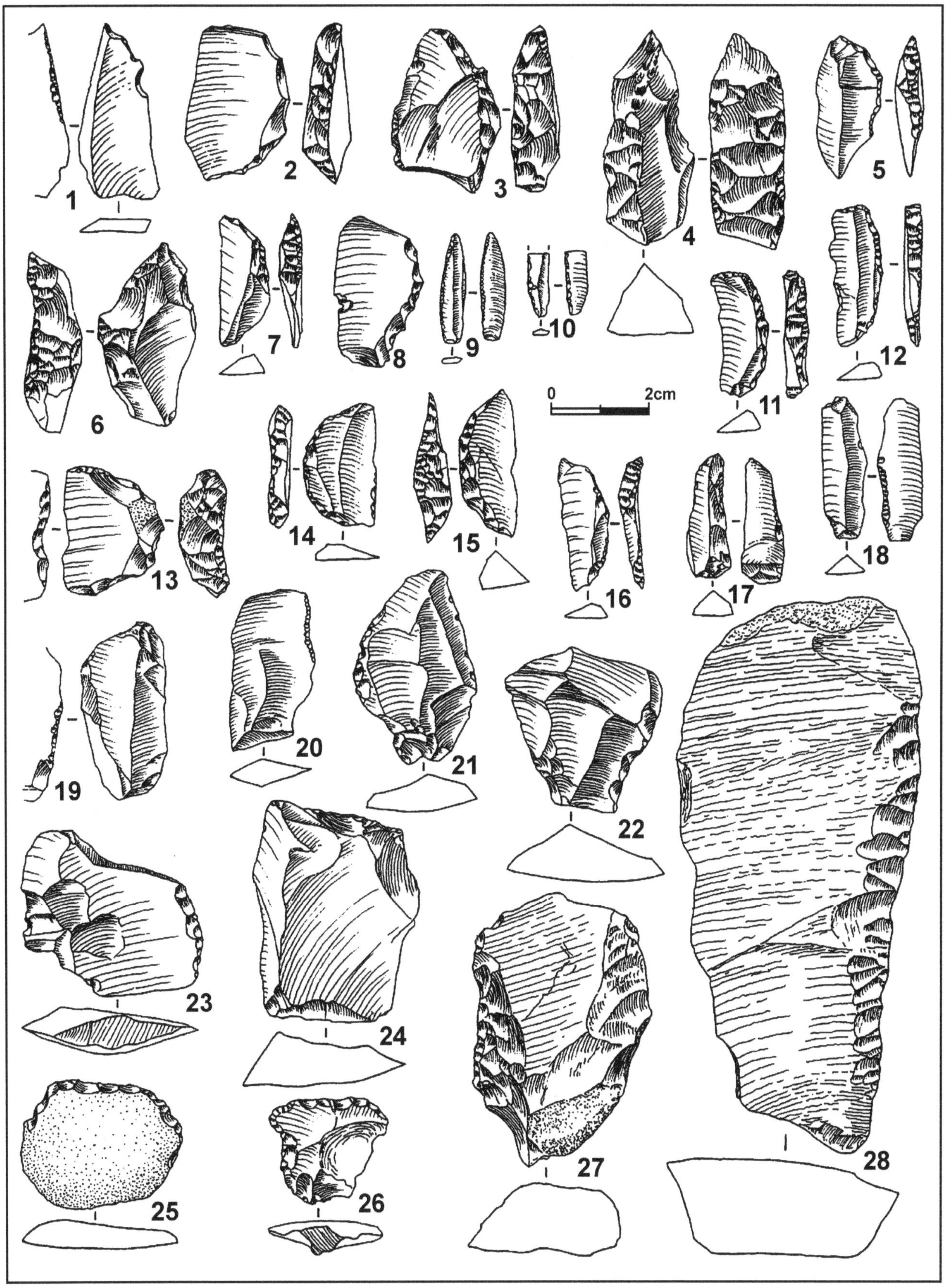

Fig. 5.10. Industry from the Uluzzian layers of Grotta di Castelcivita (from Gambassini 1997).

Table 5.3. Debitage distributions for the industry from Layer 2 of Grotta della Fabbrica.

	Local raw materials			Exotic raw materials	
	Jasper	Flint, quartzite, etc.	Quartz	Fine jasper	*Scaglia* flint
Formal tools	28	3	-	4	2
Burin spalls	5	-	-	-	-
Psp. and crp. cores	9	4	-	-	-
Centripetal cores	5	-	-	-	-
Shpl. and fragm. cores	33	5	5	-	-
Core-like *esquillées*	17	5	3	-	-
Esquillées on flakes, etc.	65	6	13	-	-
Dièdres	25	2	2	-	-
Predetermined flakes	42	10	4	-	-
Blades and bladelets	26	7	3	-	2
Rejuvenation flakes	5	1	-	-	-
Flakes	418	83	30	-	-
Totals	678	126	60	4	4
Raw material fragments	-	-	158	-	-

Critics might object to this assessment on the basis of the large number of *pièces esquillées* in this assemblage. While this feature is indisputable, it is worth noting that, at La Fabbrica, this category of stone implements includes several "core-like" types (e.g., figure 10, no. 6) which are basically exhausted cores reduced further through the use of bipolar percussion on anvil. Other technological products of this technique are known as "dièdres," which are also found in this assemblage (table 3). It is worth emphasizing that the "percussion on anvil" technique is found in several other Italian late Mousterian lithic assemblages. For instance, it is rather common in the Mousterian on small pebble of southern Latium such, for instance, that of Grotta Guattari (Bietti & Grimaldi 1996). The only really "new" tool in layer 2 at La Fabbrica is a worked bone awl (see Pitti et al. 1976: figure 6, no. 1), but we would argue that a single tool, in a deposit with a complex—and to a certain extent uncertain—stratigraphy is unlikely to be very significant.

Shifting our focus to the surface collections from Tuscany described by Palma di Cesnola (1993) as Uluzzian (e.g., Maroccone, San Romano, Poggio Clavello and others), we prefer not to discuss them in the present study. This is not because we espouse the extreme position that they *all* represent mixed samples of Mousterian and Aurignacian assemblages (see e.g., Gioia 1990, Mussi 1990), but rather because of their uncertain depositional context and the fact that they have been defined as Uluzzian solely on typological grounds, mainly based on the presence of the famous curved backed tools discussed above.

A very different argument holds for the Uluzzian of Castelcivita. As can be seen by the tools shown in figure 10, some of these crescents (notably nos. 5 & 16) are very similar to the classic ones found at Grotta del Cavallo. The two illustrated Dufour bladelets (figure 10, nos. 9–10) come from cut 10, so that some intrusion from the overlying Aurignacian of cut 9 cannot be excluded (P. Gambassini, personal communication). As well, the centripetal cores prevalent in the underlying Mousterian disappear and the vast majority of cores are shapeless (Gambassini 1997:115). Another marked difference with the Mousterian is evident in the debitage distribution for cuts 11–10, shown in the fourth column of table 2. This clearly indicates a drastic reduction in the frequency of predetermined flakes (practically no real Levallois flakes are present) and the sudden appearance of many *pièces esquillées*. More limestone artifacts are also present, and this is probably due to a change in climatic conditions, as can be inferred from Phase VI of Fumanal's sedimentary analysis (figure 7). Overall, then, the Uluzzian and Mousterian assemblages really seem to represent two entirely different worlds at Castelcivita, and we retain the Uluzzian denomination for this site to be valid, even if different in many respects from that documented at sites in the Salento.

Concluding Remarks and Comparisons with the Early Aurignacian

The first conclusion that can be drawn from these analyses is that, in Itlay, the presence of identifiable stratified "transitional" lithic industries (i.e., the Uluzzian) is practically limited to two locales: the site of Castelcivita and the cluster of sites on the Ioninan coast in the Salentine peninsula. In spite of its bone awl, the assemblage from layer 2 at Grotta della Fabbrica is much more likely to represent a late Mousterian with some affinities with the assemblage from the upper layers of Grotta Breuil. From a technological point of view, the assemblage from Via San Francesco, in Liguria, appears much more "modern"

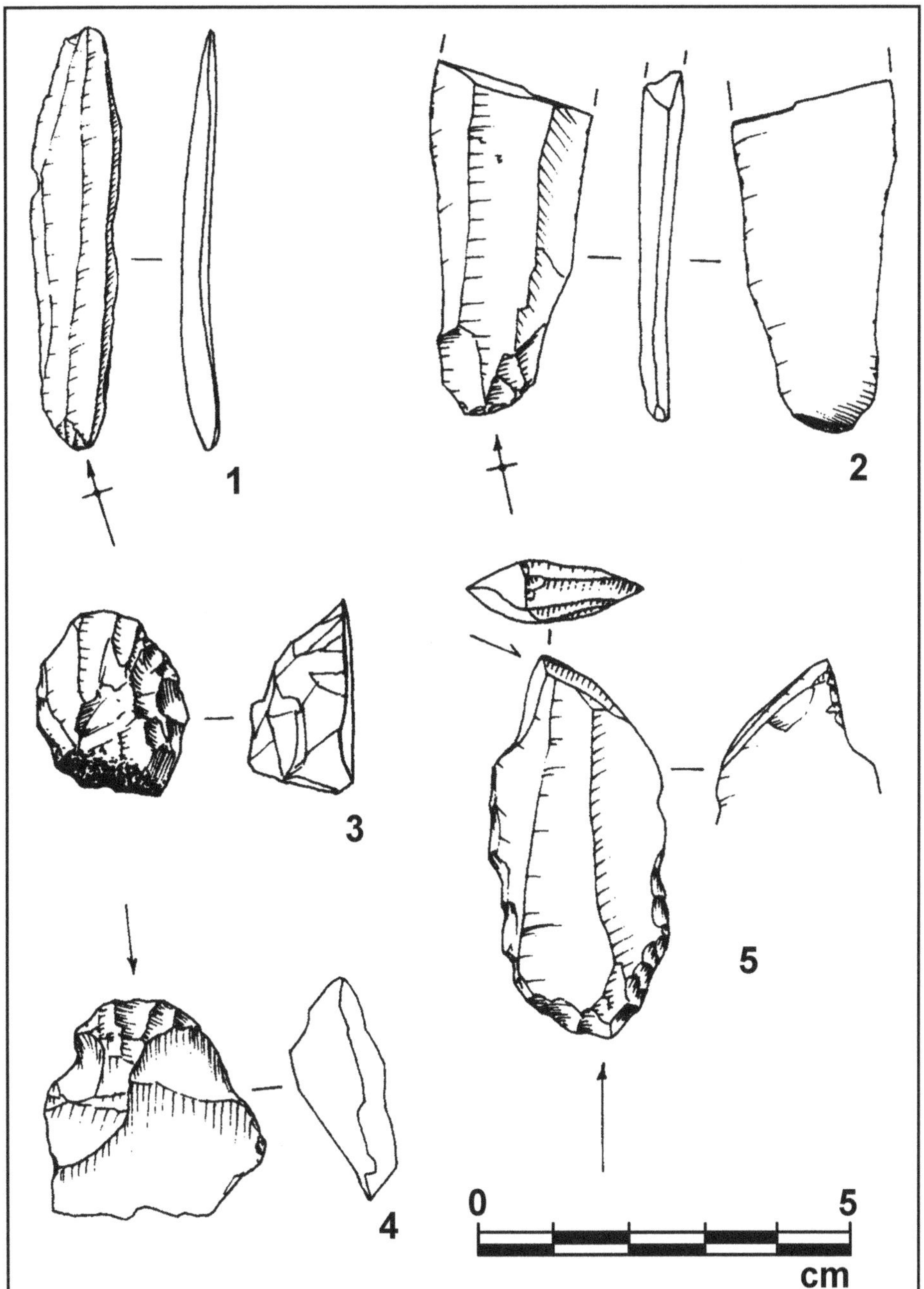

Fig. 5.11. Aurignacian industry from Riparo Mochi.
1–2: Layer G; *3–5:* Layer F (nos. 3–4 on local Ciotti flint; nos. 1, 2 and 5 on Vaucluse flint).

and "progressive" that those of Grotta Breuil or Grotta della Fabbrica. Furthermore, and always based on a perspective focused on the evaluation of efficiency and the degree of predetermination in reduction techniques, the Uluzzian assemblage from Castelcivita looks definitely more "regressive" when compared to the Levallois Mousterian assemblage of the same site, and this despite the presence of a few bone tools in the Uluzzian.

Turning to a discussion of the early Aurignacian, we have stressed repeatedly here and elsewhere that this industry is definitely older in northern Italy than elsewhere in the peninsula[1] (Bietti 1997, Kuhn & Bietti 2000). In southern Italy, the Aurignacian is younger than or occasionally contemporary with the Uluzzian (table 1; see also dates for Grotta della Cala, Grotta Paglicci & Serino in Kuhn & Bietti [2000]). Despite of differences in timing, one can nevertheless see a striking difference and "progress" both in the technology and raw material economy of early Aurignacian assemblages relative to earlier industries. As regards technology, there is the sudden appearance of a very specialized blade and bladelet technology obtained through the use of prismatic and pyramidal cores (occasionally with two opposed striking platforms), which are rather common in the assemblages from Grotta di Fumane (Bartolomei et al. 1992), layer G at Riparo Mochi (table 4; figure 2, nos. 5 & 6), layers 3–4 at Grotta della Fabbrica (table 5), and Castelcivita (Gambassini 1997).The relative frequencies of blades and bladelets (see e.g., figure 1, nos. 5–12; figure 11, nos. 1 & 2) also increases sharply, both at Castelcivita (last column of table 2) and at Grotta della Fabbrica (table 5). At these sites, the only element in common with their underlying layers is the persistence of relatively large numbers of *pièces esquillées.*

Another important difference that coincides with the early Aurignacian is in raw material procurement patterns. For instance, at Riparo Mochi, exotic flint becomes a common raw material, as has already been noted (Kuhn & Stiner 1998), and it becomes possible to determine the provenience of these flint types and their prevalence in both formal tools and debitage (table 4; note that the Ligurian radiolarite is a fine quality jasper similar to the

Table 5.4. Debitage distributions for the early Aurignacian industry from Layer G at Riparo Mochi (V.F. = very few; Pres. = present).

				Retouched and simple blades		Bladelets and "armatures"		Large formal tools (scrapers, burins, etc.)	
Riparo Mochi - Layer G		Cores	Flakes	N	%	N	%	N	%
Local *(within 20 km)*	Flint *(Ciotti)*	V.F.	V.F.	0	0	340	70.30	14	27.50
	Quartzite *(San Remo)*	-	Pres.	0	0	3	0.62	4	7.84
Exotic *(over 20 km)*	Flint *(Vaucluse, Mons, etc.)*	Rare	Pres.	3	50.00	91	18.80	24	47.10
	Riolite *(Esterel)*	-	Pres.	0	0	1	0.21	0	0
	Radiolarite *(Eastern Liguria)*	-	Pres.	3	50.00	20	4.13	3	5.88
	Scaglia flint *(Central Italy?)*	-	Pres.	0	0	10	2.07	0	0
Other flint *(uncertain provenance)*		-	Pres.	0	0	19	3.92	6	11.80

Table 5.5. Debitage distributions for the industry of Layers 3–4 from Grotta della Fabbrica.

	Local raw materials			Exotic raw materials	
	Jasper	Flint, quartzite, etc.	Quartz	Fine jasper	*Scaglia* flint
Formal tools	2	6	-	6	6
Burin spalls	1	-	-	-	-
Psp. and crp. cores	3	2	-	-	-
Centripetal cores	-	-	-	-	-
Shpl. and fragm. cores	16	5	1	-	-
Core-like *esquillées*	6	-	-	-	1
Esquillées on flakes, etc.	29	1	4	-	-
Dièdres	5	1	4	-	1
Predetermined flakes	10	2	0	1	1
Blades and bladelets	17	2	-	-	7
Rejuvenation flakes	3	-	-	-	1
Flakes	163	44	11	-	8
Total	255	63	20	7	25
Raw material fragments	-	-	85	-	-

one found at La Fabbrica). Comparing the geographical extent of raw material procurement patterns of the three kinds of industries studied in this paper (see figure 12), one can readily observe the striking differences between the relatively similar patterns of the late Mousterian and Uluzzian assemblages, on one hand, and that of the Aurignacian assemblages, on the other. It is also interesting to note that there is practically no imported raw material in the sites associated with sources of good quality local raw material (i.e., Fumane, Castelcivita, La Cala and Paglicci), whereas areas with poor quality material, such as the Monte Circeo sites, only see the importation of exotic material starting with the Aurignacian (e.g., Fossellone).

In closing, we will take a few words to discuss the "future" of the "transitional" assemblages from southern Italy. In our view, there does not appear to be an evolution from the Mousterian to the Uluzzian nor from the Uluzzian to the Aurignacian. It is well known that Palma di Cesnola (1993) believes in an evolutionary within the Uluzzian exemplified by the three "phases" identified at Grotta del Cavallo, and even of the Uluzzian into an "Uluzzo-Aurignacian" as documented in the deposits of nearby Grotta della Serra Cicora (Spennato 1981). We think that it is probably too early to postulate such scenarios (see also discussion in Bietti 1997, Kuhn & Bietti 2000). After all, even though there is a broad consensus that the Uluzzian was produced by Neandertals, the relevant human fossil record is almost non-existent which means that, strictly speaking, we cannot really assess the likelihood of any biological relation between the populations of Uluzzian and Aurignacian toolmakers, let alone some form of hybridization. And for all we know, the few bone tools documented in some Uluzzian and late Mousterian assemblages may well have been the result of contacts between Neandertals and AMH. This need not necessarily have taken the form of "acculturation" understood as some form of colonization, as suggested by d'Errico and

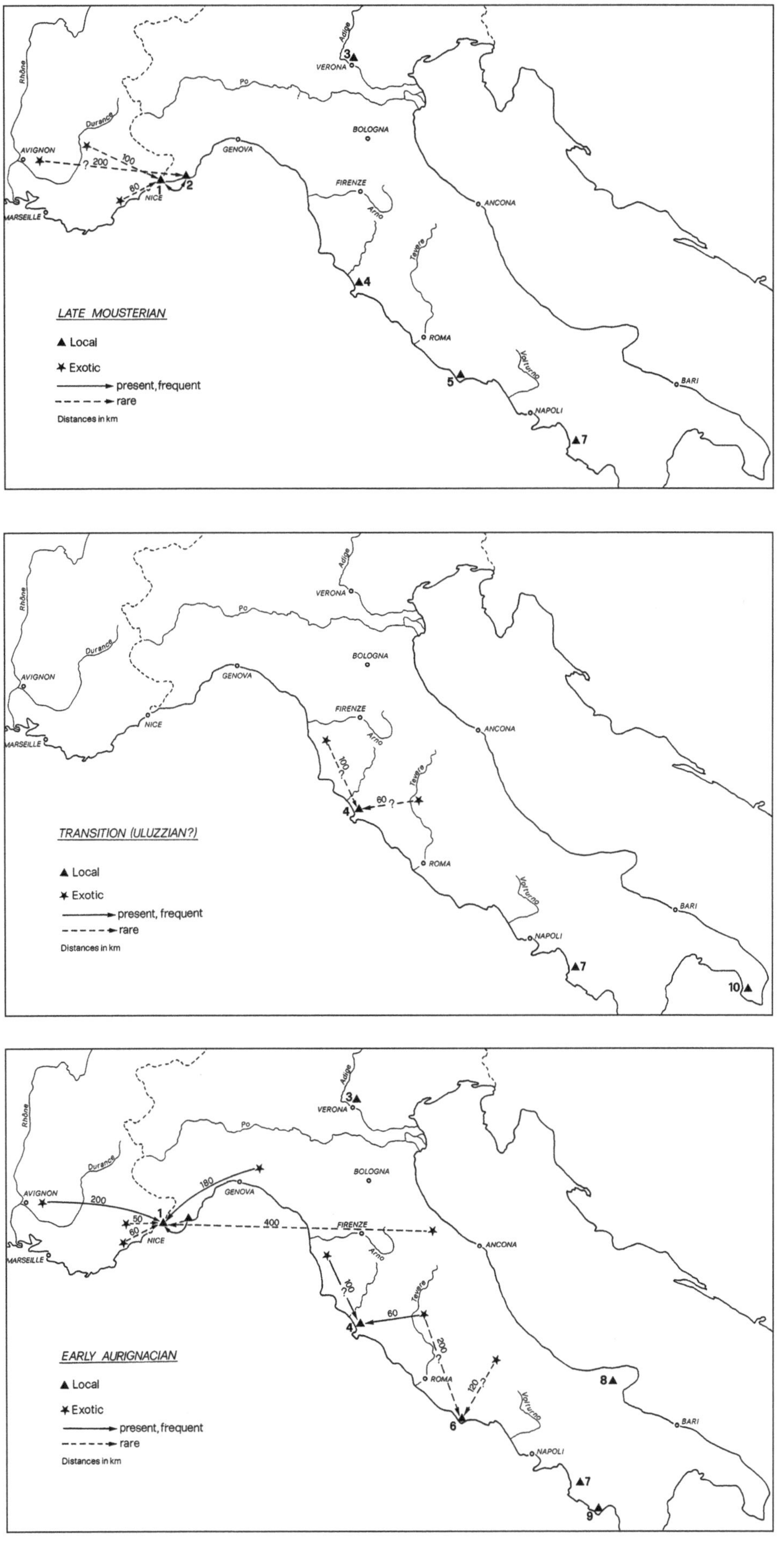

Fig. 5.12. Schematic representation of raw material procurement patterns at Late Mousterian *(top)*, "transitional" *(middle)*, and early Aurignacian sites *(bottom)*.

colleagues (1998). In fact, contacts between the two populations would not even have had to be continuous or direct for this to happen (see discussion in Kuhn & Bietti 2000). It is worth stressing again that more in-depth and sophisticated analyses should be conducted on the assemblages from Grotta della Serra Cicora. For instance, J. Riel-Salvatore and M. Barton have recently shown that at Serra Cicora, there are some surprising technological similarities between the "early Aurignacian" levels A and B and the lower Uluzzian layer D, which all diverge from patterns displayed in the intermediate Uluzzian layer C (Riel-Salvatore & Barton 2004:266). A possible local (i.e., Salentinian) evolution from the Uluzzian to the early Aurignacian is another scenario that should be tested in future studies, for instance through comparisons with the early Aurigancian of Grotta Paglicci (Palma di Cesnola 1993).

In our opinion, one main tentative conclusion is suggested by the data presented in this paper. There may well be some kind of "progressive" evolutionary process from the Mousterian to the Uluzzian in the realms of bone technology (although we are, in Italy, very far from the situation at Arcy-sur-Cure!) and of hunting efficiency (although Stiner [1994] is the only well-documented study of this behavior). Such a tendency does not, however, exist at all in the fields of lithic technology and of raw material procurement strategies. In this aspect, the early Aurignacian, especially at Riparo Mochi, really seems to indicate a completely different social organization among AMH compared to that of the previous inhabitants of the Italian peninsula.

Acknowledgments

We wish to thank the Soprintendenza Archaeologica della Liguria for their kindness and continuous assistance in examining the lithic collections of Riparo Mochi and especially those from Via San Francesco, now housed in the city museum of San Remo. For extensive discussions and exchanges of ideas on lithic raw material sources in southeastern France, we also wish to thank Dr. Didier Binder of the CNRS laboratory in Valbonne and Dr. Patrick Simon of the Museum of Monaco. One of us (A.B.) especially wishes to acknowledge the kind hospitality given by Prof. Carlo Tozzi of the Dipartimento di Scienze Archeologiche of the University of Pisa, for a reexamination of the assemblages from Grotta della Fabbrica. In the context of that work, the help given by Dr. Mario Dini with regards to the reassessment and separation of the different stratigraphic units was of paramount importance. The same co-author (A.B.) also wants to extend his acknowledgements to Prof. Paolo Gambassini of the University of Siena for his assistance during the reexamination of some samples of the assemblages from Grotta di Castelcivita. A number of discussions of the raw materials of Tuscany with Prof. A. Galiberti are also fully acknowledged. We also wish to thank Julien Riel-Salvatore for a number of fruitful discussions and suggestions.

Endnote

1. Besides the early dates from Fumane, the base of layer G at Riparo Mochi has yielded a date of 37.4 ± 1.3 kya. (Radiocarbon laboratory of the Department of Physics of the University of Rome, sample N. R-).

References

Aiello, L.
1993 The fossil evidence for modern human origins in Africa: a revised view. *American Anthropologist* 95:73–96.

Alhaique, F., A. Bietti, L. Capasso Barbato, S. Grimaldi, T. Kotskakis, S. Kuhn, C. Lemorini, G. Manzi, A. Recchi, H. P. Schwartcz, & M. Stiner
1998 Settlement patterns and subsistence strategies in the late Mousterian of Grotta Breuil (Mt. Circeo, Italy). In *Proceeding of the XIII U.I.S.P.P. International Congress,* vol. 6, workshop 5: 221–228, Abaco Edizioni, Forlì.

Alhaique, F., A. Bietti, A. del Lucchese, S. Grimaldi, G. Manzi, S. Martini, F. Negrino, & A. Recchi
2000 Biological and cultural variability at the Neandertal/modern human transition in Italian archaeological sites. *Rivista di Antropologia* 78:105–116

Bartolomei, G., A. Broglio, P. F. Cassoli, L. Castelleti, L. Cattani, M. Cremaschi, G. Giacobini, G. Malerba, A. Maspero, M. Peresani, A. Sartorelli, & A. Tagliacozzo
1992 La Grotte de Fumane. Un site aurignacien au pied des Alpes. *Preistoria Alpina* 28:131–179.

Bietti, A.
1997 The transition to anatomically modern human: the case of peninsular Italy. In *Conceptual Issues in Modern Human Origins Research* (G. A. Clark & C. M. Willermet, eds.), pp. 132–147. Aldine de Gruyter, New York.

Bietti, A., & S. Grimaldi
1996 Small flint pebbles and Mousterian reduction chains: the case of southern Latium (Italy). *Quaternaria Nova* 6:237–260.

Boëda, E.
1994 *Le Concept Levallois: Variabilité des Méthodes.* Monographie du CRA 9. Editions du CNRS, Paris.

Bordes, J. G.
2003 Lithic taphonomy of the Chatelperronian/Aurignacian interstratifications in Roc de Combe and Le Piage (Lot, France). In *The Chronology of the Aurignacian and Transitional Technocomplexes: Dating, Stratigraphies, and Cultural Implications,* (F. d' Errico & J. Zilhao, eds.), pp. 223–245.

Trabalhos de Arqueologia 33. Instituto Português de Arqueologia, Lisbon.

d'Errico, F., J. Zilhão, M. Julien, D. Baffier, & J. Pelegrin
1998 Neandertal acculturation in western Europe? A critical review of the evidence and its interpretation. *Current Anthropology* 39:S1–S44.

Fumanal, M. P.
1997 La Cueva de Castecivita. Estudio sedimentologico. In *Il Paleolitico di Castelcivita, Culture e Ambiente* (P. Gambassini, ed.), pp. 19–32. Materiae 5. Electa, Naples.

Gambassini, P. (editor)
1997 *Il Paleolitico di Castelcivita, Culture e Ambiente.* Materiae 5. Electa, Naples.

Gioia, P.
1990 An aspect of the transition between the Middle and Upper Palaeolithic in Italy: the Uluzzian. In *Paléolithique moyen recent et Paléolithique supérieur ancien en Europe* (C. Farizy, ed.), pp. 295–303. Musée Préhistorique de l'Île-de-France, Nemours.

Isetti, G.
1961 Una stazione paleolitica nel centro di Sanremo. *Rivista di Studi Liguri* 28: 5–44.

Kuhn, S. L.
1995 *Mousterian Lithic Technology: An Ecological Perspective.* Princeton University Press, Princeton.

Kuhn, S. L., & A. Bietti
2000 The late Middle and early Upper Paleolithic in Italy. In *The Geography of Neandertals and Modern Humans in Europe and the Greater Mediterranean* (O. Bar-Yosef & D. Pilbeam, eds.), pp. 49–76. Peabody Museum Bulletin no. 8. Harvard University, Cambridge.

Kuhn, S. L., & M. C. Stiner
1992 New research on Riparo Mochi, Balzi Rossi (Liguria): preliminary results. *Quaternaria Nova* 2:77–90.
1998 The earliest Aurignacian of Riparo Mochi (Liguria, Italy). *Current Anthropology* 39: S175–S189.

Mellars, P.
1996 Models for the dispersal of anatomically modern populations across Europe: theoretical and archaeological perspectives. In *The Lower and Middle Palaeolithic* (O. Bar-Yosef, L. Cavalli-Sforza, R. March & M. Piperno, eds.), pp. 225–235. Abaco Edizioni, Forlì.
2000 The archaeological records of the Neandertal-modern human transition in France. In *The Geography of Neandertals and Modern Humans in Europe and the Greater Mediterranean* (O. Bar-Yosef & D. Pilbeam, eds.), pp. 35–47. Peabody Museum Bulletin no. 8. Harvard University, Cambridge.

Milliken S.
1999–2000 The Neanderthals in Italy. *Accordia Research Papers* 8:11–82.

Mussi, M.
1990 Le peuplement de l'Italie à la fin du Paléolithique moyen et au début du Paléolithique supérieur. In *Paléolithique moyen recent et Paléolithique supérieur ancien en Europe* (C. Farizy, ed.), pp. 251–262. Musée Préhistorique de l'Île-de-France, Nemours.

Palma di Cesnola, A.
1966 Il Paleolitico superiore arcaico (facies uluzziana) della Grotta del Cavallo, Lecce. *Rivista di Scienze Preistoriche* 20:33–62.
1967 Il Paleolitico superiore arcaico (facies uluzziana) della Grotta del Cavallo, Lecce. (Continuazione). *Rivista di Scienze Preistoriche* 21:3–59
1993 *Il Paleolitico superiore in Italia: Introduzione allo studio.* Garlatti e Razzai Editori, Florence.

Peresani, M., & A. Sartorelli
1998 The lithic assemblages at the cave of Fumane. New evidence of technological variability in the Middle Paleolithic of northern Italy. In *Proceedings of XIII U.I.S.P.P. International Congress,* vol. 2, workshop 5:269–278. Abaco Edzioni, Forlì.

Pitti, C., C. Sorrentino, & C. Tozzi
1976 L'industria di tipo Paleolitico superiore arcaico di Grotta della Fabbric (Grosseto): nota preliminare. *Atti della Società Toscana di Scienze Naturali* 83:174–203.

Ricchetti G., & P. Pieri (editors)
1999 *Guida geologica regionale della Puglia e Monte Vulture.* Società Geologica Italiana, Bari.

Riel-Salvatore J., & C. M. Barton
2004 Late Pleistocene Technology, Economic Behavior, and Land-Use Dynamics in Southern Italy. *American Antiquity* 69:257–274.

Rossetti, P., & G. Zanzi
1990–1991 Technological approach to reduction sequences of the lithic industry from Grotta Breuil.

Quaternari Nova 1:351–365.

Spennato, A.
1981 I livelli protoaurignaziani della grotta di Serra Cicora (Nardò, Lecce). *Studi per l'Ecologi del Quaternario* 3:61–76.

Stiner, M. C.
1994 *Honor among Thieves: A Zooarchaeological Study of Neandertal Ecology.* Princeton University Press, Princeton.

Stringer, C. B.
1994 Out of Africa—a personal history. In *Origins of Anatomically Modern Humans* (M. H. Nitecki & D. V. Nitecki, eds.), pp.149–172. Plenum Press, New York.

Tavoso, A.
1988 L'outillage du gisement de San Francesco a San Remo (Ligurie, Italie): nouvel examen. In *l'Homme de Néandertal 8: La Mutation* (J. K. Kozlowski, ed.), pp. 193–210. ERAUL No.36, Liège.

Wolpoff, M., & R. Caspari
1996 Why aren't Neandertals modern humans? In *The Lower and Middle Palaeolithic* (O. Bar-Yosef, L. Cavalli-Sforza, R. March & M. Piperno, eds.), pp. 133–156. Abaco Edizioni, Forlì.

6

New Quantitative Perspectives on the Middle–Upper Paleolithic Transition: The View from the Northern Mediterranean

Julien Riel-Salvatore
Arizona State University

C. Michael Barton
Arizona State University

Abstract

In this paper, we present a method to analyze lithic assemblages that can yield detailed information about Paleolithic technological organization patterns and land-use strategies. Applying it to series of Late Pleistocene assemblages from sites distributed along the northern Mediterranean (i.e., Gibraltar, eastern Spain and southeastern Italy) and spanning the Middle–Upper Paleolithic Transition, we show how this approach can be used to track diachronic behavioral change for Paleolithic hunter-gatherers. We demonstrate that the approach has wide applicability, that it can be used to reconstruct prehistoric land-use practices provided sedimentation rates can be accounted for, and that it provides a better explanation for diachronic lithic variability than schemes that simply classify assemblages in poorly defined "archaic" to "evolved" sequences. This analysis strongly suggests that no marked shift in the flexibility of technological organization or land-use strategies coincides with the Middle–Upper Paleolithic Transition. The implications of these findings and the apparent disjunction of behavioral and biological change known to have taken place during that interval lead us to outline new kinds of research questions which may better enable researchers to comprehend the mechanisms of the Transition process.

The Middle–Upper Paleolithic transition (hereafter, the Transition) continues to generate considerable popular and scientific interest. As the interval that saw the disappearance of Neanderthals as a discrete population and the putative widespread diffusion of 'modern behavior' outside of Africa, the Transition provides continued impetus for research across an ever-widening array of disciplines, including most prominently archaeology, human paleontology, ecology, geology, and genetics. Despite the number of studies centered on this topic, resolution of the fundamental questions associated with the Transition still eludes us, as the lively debates fueled by its empirical record demonstrate (see papers in Bar-Yosef & Pilbeam 2000, Hays & Thacker 2001, Zilhão & d'Errico 2003, Brantingham et al. 2004a, Straus 2005). In part, this situation is due to a generalized lack of interest in formulating widely accepted theoretical scaffoldings to frame the debate and explicitly define the expectations of researchers on both sides of the issue (see Clark & Riel-Salvatore 2006, Riel-Salvatore & Clark 2001).

To a degree, this is inevitable in a field of research as discovery driven as paleoanthropology (but see Brantingham et al. 2004b for a welcome exception). Perhaps more problematic, however, is the lack of appropriate methodologies to effectively capture and coherently present the dynamics of the Transition empirical record from an empirical record whose most abundant and best preserved components are lithic assemblages. As we have argued elsewhere (e.g., Riel-Salvatore & Barton 2004), this is in part due to the influence of typological approaches, so formative to our understanding of the Pleistocene archaeological record, that have imposed an *a priori* distinction between the Middle and Upper Paleolithic records (Simek and Price 1990, Grayson and Cole 1998, Marks et al. 2001). Theses systematics require that assemblages first be classified as either Middle *or* Upper Paleolithic. They are subsequently classified in typologies that differ fundamentally with regards to the types that make up each system. Furthermore, this makes these typologies inherently inadequate tools to classify and study the *transition* in lithic assemblages that span the two periods (Grayson & Cole 1998).

This paper is an effort to draw together the results of several past studies in order to demonstrate the broad usefulness of an approach to lithic analysis (originally proposed by Barton [1998]) that can be applied equally to Middle and Upper Paleolithic assemblages and its potential for revealing changing behavioral dimensions of the Transition. Here, we summarize the results of three studies that have employed this methodology to analyze Late Pleistocene assemblages from Gibraltar, eastern Spain and southeastern Italy. We use these results as a basis for comparing the flexibility of Mousterian and early Upper Paleolithic technocomplexes and the ways in which the makers of these artifacts exploited their landscapes. Our results, drawn from a dozen geographically dispersed sites, situated across a variety of ecological contexts, provide compelling evidence that the Transition does not coincide with a dramatic shift in landscape use. We conclude by considering the implications of our findings for clarifying our understanding of fitness-enhancing mechanisms that may have been at play during that crucial interval of recent human evolution, linking our behavioral observations to some of the biological dimensions of the Transition.

Methodology

The method used in this paper was originally proposed by Barton (1998) and subsequently developed in greater detail by the authors (Riel-Salvatore & Barton 2004). We present a brief overview of it here and refer the reader to the earlier publications for a more detailed discussion.

Our approach begins with plotting the frequency of retouched tools in an assemblage against its total lithic artifact volumetric density, defined as the quantity of chipped stone per cubic meter of sediment from which the assemblage was derived under conditions of similar depositional rates and archaeological recovery methods. The resulting relationship is predicted to be negative; assemblages with high frequencies of retouched pieces will tend to have comparatively low densities of artifacts while, in contrast, assemblages with low amounts of retouch will have higher densities. This pattern, we argue, reflects accumulated artifacts deriving from primarily curated to primarily expedient artifact use respectively. These differences are best distinguished when all the assemblages from a site or a series of sites are plotted on the same graph, and both axes are expressed as log scales (see figure 1). It is important to emphasize that we are not here arguing that the terms "expedient" and "curated" reflect individual site-occupation events. Rather, in an archaeological context, these labels refer to time-averaged suites of strategies resulting from a palimpsest of occupations, the predominant character of which will dominate the signature of a given archaeological assemblage.

Expedient assemblages often accumulate at intensively occupied sites, where raw material was *effectively abundant* as a result of any combination of natural availability, stockpiling and/or other behavioral mechanisms. Curated assemblages, in contrast, more commonly derive from relatively short occupations of a site under conditions of raw material *effective scarcity,* due either to natural unavailability or to behavioral mechanisms, such as high mobility, that restrict raw material use.

These characteristics of assemblages also can be linked to the prevalent land-use strategy employed by the hunter-gatherer groups responsible for their manufacture, use, maintenance and deposition as originally defined by Binford (1979, 1980; see also Nelson 1991). Expedient assemblages are often a consequence of logistical mobility, whereby a central site is occupied for prolonged periods of time while task-groups come and go from it to procure various non-local resources. In contrast, curated assemblages are by and large expected in cases of residential mobility whereby a group of hunter-gatherers moved their home base camp frequently to go from resource patch to resource patch and artifact portability is important.

Thus, it becomes possible to link lithic assemblages to generalized patterns of land-use and landscape exploitation. Importantly, since this analytical method does not depend on typologies specific to either Middle or Upper Paleolithic assemblages, it enables the comparison of behavioral modalities across time and space. In the context of Transition research, it therefore provides a powerful methodological tool to assess whether the makers of different industries appear to have exploited their landscapes differently, or whether they display comparable ranges of behavioral flexibility. As will be discussed below, the dominance of one or the other mode in given technocomplexes may also have significant implications about how 'behavioral modernity' may be defined as based on lithic assemblages.

This method addresses questions related to the use-life of lithic assemblages and to landscape-use; it provides no information about the typological or technological make-up of given assemblages. Our approach is not designed to answer all the questions pertaining to lithics in the context of the Transition. What it does allow, however, is the direct comparison of assemblages argued on technotypological grounds to be different. This therefore enables researchers to address questions which traditional typotechnological systematics simply are not suited to tackle, such as the potential behavioral connotations of technotypological differences.

As we have previously noted (i.e., Barton 1998, Riel-Salvatore & Barton 2004), sedimentation rates and diagenesis can influence the results obtained through this approach. Fine-grained chronological data can provide estimates of deposition rates for given assemblages, while sedimentary analysis can indicate the effects of postdepositional forces on artifact counts and sediment volume. Such information significantly helps to clarify the original depositional regime under which an assemblage entered the archaeological record (Karkanas et al. 2001, Weiner et al. 2002). In the absence of such information, the inferences drawn from our method can be admittedly tentative in cases where non-cultural depositional processes appear to have varied across assemblages. This further emphasizes the need for careful geoarcheological studies in contemporary prehistoric research of all sorts. However, deviation from expected patterning can equally serve to identify contexts of variable depositional and postdepositional processes.

Case Studies

The case studies used in this paper extend around the western and northern margins of the Mediterranean Sea: Gibraltar, eastern Spain, and southeastern Italy. Because these sites are distributed over a wide area, they permit us to test this method beyond the influence of local conditions on Late Pleistocene technological behavior. As well, each of the three regions contains several technocomplexes, including the Mousterian, the Aurignacian, and the Uluzzian in the case of southeastern Italy. This offers the potential to take an interregional view of the process of the Middle–Upper Paleolithic Transition and to highlight its common aspects as well as its regional peculiarities.

1. Gibraltar

The site of Gorham's Cave is located on the Gibraltar peninsula and has a long history of research. The results of our analysis are based on a sample of the material from Waechter's original excavations at the site between 1948 and 1954, reported in detail by Barton (1988). Our model was applied first to the Gorham's Cave assemblages and they closely match its expectations (figure 1). The data suggest that Mousterian occupations at Gorham's Cave did not conform to a single land-use modality. Rather they show that site-use varied, suggesting possible shifts in overall land-use strategy, with the site alternating between a ephemeral residential site and a 'home base' from which logistical forays were conducted to procure other resources. Layers G, K, M and P appear to represent a prolonged use of the site as a 'home base' reflected by higher densities of artifacts and lower frequencies of retouched pieces, indicating that the site likely was provisioned with

Fig. 6.1. Graphical depiction of patterns for Paleolithic assemblages form Gorham's Cave, predicting a negative relationship (r = -.924, p = 0.000, N = 10) between artifact volumetric density and frequency of retouched pieces, along with associated technological strategies (after Barton 1998).

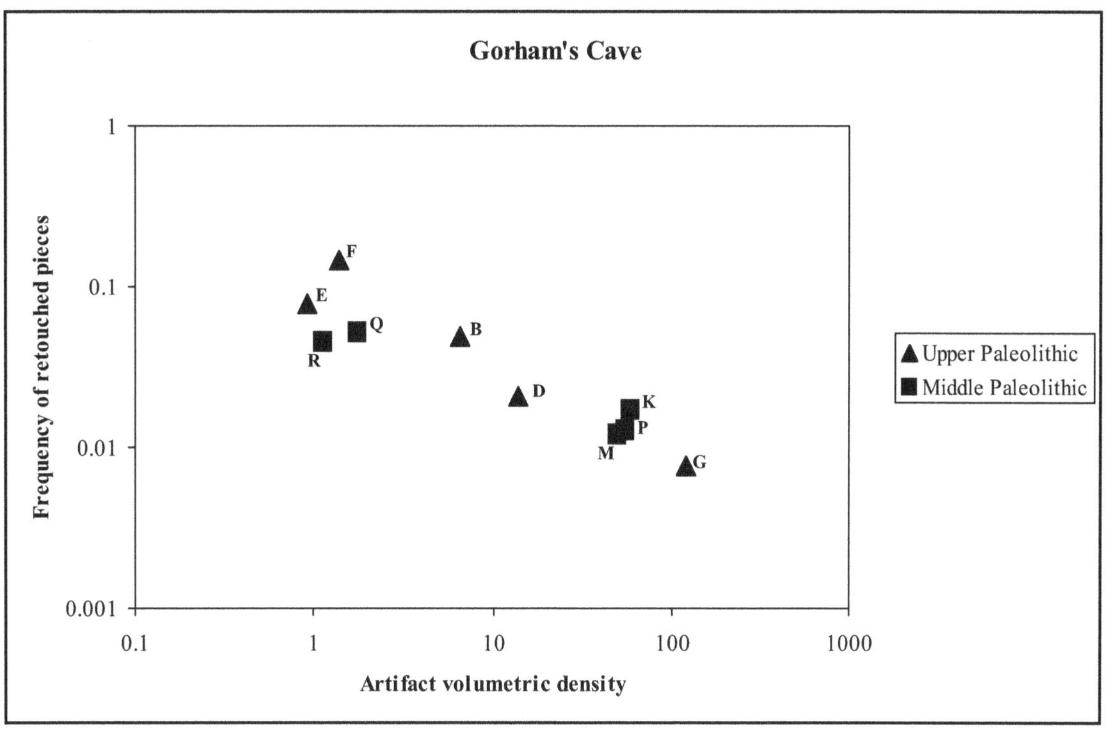

raw material and that expedient lithic production was the dominant strategy at the site. In contrast, layers Q and R document more intensive use of lithic resources, resulting in lower artifact densities and higher frequency of retouch. These assemblages probably reflect the use of Gorham's Cave as short-term residential sites by highly mobile foragers who discarded exhausted (i.e., highly retouched) components of their lithic assemblages. Based on currently available data, it is difficult to assess whether layers Q and R could reflect repeated use of the site as specialized activity area (cf. Riel-Salvatore & Barton 2004:259).

In light of a replacement model which often predicts large-scale discontinuities in land-use patterns between the Middle and Early Upper Paleolithic (e.g., Bar-Yosef 2002, Mellars 1996), it is striking that the upper levels of Gorham's Cave provide clear evidence of the continuation of this alternation in site occupation modality during the Upper Paleolithic. For this time period, layers B and D appear to document more prolonged use of the site with a concomitant increased reliance on expedient lithic technology while layers E and F document short-term site occupations with discard of more curated material. Also of interest is that there appears to be no unilineal trend in the adoption of one or the other strategy during the Upper Paleolithic. Again, currently available data make it hard to assess whether layers E and F reflect specialized activity or short-term residential occupations.

Thus, it cannot be excluded that curated Upper Paleolithic assemblages represent special activity occupations by logistical task groups in contrast to Mousterian short-term residential occupations (or vice-versa). An examination of paleoclimatic data, however, indicates a potential common thread linking Middle and Upper Paleolithic curated assemblages from Gorham's Cave. Figure 2 shows that expedient assemblages for both periods correspond to

Fig. 6.2. The correlation between OIS-scale climatic fluctuations and the density of lithic artifacts and incidence of cores and retouched tools in the Gorham's Cave assemblages (after Barton 1998).

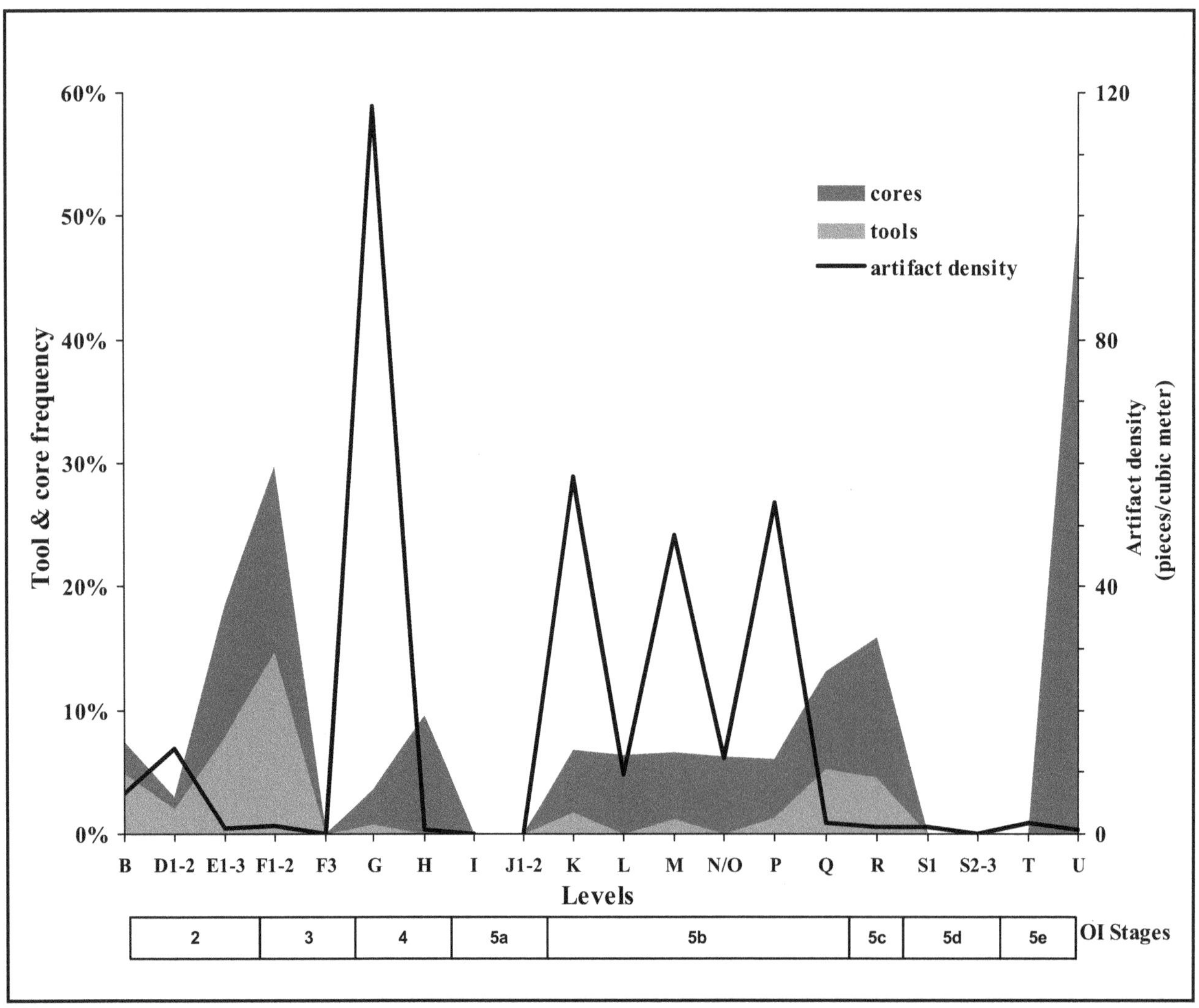

phases of climatic deterioration, while curated assemblages indicate milder climatic conditions. This suggests that, in both periods, the cave's human occupants had very similar land-use patterns which emphasized use of the site as a home-base from which logistical forays could be organized during colder periods, while warmer periods saw an increase in mobility reflected by the use of the site as an ephemeral (and probably repeatedly used) residential site.

It would be quite interesting to see whether these conclusions, based on material from the original excavations at Gorham's Cave (Waechter 1951, 1964), find support in the data resulting from the recent, multidisciplinary reexcavation of the site (see papers in Stringer et al. 2000), but the complementary and self-reinforcing conclusions drawn from lithic technology and paleoenvironmental reconstruction do suggest that we are dealing with robust, meaningful patterns at Gorham's Cave.

2. Valencia, eastern Spain

Northwest of Gibraltar along the Mediterranean coast are three Late Pleistocene sites in the region of Valencia, in southeastern Spain: the Mousterian sites of Cova Negra and Cova del Salt, and the Middle–Upper Paleolithic site of Cova Beneito (figure 3). The Mousterian of Cova Negra adheres closely to the theoretical expectations of our model (r = -.851), with low volumetric artifact densities (2–14 artifacts/m^3) and high retouch frequencies (44–77%). While these patterns have previously been interpreted as reflecting increased artifact curation due to exhaustion of local lithic resources over the course of long occupations (e.g., Barton 1988, 1990; Villaverde 1984), the methodology used here suggests rather that, during the Mousterian, the site was used for short periods during which mainly exhausted lithics were discarded (Villaverde et al. 1998:146). The interpretation of the Cova Negra

Fig. 6.3. The relationship between absolute artifact density and frequency of retouched pieces for Paleolithic assemblages from eastern Spain (after Villaverde et al. 1998).

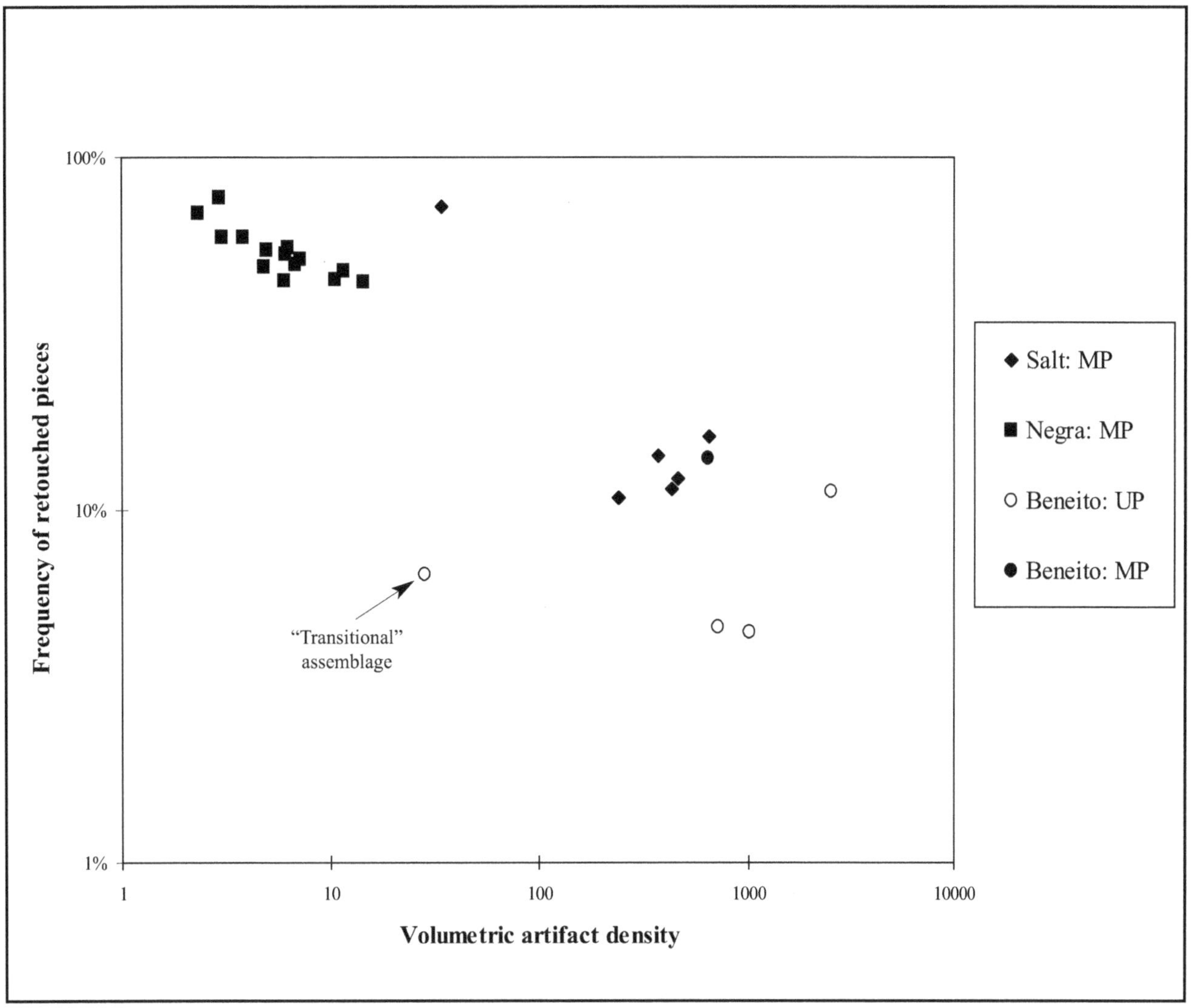

Mousterian layers as the results of short-lived occupations is further supported by the site's faunal assemblages. Microfauna indicate that the site was occupied by bats and raptors that accumulated lagomorphs bearing no traces of human processing (Villaverde et al. 1996) while most gnaw marks on the remains of medium-sized ungulates (mainly ibex and thar) suggest that wolves and dholes were responsible for their accumulation at the site (Villaverde 1984). Humans appear to be mainly responsible for the accumulation of large herbivore remains found at the site, however, indicating alternating use of the site by human and non-human carnivores, which strengthens the interpretation of the site as being ephemerally occupied by Mousterian toolmakers. While it is difficult at present to establish whether the Cova Negra assemblages represent residential as opposed to specific-activity occupations, studies of artifact assemblage richness combined with microwear studies would likely yield useful information in that regard (see Kuhn 1989).

Overall, the Cova Negra material also indicates a consistent use of the site in Mousterian times, which differs from the shifting land-use pattern identified at Gorham's Cave. This is all the more striking in light of the patterns documented by the material from Cova Beneito and Cova del Salt (figure 3). In stark opposition to the Cova Negra pattern, the chipped stone assemblages from the Mousterian levels of those two sites (with the exception of one layer from Cova del Salt) attest mainly to an expedient technological strategy that can be referred to a use of those sites as home bases in Mousterian times. While it is impossible to determine whether the single Mousterian assemblage from Cova Beneito fits our model's theoretical expectations, the material from Cova del Salt agrees with them closely (r = -.893).

Studies of the sedimentary history at Cova del Salt and Cova Negra (i.e., Barton 1988, Barton & Clark 1993) indicate that the significant difference in artifact densities at the two sites is not merely the result of differences in sedimentation rates and depositional environments. This warrants the possible interpretation that the sites served different functions within a broader regional land-use system used by Neanderthals in the region, especially in light of their probable contemporaneity (Barton 1988). Barton (1990) has proposed that altitudinal differences between the two sites might be a significant dimension to explain the variability displayed in their lithic assemblages: While Cova del Salt and Cova Beneito are upland sites, Cova Negra lies under 150 m asl. in a coastal setting *sensu lato.* Cova del Salt and Cova Beneito are also located at ecotones at the edge of the valleys onto which they open, endowing them with strategic positions from which to coordinate activities such as the tracking and logistical exploitation of migrating herds of animals in the valley below and the procurement of other resources in the neighboring highlands. Given this, it is tempting to hypothesize that these sites might have formed a single socio-ecosystem in which Cova Negra—located slightly over 30 km away—could be interpreted as a site visited during targeted resource exploitation forays. The sedimentary studies mentioned above suggest the possibility of comparing retouch frequencies and artifact densities for the assemblages of all on a single graph (figure. 3). The resulting pattern conforms almost perfectly to the model's expectation (r = -.937) while nonetheless clearly distinguishing between the ephemeral and long-term occupations. This reinforces the impression that these sites might well have articulated together within a single land-use system that can be defined as broadly logistical.

Shifting our attention to the Upper Paleolithic sequence of Cova Beneito, all the pertinent assemblages show lower frequencies of retouch and higher artifact densities than the does the site's Mousterian assemblage (see also Iturbe et al. 1993). This suggests that the site was used for longer periods during the Upper Paleolithic, in agreement with a pattern of generalized forager mobility in Mediterranean Spain during that period (Villaverde et al. 1998). This interpretation is also corroborated by the faunal record, which is dominated by the remains of lagomorphs and locally available medium-sized ungulates (especially ibex), many of which bear traces of human modification (Villaverde et al. 1996). This depart from the Mousterian pattern at the site where human modification is concentrated on the bones of large-bodied ungulates (horse, aurochs) and infrequent on lagomorph remains (Iturbe et al. 1993). Thus, lithic data complement faunal data to indicate that the earliest Upper Paleolithic of the region witnessed a shift towards decreasing mobility and increased exploitation of more immediately available resources than was the case during the preceding Mousterian (Villaverde et al. 1998). However, Cova Beneito is only one site and there are no other excavated sites of the same age in its vicinity. It is possible that additional Upper Paleolithic sites might show a similar range of occupational patterns seen in the Middle Paleolithic assemblages of Cova del Salt and Cova Negra. The site from southern Italy, discussed below, provide such data.

The alleged "transitional" assemblage at Beneito reported by Iturbe et al. (1993) is also interesting in the context of using a whole assemblage analysis approach to investigate the Transition. This assemblage was originally argued to document the transition between Middle and Upper Paleolithic but has recently been shown to be the result of mixing of the Mousterian and Aurignacian assemblages that sandwich it (Domenèch Faus 2004). For this reason, it has been omitted from this study (in terms of deriving r-values for the site as a whole), although it was discussed in an earlier publication that used the method presented here to interpret it as representing only "sporadic occupation of the site" (i.e., Villaverde et al. 1998:148). Interestingly, that study also unambiguously showed that assemblage as departing from the expected pattern based on overall sedimentation rates at the site (see Villaverde et al. 1998: figure 5). That this problematic assemblage is reported by Domenèch Faus (2004) to be the result of post-depositional taphonomic processes demonstrates our methodology's potential usefulness as a tool to investigate site formation processes (see Riel-Salvatore & Barton 2004), since it had clearly highlighted that this assemblage as a depositional "odd man out" in the earlier study.

3. Salento, southeastern Italy

In order to further test the applicability of our methodology and to stress its usefulness in highlighting behavioral patterns that can help clarify our understanding of the Transition, we also applied it to a series of eight cave and rockshelter sites located on the western coast of the Salento peninsula in southeastern Italy (for a full discussion, see Riel-Salvatore & Barton 2004). An important criterion for site selection was the presence of a well-implanted 'transitional' industry in the region—the Uluzzian—along with Mousterian, proto-Aurignacian and Epigravettian levels, as well as the close geographical clustering of the sites and their comparable depositional settings that, despite the absence of detailed sedimentary studies, nonetheless allows for more or less direct comparisons of their deposits.

Figure 4 presents the results of this effort. Broadly speaking, most sites display a general adherence to the expectations of the model, although the r-values tend to be low. While this may be in part the result of the problems inherent in estimating excavated volumes from published reports, it is interesting to note that in most cases this rather appears to be due to the presence of one or a few anomalous assemblages within a site's sequence. As suggested by Barton (1998) in the case of Gorham's Cave, the position of these assemblages may reflect sedimentary vagaries rather than the result of long-term patterned human behavior, though the available published data does not allow us to satisfactorily resolve the question at this time.

Another factor likely to be of greater importance in explaining these patterns in the specific case of these assemblages from the coastal Salento is raw material availability. Milliken (1998, 2000) has argued that the Salento does not contain sources of fine-grained siliceous material, although Bietti and Negrino (this volume; see also Richetti & Pieri 1999: figs 10 & 11) use geological reports to suggest that the region likely contains some outcrops of good-quality cryptocrystalline rock. Nonetheless, while there may be some outcrops of usable lithotypes in the Salento, none are currently known in the immediate vicinity of the cluster of eight sites. Cryptocrystalline raw material therefore probably should not be considered "local" in the sense usually attributed to it in most studies of lithic procurement (e.g., Féblot-Augustins 1997). As such, the pattern of raw material utilization is quite striking, in that non-local (and perhaps even exotic, according to Milliken) lithotypes appear to have been systematically targeted for retouch and resharpening (Palma di Cesnola 1996).

Multiple scenarios can be invoked to explain the presence of distinct types of raw materials within archaeological lithic assemblages, most of which suggest that non-local stone was procured and maintained differentially from local lithotypes (see Riel-Salvatore & Barton 2004:265). For this reason, it may be useful to separate given assemblages into their local and non-local components to better reconstruct patterns of prehistoric technological behavior. Doing so for the four sites for which the published information is adequate results in their assemblages displaying a very strong adherence to the predicted negative relationship between retouch frequency and lithic artifact volumetric density predicted by our model (figure 5). These results suggest two main observations. First, and unsurprisingly, it is important to account for raw material variability in whole assemblage analyses of prehistoric assemblages. Second, it appears that the Mousterian and Uluzzian display equally flexible technoeconomic strategies involving both residential and logistical land-use patterns. It is, however, important to note that fine-grained, non-local material represents a commodity embedded in the context of time-specific procurement systems and that it was not necessarily always heavily curated simply *because* it is non-local. For instance, in the Epigravettian assemblages from our sample, flint is the dominant—and some times the only—exploited lithotype. Despite this, these assemblages show expedient management strategies, suggesting that non-local material had become effectively abundant after the LGM (see also Milliken 1998).

It is also possible to make somewhat more subjective observations about some of the sites for which information about raw material was insufficient to subdivide the assemblages into distinct components. For instance, at Serra Cicora A, where a full "transitional" sequence is found, studying the Uluzzian and proto-Aurignacian assemblages separately reveals an extremely strong adherence to the pattern predicted by the model, even in the absence of raw material differentiation ($r = -.991, p = .009, N = 4$). Spennato (1981) reports that limestone is abundant throughout the sequence, but that it only dominates in the lowermost Uluzzian assemblage from layer D. Despite this, that assemblage is tightly clustered with the proto-Aurignacian ones from layers A and B, all of which are more curated than the upper Uluzzian from layer C which documents a more expedient strategy. The Salento assemblages demonstrate that, over the course of the Early Upper Paleolithic in southern Italy, we cannot speak of a simple unidirectional switch from expedient to curated technoeconomic

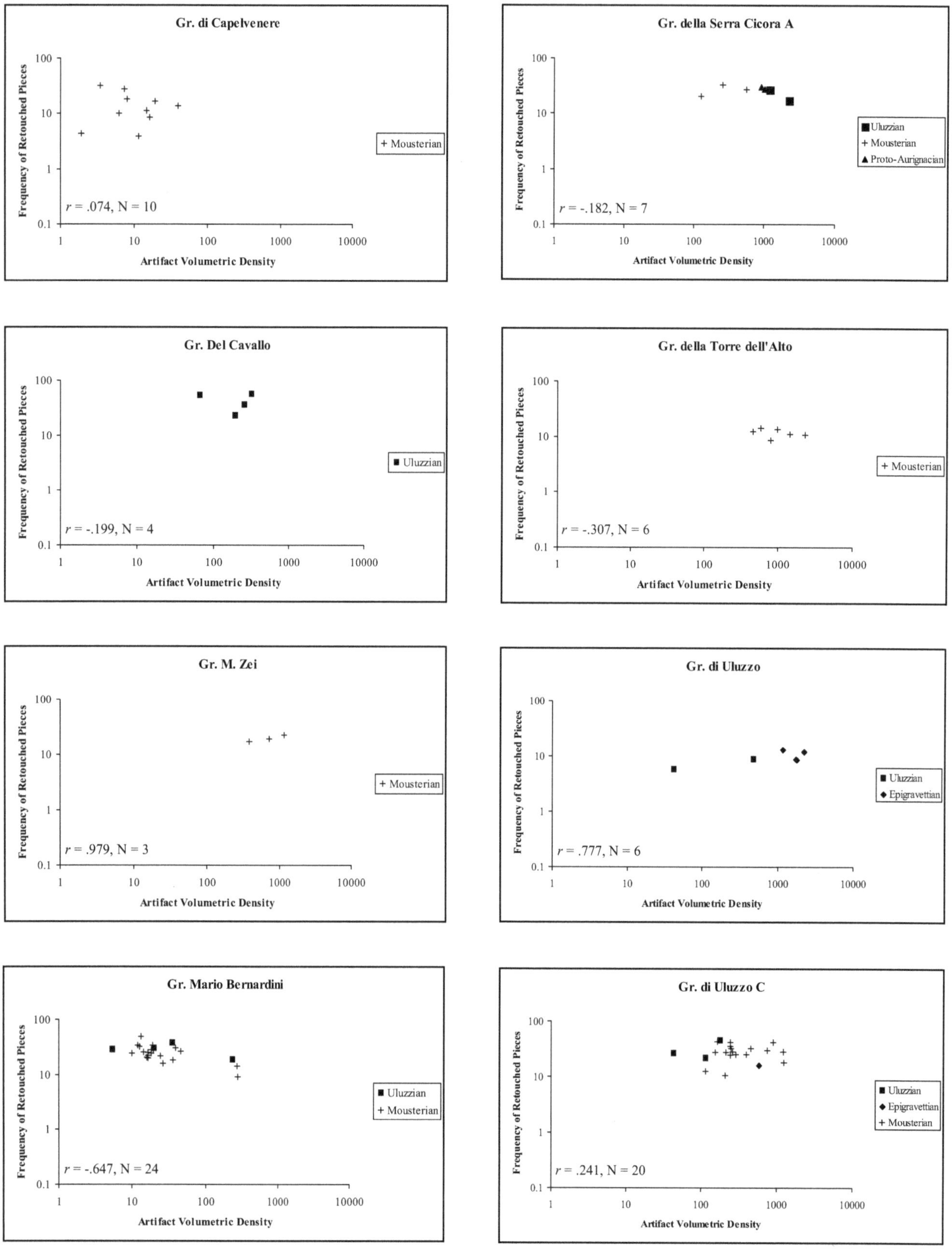

Fig. 6.4. The relationship between absolute artifact density and frequency of retouched pieces for the eight sites from the Salento (after Riel-Salvatore and Barton 2004).

Fig. 6.5. The relationship between absolute artifact density and frequency of retouched pieces after differentiating between flint (i.e., high quality raw material) and limestone (i.e., low quality raw material) sub-assemblages, for the four sites from the Salento for which adequate information is available.

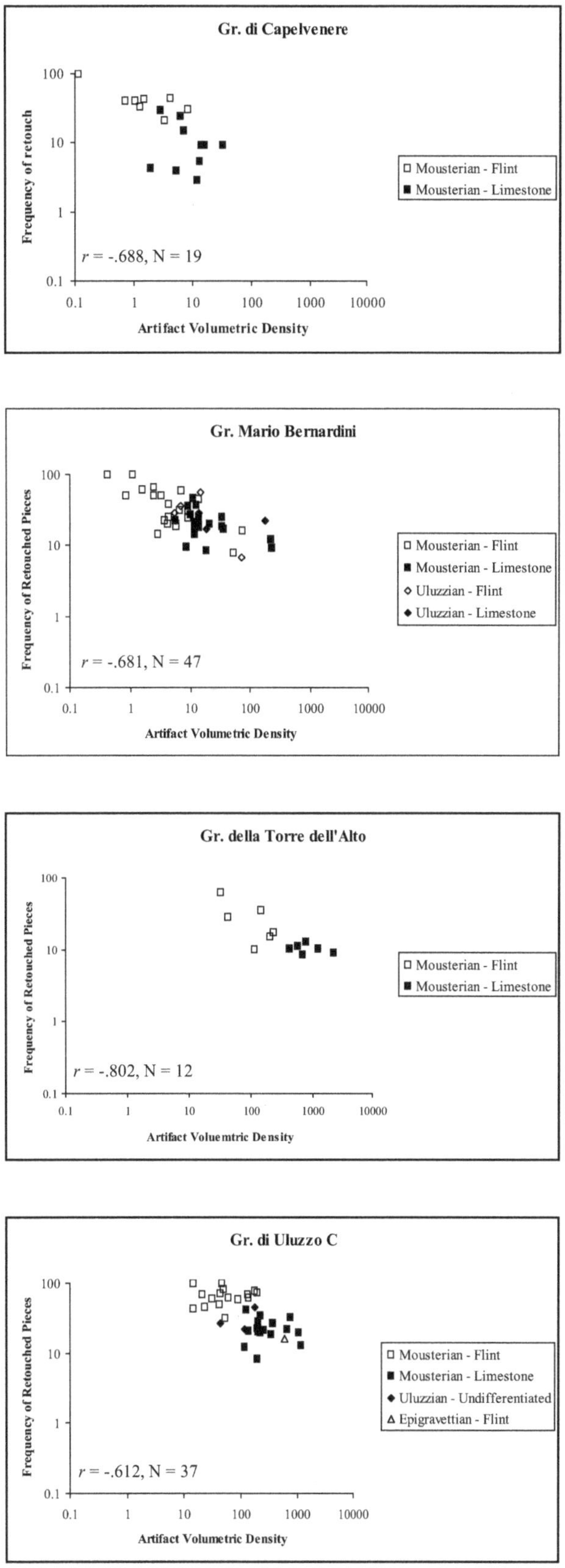

strategies. This analysis offers researchers a more nuanced perspective on the variability of behavioral systems at that time, one which must incorporate detailed discussions of the context for the adoption of alternative adaptive modes.

The linkages between assemblage composition and human ecology also provide the basis for a new interpretation of lithic variability within the transitional Uluzzian industry. Because of its relatively crude tools and reliance on large amounts of local siliceous limestone, the expedient Uluzzian assemblage of Grotta Mario Bernardini (i.e., layer A-IV) has been interpreted as representing an "archaic" phase of that technocomplex, as defined by Palma di Cesnola (1966, 1967, 1989, 1993). In contrast, the "evolved" and "final" phases of the Uluzzian are characterized by a lesser dependence on local lithotypes and by a more refined tool typology, something which Palma di Cesnola (1993) has interpreted as a gradual evolution of the Uluzzian through time marked by increasing technotypological "refinement." The whole assemblage analysis approach used here, however, interprets the site's "final" Uluzzian assemblage (i.e., layer A-I-II) as reflecting a comparatively curated technological organization. Here, we would argue that the various "phases" of the Uluzzian probably represent different land-use strategies used by its makers, imbuing Palma di Cesnola's keen observation of differences within the Uluzzian with behavioral as opposed to 'evolutionary' meaning. This perspective also has the advantage of accounting for disparities in the placement of various Uluzzian assemblages in an "evolutionary" framework, for instance as at Serra Cicora A where the most expedient Uluzzian assemblage is not the lowermost one (Spennato 1981, Palma di Cesnola 1989, 1993). It also provides an explanation for why Palma di Cesnola's evolutionary sequence does not seem to apply to stratified Uluzzian assemblages outside of the Salento, since little attention has been paid to the behavioral significance of varying retouch frequencies and of shifting modalities of raw material use. The whole assemblage analysis approach also allows the incorporation of assemblages with too few tools to permit a secure typological attribution (e.g., Grotta Mario Bernardini, layer A-II) into broader models of behavioral variability within the Uluzzian in this particular case, and prehistoric assemblages more generally.

Discussion and Conclusions

This overview of work conducted in three distinct regions of the northern Mediterranean offers new insights about the Transition. On a methodological level, the whole assemblage analysis approach used here and in other papers (i.e., Barton 1998; Riel-Salvatore & Barton 2004; Villaverde et al. 1998) permits us to directly compare assemblages assigned to different technocomplexes and otherwise analyzed using distinct typological systems. Doing so reveals that, across a broad geographic area, the Mousterian evidences a considerable range of technoeconomic flexibility and of concomitant land-use strategies. This is equivalent to the range of behavioral flexibility apparent in industries largely thought to have been manufactured by modern humans from Gibraltar, eastern Spain and the Salento. Perhaps most interesting in the context of a discussion of "transitional" behavioral dynamics is that the makers of a transitional industry (the Uluzzian) employ the range of land-use and associated lithic management strategies seen in the Middle and Upper Paleolithic but do not show trends toward any particular strategy. The ultimate impetus behind these shifts remain to be fully elucidated, but paleoenvironmental data from Gorham's Cave suggest that changes in ecological conditions or climatic regimes might have been major catalysts of behavioral dynamics throughout the Late Pleistocene (Barton 1998).

This study joins a growing body of work (e.g., Grayson & Cole 1998, Simek & Price 1990) in establishing that the use a single framework to study lithic assemblages across the Transition interval usually does not bear out the differences implicitly embedded in typological systematics. As a result, we would argue that a large part of the distinction between the Middle and Early Upper Paleolithic in Eurasia may itself be an artifact of different typologies traditionally having been used to characterize lithic variation across this analytical boundary (see also Barton et al. 1996; Riel-Salvatore & Clark 2001, this volume). As suggested by the results from the Salento and studies based on other lines of evidence (e.g., art [Barton et al. 1994]), the most significant changes in land-use strategies in Eurasia seem to coincide with the LGM, some 20–25,000 years *after* the Transition. There is no question that sometime between 45–25 kya, Neanderthals as a distinct morphological population disappeared from Eurasia (Trinkaus 2005, Wolpoff et al. 2004). However, the disjunction between this process and some of the fundamental ecological and economic adaptations that might best account for it forces us to recast the questions we should be asking about the Transitions itself. In fact, we would argue that before attempting to delineate the exact process by which 'modernity' spread across Eurasia, it might be more illuminating to figure out what were the precise selective forces that would have driven such a spread (Riel-Salvatore & Barton 2004:269). Only thus will we truly be able to tackle an evolutionary problem using appropriate conceptual frameworks that adequately direct inquiry rather than simply applying an evolutionary veneer to the inductive pattern searches that still characterize much paleoanthropological research (e.g., Mellars 2004, 2005).

In sum, the data presented in this paper suggests that, in terms of the organizational strategies employed to create lithic assemblages, the Neanderthals responsible for manufacturing Mousterian and Uluzzian assemblages were able to adapt their technological responses to changing local conditions to a degree comparable to that of the makers of the Upper Paleolithic, often assumed to be behaviorally and morphologically modern humans. Along with other lines of evidence for 'modern' Neanderthal behavior such as fully efficient large-game hunting (e.g., Marean & Kim 1998, Marean & Assefa 1999) and symbolic behavior (e.g., d'Errico 2003), this observation contradicts the widespread assumption that the two hominin populations were somehow behaviorally unequal (e.g., Bar-Yosef 2002, 2003; Mellars 2004, 2005). We certainly do not pretend that there was no behavioral change over the course of the Late Pleistocene or over the 50–30 kya interval, for that matter. However, while novel behaviors are certainly noteworthy topics of investigation, we advocate that they can only be fully understood when studied in the broader adaptive context within which they emerge. That this context, over the Transition interval, is largely unchanging is, to us, just as important to explain (if not more!) as is the nature of any behavioral changes that did manifest themselves at that time.

The methodology employed in this study offers the potential to directly and objectively compare Middle Paleolithic, "transitional" and Upper Paleolithic assemblages within a single, unified conceptual framework that sheds light on the behavior of the hominins responsible for their manufacture. Its simplicity makes it easily applicable across a large range of archaeological contexts, and we are heartened to see that others who have employed it (e.g., Sandgathe 2005) have found it a provocatively informative source of information on Paleolithic lifeways. As we have shown, this approach also has the advantage of helping to monitor depositional regimes and even offers an alternative explanation to putative pseudo-evolutionary trends within a single technocomplex, which are an integral part of most traditional typological classificatory schemes. The results of applying this common analytical framework to a dozen sites from the northern Mediterranean also strongly suggests a pronounced disconnect between the biological and behavioral aspects of the Transition, thus enabling us to redirect the attention of researchers away from painstaking description of the archaeological manifestations of the Transition and towards the nature of the link between the two kinds of change, behavioral and morphological. As such, this study offers new avenues to explore (and new ways to re-explore old ones) in the ongoing effort to better understand this fascinating segment of recent human evolution.

Acknowledgments

We would like to thank Geoff Clark and Frank Harrold for constructive comments on the version of this paper presented in the original SAA symposium from which this volume derives. Our understanding of the Uluzzian has greatly benefited from discussions with Amilcare Bietti, Paolo Gambassini and Fabio Negrino. Support for Riel-Salvatore during this research was provided in part by the Social Sciences and Humanities Research Council of Canada.

References

Barton, C. M.
1988 *Lithic variability and Middle Paleolithic behaviour. New evidence from the Iberian Peninsula.* BAR International Series, 408, Oxbow Press, Oxford.
1990 Stone tools and Paleolithic settlement in the Iberian peninsula. *Proceedings of the Prehistoric Society* 56: 15–32.
1998 Looking back from the world's end: Paleolithic settlement and mobility at Gibraltar. In *Las culturas del Pleistoceno superior en Andalucía* (J. L. Sanchidrián Torti & M. D. Simón Vallejo, eds.), pp. 13–22. Patronato de la Cueva de Nerja, Cordoba, Spain.

Barton, C. M., & G. A. Clark
1993 Cultural and natural formation processes in late Quaternary cave and rockshelter sites of western Europe and the Near East. In *Formation Processes in Archaeological Context,* (P. Goldberg, D. T. Nash, & M. D. Petraglia, eds.), pp. 33–52. Prehistory Press, Madison.

Barton, C. M., G. A. Clark, & A. E. Cohen
1994 Art as information: Upper Palaeolithic art in Western Europe. *World Archaeology* 26:185–207.

Bar-Yosef, O.
2002 The Upper Paleolithic revolution. *Annual Review of Anthropology* 31:363–393.
2003 Reflections on selected issues of the Upper Paleolithic. In *More than Meets the Eye: Studies on Upper Palaeolithic Diversity in the Near East* (A. N. Goring-Morris & A. Belfer-Cohen, eds.), pp. 265–273. Oxbow Books, Oxford.

Bar-Yosef, O., & D. Pilbeam (editors)
2000 *The Geography of Neanderthals and Modern Humans in Europe and the Greater Mediterranean.* Peabody Museum Bulletins, Harvard University Press, Cambridge.

Binford, L. R.
1979 Organization and formation processes: looking at curated technologies. *Journal of Anthropological Research* 35:255–273.
1980 Willow's smoke and dogs' tails: hunter-gatherer settlement systems and archaeological site formation. *American Antiquity* 45:4–20.

Brantingham, P. J., S. L. Kuhn, & K. W. Kerry (eds.)
2004a *The Early Upper Paleolithic Beyond Western Europe.* University of California Press, Berkeley.
2004b The Early Upper Paleolithic and modern human behavior. In *The Early Upper Paleolithic Beyond Western Europe* (P. J. Brantingham, S. L. Kuhn, & K. W. Kerry, eds.), pp. 242–248. University of California Press, Berkeley.

Clark, G. A., & J. Riel-Salvatore
2006 Observations on systematics in paleolithic archaeology. In *Transitions Before the Transition: Evolution and Stability in the Middle Paleolithic and Middle Stone Age* (E. Hovers & S. Kuhn, eds.), pp. 29–56. Springer, New York.

d'Errico F.
2003 The invisible frontier: A multiple species model for the origin of behavioral modernity. *Evolutionary Anthropology* 12:188–202.

Domenèch Faus, E.
2004 Le Paléolithique moyen et supérieur dans le Levant espagnol: la séquence de La Cova Beneito (Muro, Alicante, Espagne). In *Acts of the XIVth UISPP Congress, University of Liège, Belgium, 2–8 September 2001, Section 6: Le Paléolithique Supérieur/The Upper Palaeolithic, General sessions and posters* (eds.), pp. British Archaeological Reports no. 1240. Archaeopress, Oxford.

Féblot-Augustins, J.
1997 *La circulation des matières premières au Paléolithique.* ERAUL 75, Liège.

Grayson, D. K., & S. C. Cole
1998 Stone tool assemblage richness during the Middle and Early Upper Palaeolithic in France.

Journal of Archaeological Science 25:927–938.

Hays, M. A., & P. T. Thacker (editors)
2001 *Questioning the Answers: Re-Solving Fundamental Problems of the Early Upper Paleolithic.* British Archaeological Reports 1005. Archaeopress, Oxford.

Iturbe, G, M.P. Fumanal, J.S. Carrion, E. Cortell, R. Martinez, R. M. Guillem, M. D. Garralda, & B. Vandermeersch
1993 Cova Beneito (Muro, Alicante): una perspective interdisciplinar. *Recerques del Museu d'Alcoi* 2: 23–88.

Karkanas, P., O. Bar-Yosef, P. Goldberg, & S. Weiner
2000 Diagenesis in prehistoric caves: the use of minerals that form in situ to assess the completeness of the archaeological record. *Journal of Archaeological Science* 27:915–929.

Kuhn, S. L.
1989 Hunter-gatherer foraging organization and strategies of artifact replacement and discard. In *Experiments in Lithic Technology* (D. S. Amick & R. P. Mauldin, eds.), pp. 33–48. British Archaeological Reports, 528. BAR International Series, Oxford.

Marean, C. W., & S. Y. Kim
1998 The Mousterian faunal remains from Kobeh Cave: Behavioral implications for Neanderthals and early modern humans. *Current Anthropology* 39:S79–S114.

Marean, C. W., & Z. Assefa
1999 Zooarchaeological evidence for Neandertal and Early Modern Human faunal exploitation. *Evolutionary Anthropology* 8:22–37.

Marks, A. E., H. J. Hietala, & J. K. Williams
2001 Tool standardization in the Middle and Upper Paleolithic: a closer look. *Cambridge Archaeological Journal* 11:17–44.

Mellars, P.
2004 Neanderthals and the modern human colonization of Europe. *Nature* 432:461–465.
2005 The impossible coincidence: a single-species model for the origins of modern human behavior in Europe. *Evolutionary Anthropology* 12: 12–27.

Milliken, S.
1998 Hunter-gatherer land use in Late Glacial southern Italy. *Oxford Journal of Archaeology* 17:269–286.
2000 The Neanderthals in Italy. *Accordia Research Papers* 8:11–82.

Nelson, M. C.
1991 The study of technological organization. *Archaeological Method and Theory* 3:57–100.

Palma di Cesnola, A.
1966 Il Paleolitico Superiore Arcaico (Facies Uluzziana) della Grotta del Cavallo, Lecce. *Rivista di Scienze Preistoriche* 20:33–62.
1967 Il Paleolitico Superiore Arcaico (Facies Uluzziana) della Grotta del Cavallo, Lecce. (Continuazione). *Rivista di Scienze Preistoriche* 21:3–59.
1989 L'Uluzzien—Facies Italien du Leptolithique Archaïque. *L'Anthropologie* 93:783–812.
1993 *Il Paleolitico Superiore in Italia: Introduzione allo Studio.* Garlatti e Razzai Editori, Florence.
1996 *Le Paléolithique Inférieur et Moyen en Italie.* Jérôme Million, Grenoble.

Richetti, , G., & P. Pieri (editors)
1999 *Guide GeologicheRegionali 9: Itinerari Puglia e Monte Vulture, Prima Parte.* BE-MA Editrice, Rome.

Riel-Salvatore, J., & G. A. Clark
2001 Grave markers: Middle and Early Upper Paleolithic burials and the use of chronotypology in contemporary Paleolithic research. *Current Anthropology* 42:449–481.

Riel-Salvatore, J., & C. M. Barton
2004 Late Pleistocene technology, economic behavior and land-use dynamics in southern Italy. *American Antiquity* 69:257–274.

Sandgathe, D. M.
2005 *An Analysis of the Levallois Reduction Strategy Using a Design Theory Framework.* Unpublished Ph.D. dissertation, Dept. of Archaeology, Simon Fraser University.

Simek, J. F., & H. A. Price
1990 Chronological change in Périgord lithic assemblages. In *The Emergence of Modern Humans: An Archaeological Perspective* (P. A. Mellars, ed.), pp. 243–261. Edinburgh University Press, Edinburgh.

Spennato, A. G.
1981 I livelli protoaurignaziani della Grotta di Serra Cicora (Nardò, Lecce). *Studi per l'Ecologia del Quaternario* 3:61–76.

Straus, L. G. (editor)
2005 *Armageddon or entente? The demise of the*

European Neandertals in Isotope Stage 3. Special issue of *Quaternary International* 137: 1–89.

Stringer, C. B., R. N. E. Barton, & J. C. Finlayson (editors)
2000 *Neanderthals on the Edge*. Oxbow Books, Oxford.

Trinkaus, E.
2005 Early modern humans. *Annual Review of Anthropology* 34:207–230.

Villaverde, V.
1984 *La Cova Negra de Xativa y el Musteriense de la Region Central del Mediterraneo Espanol*. Serie de Trabajos Varies, 79, Servicio de Investigaciones Prehistoricas, Valencia.

Villaverde, V., J. E. Aura, & C. M. Barton
1998 The Upper Paleolithic of Mediterranean Spain: a review of current evidence. *Journal of World Prehistory* 12:121–198.

Villaverde, V, R. Martinez-Valle, P. M. Guillem, & M. P. Fumanal
1996 Mobility and the role of small game in the Middle Paleolithic of the Central Region of the Spanish Mediterranean: A comparison of Cova Negra with other Paleolithic deposits. In *The Last Neandertals, the First Anatomically Modern Humans: A Tale About the Human Diversity. Cultural Change and Human Evolution: The Crisis at 40 KA BP* (E. Carbonell & M. Vaquero, eds.), pp. 267–288. Universitat Rovira i Virgili, Tarragona.

Waechter, J. d'A.,
1951 Excavations at Gorham's Cave, Gibraltar. *Proceedings of the Prehistoric Society* 17:83–92.

Waechter, J. d'A.
1964 The excavation of Gorham's Cave (Gibraltar): 1951–54. *Bulletin of the Institute of Archaeology* 4:189–222.

Weiner, S., P. Goldberg, & O. Bar-Yosef
2002 Three-dimensional distribution of minerals in the sediments of Hayonim Cave, Israel: diagenetic processes and archaeological implications. *Journal of Archaeological Science* 29:1289–1308.

Wolpoff, M.H., B. Mannheim, A. Mann, J. Hawks, R. Caspari, K. R. Rosenberg, D. W. Frayer, G. W. Gill, & G. A. Clark
2004 Why not the Neandertals? *World Archaeology* 36:527–546.

Zilhão, J., & F. d'Errico (editors)
2003 *The Chronology of the Aurignacian and the Transitional Technocomplexes: Dating, Stratigraphies, Cultural Implications*. Instituto Portugêse de Arqueologia, Lisbon.

7

The Middle/Upper Paleolithic Interface in Croatia

Ivor Karavanić
University of Zagreb

Abstract

Recent publications have discussed Neanderthal remains found associated with a possible "transitional industry" (the Olschewian) at Vindija Cave, in northwestern Croatia. The Middle/Upper Paleolithic interface from the Adriatic region of Croatia, in contrast, is not well known. In addition to the sites of Vindija and Velika Pećina (both in northwestern Croatia), this paper presents two sites from the Eastern Adriatic region: Mujina Pećina, in Dalmatia (late Middle Paleolithic,) and Šandalja II, in Istria (early Upper Paleolithic). We describe some differences between the Mousterian industries of Vindija and Mujina Pećina and note that the Initial Upper Paleolithic industries of Vindija and Velika Pećina do not conform to the "classic" Aurignacian pattern while a typical Aurignacian industry is present at Šandalja II. The differences between the archaeological remains from these four sites may be interpreted as representing different adaptations by Paleolithic populations to different environments. The absence of a "transitional industry" in the Eastern Adriatic provides further evidence of the complex pattern of adaptive change during the Middle/Upper Paleolithic interface in Europe.

Croatia is known for a number of Paleolithic sites significant for our understanding of the evolution of Upper Pleistocene hominids and of their material cultures. In the Hrvatsko Zagorje region (northwestern Croatia), the most important sites to have yielded human remains and/or Paleolithic assemblages are Krapina, Vindija, Velika Pećina and Veternica (near Zagreb). The Neanderthal remains from Krapina and Vindija have been studied from a number of perspectives and these studies have been published in several venues (Smith 1976, 1982; Radovčić et al. 1988; Wolpoff 1996). As well, the Mousterian artifacts from Krapina (Malez 1979; Šimek 1991; Šimek & Smith 1997) and Velika Pećina (Malez 1967, 1979; Karavanić & Smith 1998) have been described by several analysts while the Middle Paleolithic material from Vindija has been only partially presented (Malez 1978, 1979; Karavanić & Smith 1998). Besides Mousterian assemblages, Vindija and Velika Pećina have also yielded early Upper Paleolithic assemblages. The initial Upper Paleolithic industry documented at Vindija and Velika Pećina differs from the typical Aurignacian and are discussed in recent publications (Miracle 1998; Straus 1999; Karavanić 2000b; Smith & Karavanić 1998, 2000). Importantly, Neanderthal remains were found in level G1 at Vindija where they were associated with stone tools combining Middle and Upper Paleolithic typological characteristics and with bone artifacts of Upper Paleolithictype. In contrast, the Middle/Upper Paleolithic interface from the Adriatic region of Croatia is not well known. Hence, this paper presents data from two sites located in the Eastern Adriatic region, Mujina Pećina, in Dalmatia (figure 1, no. 3), and Šandalja II, in Istria (figure 1, no. 4) ,in addition to a discussion of data about Vindija and Velika Pećina (figure 1, nos. 1 & 2). Mujina Pećina contains a well-dated Mousterian industry and is important for reconstructions of late Mousterian adaptations in the Eastern Adriatic, while Šandalja II is the only systematically excavated site to have yielded an Aurignacian industry in that region.

The Hrvatsko Zagorje Landscape and Sites

The Hrvatsko Zagorje is a hilly part of the Pannonian valley in northwestern Croatia. That lowland area with few isolated mountains is bordered by the Sutla river to the west, Medvednica mountain (1,035 m above sea level) tothe south, the Kalnik hills (Kalničko Gorje; 643 m above sea level) to the southeast, and Ravna Gora to the north. The Hrvatsko Zagorje is divided in two parts by the Ivanščica mountain (30 km long, 9 km wide and 1,061 mabove sea level [Crkvenčić et al. 1974]): a smaller one around the Bednja river on the north (i.e., the Podravina area), and a larger one between the Ivanščica and Medvednica mountains. Except for Medvednica, all the mountains of the Hrvatsko Zagorje are part of the Eastern Alps. During the Upper Pleistocene, the whole area would have been in the periglacial region.

Fig. 7.1. Important late Middle and/or early Upper Paleolithic sites of Croatia. 1. Vindija (late Middle and early Upper Paleolithic), 2. Velika Pećina (Late Middle and early Upper Paleolithic), 3. Mujina Pećina (Late Middle Paleolithic), 4. Šandalja II (early Upper Paleolithic).

Vindija

The site is located 2 km west of village of Donja Voča, and 20km west of Varaždin (figure 1, no.1). S. Vuković, who visited the site in 1928 for the first time, excavated there off and on for over thirty years, while M. Malez started systematic excavations at Vindija in 1974, with subsequent yearly field seasons until 1986. It is during this second period that most of the lithic and faunal material—and all of the fossil human remains—were recovered. The cave is more than 50 m. deep, up to 28 m wide, and over 10 m. high. The stratigraphic profile is about 9m. high and comprises about twenty strata which, according to Malez and Rukavina (1979), were deposited over an interval spanning the onset of Riss glaciation (OxygenIsotope Stage 6 or perhaps earlier) through the Holocene.

Neanderthal remain from level G3 were recently dated by AMS ^{14}C (Ua-13873) to over 42 ka B.P. (Krings et al. 2000), while a cave bear longbone from level G1 yieldedan age of 33,000 ± 400 (ETH-12714) (Karavanić 1995). Additionally, AMS ^{14}C dates were obtained directly from two Neanderthal skeletal specimens from level G1: a right ramus and posterior mandibular corpus (Vi-207) yielded an age of 29,080 ±400 B.P. (OxA-8296) while ana nterior superior fragment of a left parietal (Vi-208) was dated to 28,020 ± 360 B.P.(OxA-8295) (Smith et al.1999). Animal long bones from next younger level (Fd/d) have been dated by radiocarbon to 26,600 ± 900B.P.(Z-2443) (Obelić et al.1994), while a sample of charcoal from level F/d yielded an age of 27,000 ±600 B.P.(Srdoč et al. 1979).

The association of the Vindija hominid remains with indus-tries combining Middle and Upper Paleolithic features creates the most interesting and puzzling problems per-taining to this site. The human remains can be stratigraphi-cally separated into three groups (Smith 1982:634; Smith & Ahern 1994): 1) agroup consisting of Neanderthal remains from level G3 associated with a late Mousterian industry; 2) a group comprising the hominid remains from level G1 (a right ramus and posterior mandibular corpus, teeth, an anterior superior fragment of a left parietal, a zygomatic, a supra-orbital torus), associated with a Mousterian industry and "Upper Paleolithic" bone points—three isolated teeth from level Fd and a posterior fragment of a left parietal from the contact of levels Fd and Fd/d also belong to this group; and 3) a group comprising Homo sapiens sapiens remains from Epigravettian level D. Recently, a Nean-derthal specimen was also found in level I (Ahern et al. 2004).

The Upper Paleolithic bone tools and lithics from Vindija have been analyzed (Malez 1978, 1988; Karavanić1995) while the Middle Paleolithic material has been presented only in part (Malez 1978; Montet-White 1996; Karavanić & Smith 1998). The lithic assemblage of level G3 is dom-

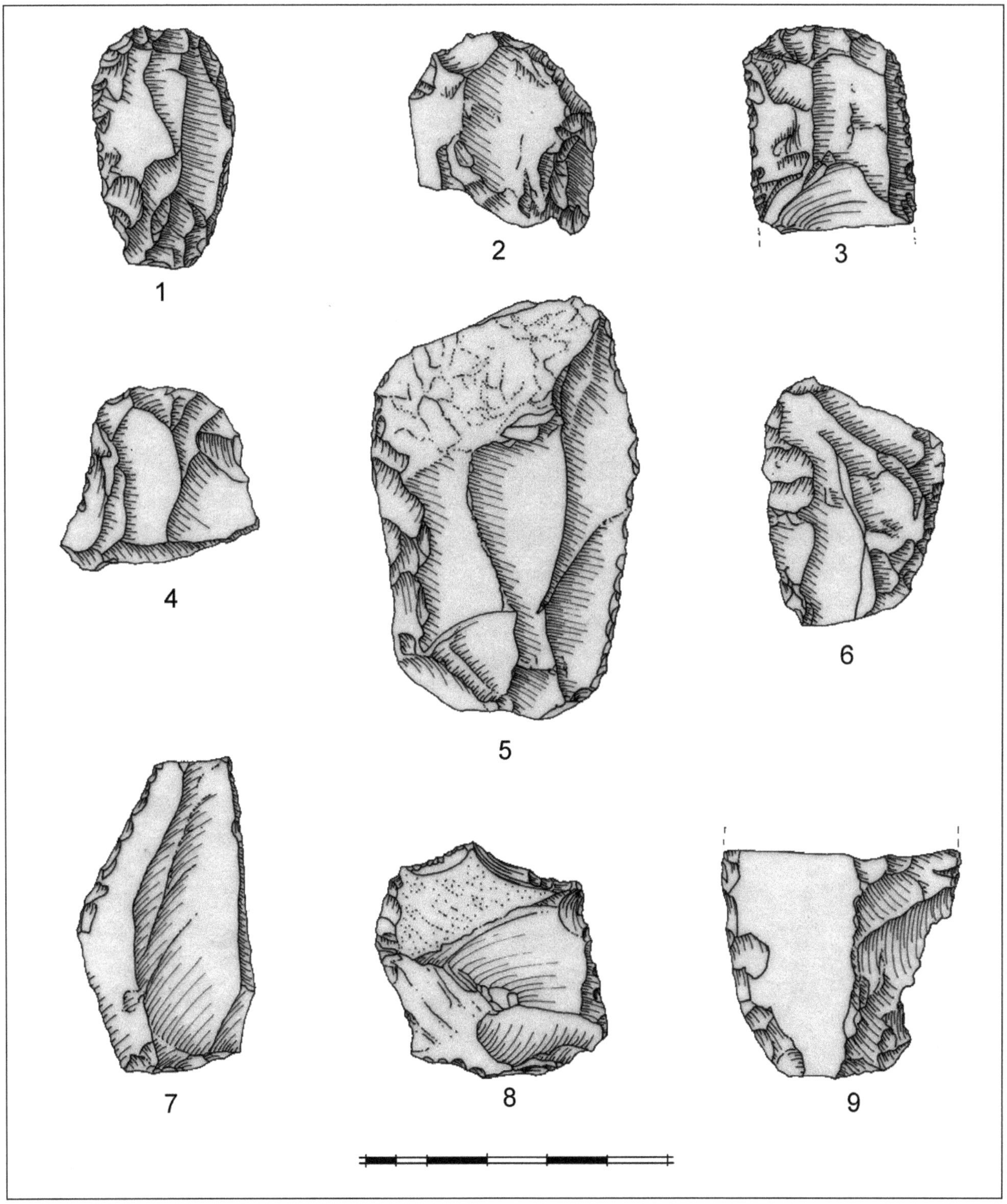

Fig. 7.2. Selected artifacts from level G3, Vindija.1, 2 and 4. endscrapers on flake, 3. endscraper on broken blade, 5. double straight convex sidescraper, 6. alternate retouched sidescraper, 7. retouched piece, 8. denticulated piece, 9. single convex sidescraper (redrawn and modified by M. Bezić after Karavanić and Smith 1998:Figure 3 and 4). Scale is in cm.

inated by flake technology, although Levallois technology is absent altogether. Among tools, the most frequent type are retouched pieces (some of them blades), followed in decreasing order of importance by notches and denticulated pieces (figure 2, no. 8), and by various types of sidescrapers (figure 2, nos. 5, 6, 9). Four end scrapers were also found (figure 2, nos. 1–4), and they are smaller and more finely made than typically Mousterian tools. All of the raw materials used in tool production in this level come from the cave's vicinity, and there is no evidence to

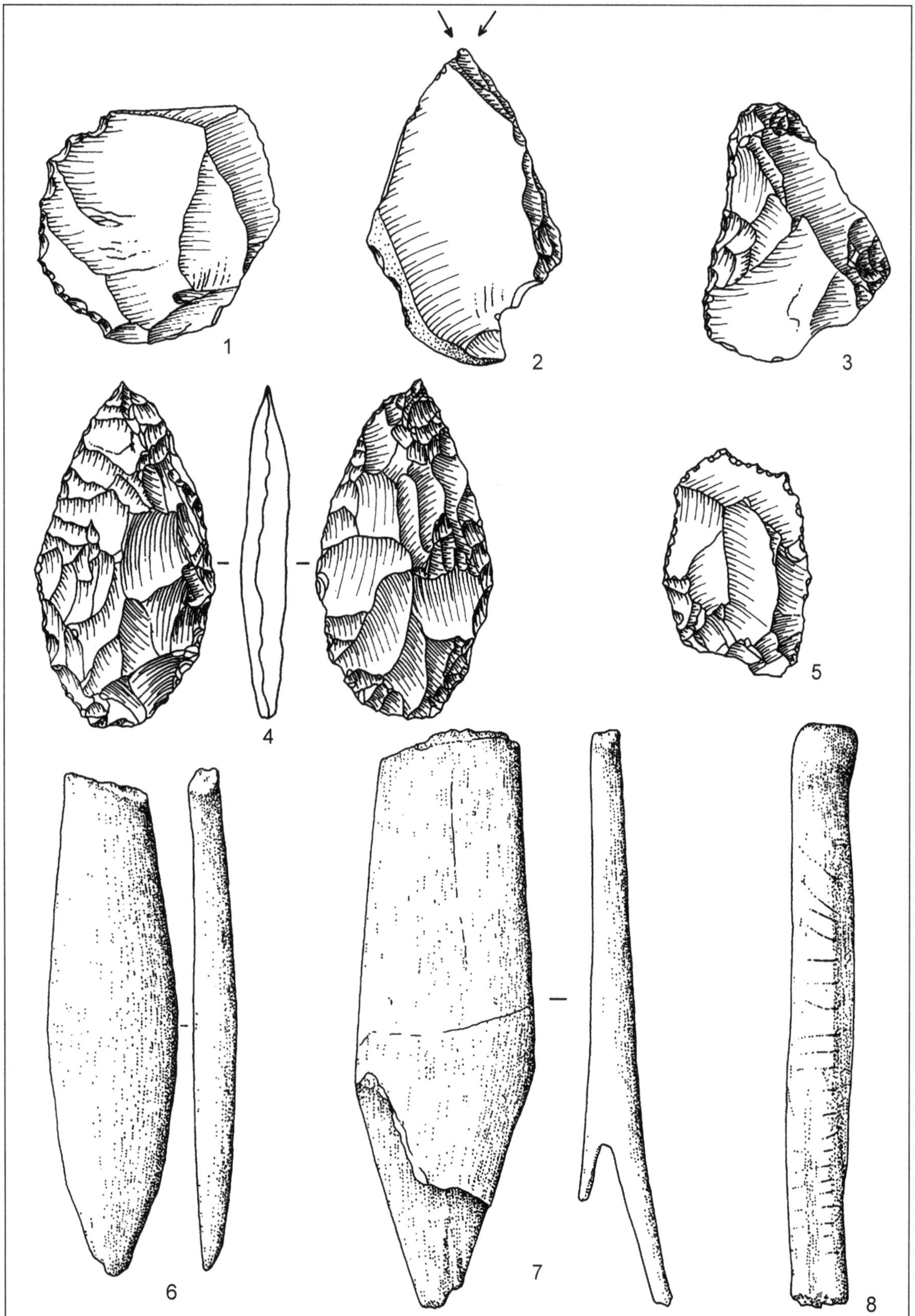

Fig. 7.3. Selected artifacts from level G1, Vindija. 1. denticulated piece, 2. straight dihedral burin, 3. sidescraper, 4. leaf-shaped bifacial piece, 5. endscraper on a flake, 6. massive-base bone point, 7. split-base bone point, 8. bear baculum bone with carved circular markings (redrawn and modified by M. Bezić after Karavanić 1995:Figure 3). Scale is in centimeters..

suggest that material from any other source was used (Kurtanjek & Marci 1990:235). The majority of debitageis made on quartz, and other raw materials include chert, tuff and sandstone (Kurtanjek & Marci 1990, Blaser et al. 2002). The stone artifacts from this level can securely be attributed to a late Mousterian industry.

Level G2 is only present in some parts of the cave, sandwiched between levels G3 and G1, and there are thus sections of the cave where G1 and G3 are in direct stratigraphic contact. Level G2 contained only nine pieces of debitage and four tools—too few for a secure cultural attribution.

The stone tools from level G1, where bone points where also found (Karavanić 1995), combine Middle and Upper Paleolithic typological characteristics. Typologically, sidescrapers (figure 3, no. 3) and denticulates (figure 3, no. 1) are the most common tools, but we also find some Upper Paleolithic tool types in this level, like burins (figure 3, no. 2) and endscrapers on flake (figure 3, no. 5) although only one tool (i.e., an endscraper on Aurignacian blade) may be said to be of Aurignacian type. A leaf-shaped bifacial piece (figure 3, no. 4), made on non-local red radiolarite has also been recovered from this level, and it appears that this piece was imported from elsewhere (Montet-White 1996). J. Kozlowski (1996:211) has attributed the lithics from level G1 to a "Moustero-Levalloisian" assemblage but, while the dominant tool types (i.e., sidescrapers and denticulated pieces) are typical of the Mousterian sensu lato, the Levallois method was not used at all in this level, casting doubt on that attribution. Overall, the lithic industry of this level strongly suggests affinities to a non-Levallois Mousterian technological and typological tradition, while the bone tools are instead more typical of the Upper Paleolithic. The bone points (actually antler points) include a split-base point (figure 3, no. 7) while the other three are massive-base points (figure 3, no. 6). There are also five other bone tool fragments securely associated with level G1, and an additional piece—a bear baculum bone engraved with circular markings (figure 3, no. 8)—probably also originated from this layer, although it is possible that it originated from level G3, which would make it even older.

The sedimentary interface between complexes G and F caps the Vindija sequence and contains a lithic assemblage comprising retouched blades and endscrapers, as well as massive-base bone points. A keeled endscraper, a flat-nosed endscraper and an endscraper on an Aurignacian blade are all atypical in form, but nonetheless generally associated with the Aurignacian. Differently from level G1, the presence of few Aurignacian stone tool types and of blade technology at the G/F interface may suggest aregional variant of what is generally called Central European Aurignacian.

Velika Pećina

Velika Pećina—another important northwest Croatian Paleolithic site—is situated between the sites of Krapina and Vindija, near the village of Goranec on the Ravna Gora (figure 1, no. 2). The cave is 25 m deep, and excavations were initially conducted by M. Malez in 1948. Subsequent excavations started in 1957 and continued, with some interruptions, until 1979. The stratigraphy can be over 10 m deep in some parts of the cave and it comprises 16 distinct levels, encompassing an interval of time ranging from the end of the Riss glaciation (OxygenIsotope Stage 6 or earlier) through the Holocene. Radiocarbon dating for Velika Pećina's early Upper Paleolithic layers yielded a date of 33,850 ± 520 B.P. (GrN-4979) for level i (Malez & Vogel 1970) and a date of 27,300 ± 1200 B.P. (Z-189) for level g (Srdoč et al. 1973). The human frontal bone that had been found inlevel j, which was generally considered as one of the earliest modern human fossil in Europe, has been recently directly dated by AMS radiocarbon to 5,045 ± 40 B.P. (OxA-8294) (Smith et al. 1999). This Holocene age has removed the Velika Pećina frontal from the list of early modern human remains in Europe.

Velika Pećina's lower levels (levels p to k) have yielded Mousterian assemblages (Malez 1979), although Malez (1967:28) attributed the artifacts from the lower part of level k to the Mousterian and tentatively attributed those from the upper part to the proto-Aurignacian. A recent reanalysis of the artifacts from level k suggests however that there is no convincing reason to recognize two different industries (Karavanić & Smith 1998). Rather, it seems likely that the lower part of level k contains credible Mousterian lithics while the "tools" from the upper part are only pseudo-tools. All Mousterian tools are small, reminiscent of those found in so-called Micromousterian assemblages, and the small numbers of artifacts in all Mousterian levels of Velika Pećina suggests several very short occupation episodes.

Only one stone tool was found in level j (a blade with two continuously retouched edges with a notch), while the first level to yield bone tools—again, actually antlerpoints—is level i, which contains six lithics, four bone tools and a bladelet core. Of the bone tools, three are points that probably had split bases, while one is a massive-base point. While the lithics are few, the presence ofbone points strongly suggests an Upper Paleolithic designation for this industry.

The Eastern Adriatic Landscape and Sites

The physical geography of the Eastern Adriatic area is shaped by the Dinarid Mountains to the east and the

Adriatic Sea to the west. The Dinarids are a southeastern extension of the Alps and they extend northwest–southeast along the Adriatic. While the highest mountains in northwestern Croatia are about 1000 m high, the Dinarids are over 2000 m high, in addition to being very steep and rugged on their western side. Furthermore, although they are not glaciated today, the Dinarids would have supported small isolated glaciers during the Pleistocene (Marjanac et al. 1990), and they may well have acted as a barrier that would have impeded migrations between the coast and continent during the Pleistocene. Reconstructions of the Pleistocene geography thus suggest the presence of two major landscapes in the Adriatic region: the exposed Adriatic plan and the karstic hinterland. However, the sea level and coastline repeatedly rose and subsided during the Pleistocene. At the height of the last interglacial, the level of the Adriatic Sea was 1.7 m higher than it is today, while it would have been 96 m lower during the Last Glacial Maximum (Šegota 1982). It is important to consider in studies of the Middle/Upper Paleolithic interface that a large part of the Pleistocene landscape—and probably some important Paleolithic sites—is now covered by water due to rising sea levels since the beginning of Holocene.

Mujina Pećina

In the Eastern Adriatic region, Mujina Pećina is the only systematically excavated site with homogenous Mousterian deposits in clear stratigraphic context to have been well-dated chronometrically. The cave, which is about 10m deep and 8 m wide, is located in Dalmatia, north of Trogir and west of Split, in typical karstic terrain (figure1, no. 3). Lithic material was initially collected from the surface inside the cave and in front of its entrance (Malez1979) and the first excavation was undertaken in 1978 (Petrić 1979). In 1995, systematic excavations, which are still in progress, were initiated by a joint project of the Department of Archaeology, of the University of Zagreb and the Kaštela City Museum (Karavanić & Bilich-Kamenjarin 1997). Excavation levels followed the cavede posits' natural stratigraphy, all sediments were sieved, and all artifacts and ecofacts over 2 cm in size have been entered in three dimensions on site plans. The stratigraphic profile in the cave is only about 1.5 m deep, and the deposits comprise poorly sorted Quaternary sediments composed of large fragments of carbonate rock (debris), gravel and sand grains, rarely silt, and some clay.

Electron spin resonance (ESR) dating was conducted ontwo teeth from Mousterian level E1 at Mujina Pećina, while five bone and one charcoal samples coming from five different strata were dated by AMS 14C (Rink et al.2002). For level E1, mean ESR age estimates of 40 ± 7 ka (EU) and 44 ± 5 ka (LU) have been obtained (assuming 30% moisture for both the gamma and beta dose rate calculations). The interface between Level E2 and E1 has been dated by AMS to 45,170 ± 2780/2060 B.P. (GrA-9635) while the AMS age of overlying levels (D2, D1, C and B) calculated as the mean of the five dates from these levels is 39,222 ± 2956 B.P. (for individual dates and lab numbers see Rink et al. 2002, table 3), and the true (calibrated) age of this mean age is about 42 kya. These dates suggest that entire the Mujina Pećina Mousterian deposits were deposited very rapidly during Oxygen Isotope Stage 3 and they indicate that the radiocarbon results are in close agreement with the ESR ages calculated at 30% moisture (Rink et al. 2002).

Starting at the bottom of the Mujina Pećina sequence, the oldest levels (E3, E2, E1) are the richest in archaeological material, which indicates much more intensive use of the site by humans than in more recent levels where artifacts are less dense. The richness of levels E3, E2, and E1 may represent long-term occupations of the cave (Karavanić 2000a), but may also have been deposited as the result of the site being occupied for many brief occupations (see Conard 1996). In levels D1 and D2 are characterized by a significant presence (about 20% of all debitage) of Levallois debitage (see e.g., figure 4, nos. 10 & 11). In contrast, there is a modest representation of Levallois debitage in overlying level B—including one small Levallois core which suggests that the "micro-Levallois" technique may have been used—but regular flakes, flakes <2 cm in size, and debris are all more frequent. Over 50% of all primary, secondary and regular flakes in level B were transformed into tools, including small tools similar to those of Micromousterian type. Such small tools were found in different levels of Mujina Pećina along with "typical" Mousterian tools, the majority of which are retouched flakes, notched pieces (figure 4, no. 4) and denticulates (figure 4, nos 1 & 7). Sidescrapers are also present in all levels (figure 4, nos. 2, 3, 8), but we only note a significant presence of "Upper Paleolithic" types in level B.

The small size of tools at Mujina Pećina can probably best be explained by the size of available local raw material, which consists of small chert pebbles and nodules whose diminutive size impeded the removal of much cortex on the tools and cores from level B, for instance. The low flaking quality of local cherts present in larger packages also limits tool size, since knapping experiments conducted by the author have demonstrated that it was rarely possible to detach large and regular flakes from the larger blocks of such materials and, moreover, that many of the few large flakes produced experimentally would break during retouching.

Šandalja II

Šandalja II is the only systematically excavated site in the Eastern Adriatic region with Aurignacian deposits in clear

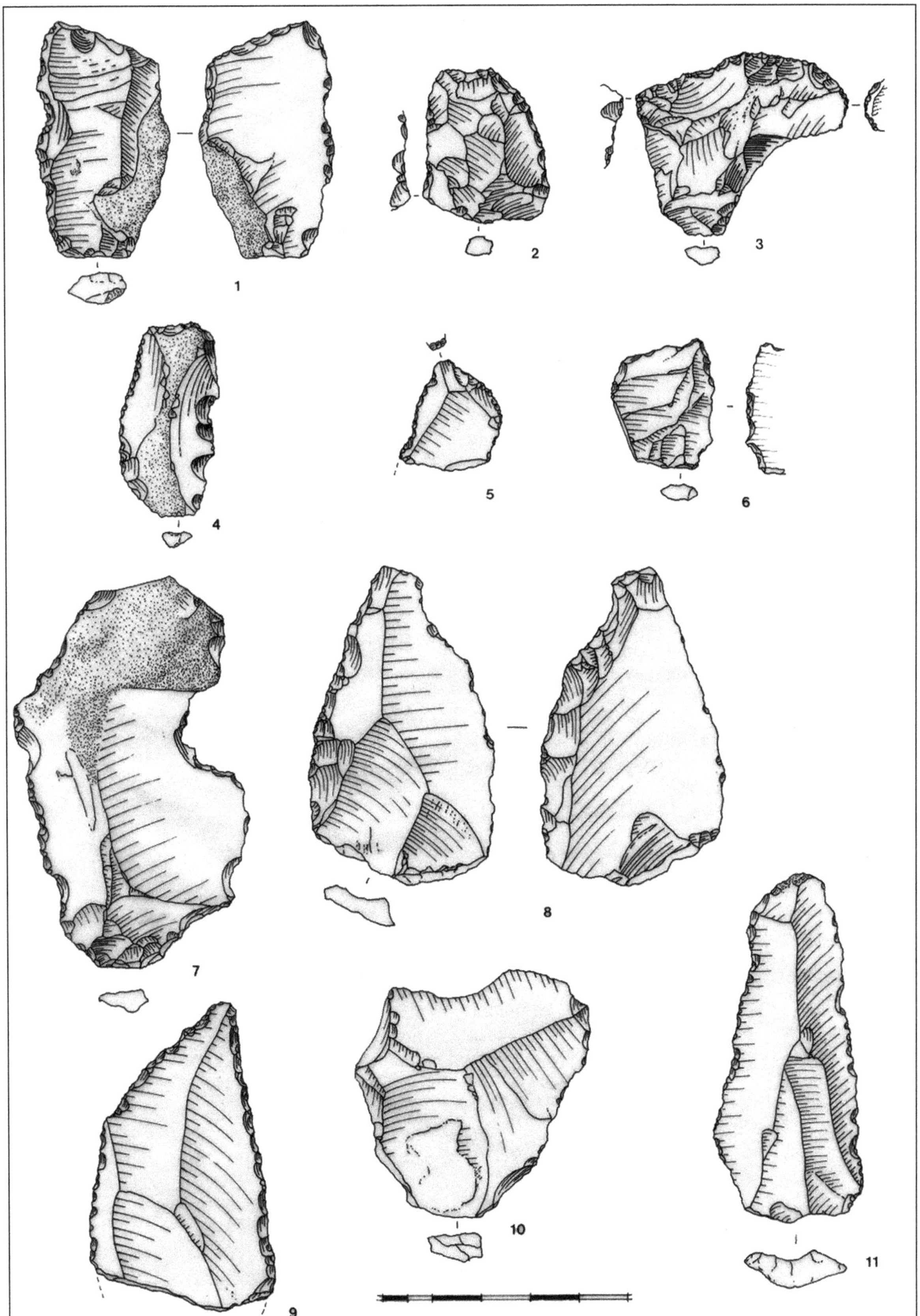

Fig. 7.4. Selected artifacts from Mujina Pećina. Level B: 1. denticulated piece, 2. single convex sidescraper, 3. transversal convex sidescraper, 4. notch, 5. alternate retouched bec, 6. drill. Level D1: 7. denticulated piece, 8. alternate scraper, 9. Mousterian point, 10. Levallois flake, 11. Levallois blade. Scale is in centimeters. Drawing: M. Bezić (modified after Rink et al. 2002: Figure 4).

stratigraphic context. The site is located in a quarry about 4 km north-east from the city of Pula on the Istrian peninsula (figure 1, no. 4). The cave is 14 m deep and 18 m wide and was discovered in 1961, during the mining of a karstic system. The initial excavations were conducted by B. Bačić and continued by M. Malez, and the site has been excavated on 22 occasions between 1962 and 1989 (Miracle 1995). The stratigraphic profile is roughly 8 m high and comprises eight strata deposited mainly during late Würm period. These levels are composed of compact or sandy clay with large or small rock fragments. They contain Aurignacian and Epigravettian industries (Karavanić 1999), and a humus level containing Bronze Age pottery is the only to have been deposited during the Holocene. Animal bones from level H were dated by radiocarbon to 17,600 ± 370 B.P. (Z-2422) (Obelić et al. 1994), while a charcoal sample from a hearth in level G yielded an age of 27,800 ± 800 B.P. (Z-536) (Srdoč et al. 1979). Level F has been dated to 25,340 ± 170 B.P. (GrN-4977) by radiocarbon on animal bones (Malez & Vogel 1969), although a charcoal sample from a hearth from that same level wasdated to 22,660 ± 460 B.P. (Z-537) (Srdoč et al. 1979). Faunal remains from next younger level (F) produced adate of 23,540 ± 180 B.P. (GrN-5013) (Malez & Vogel 1969). Overall, the radiocarbon dates obtained for these levels are young for Aurignacian levels compared to those from other Mediterranean and Central European sites.

A preliminary study of the industries from Šandalja II was published by M. Malez (1987), although they have only recently been analyzed in detail (Karavanić1999). Chert was used for tool production in all levels. Only a few stone tools were found in level H and they can be generically attributed to the Upper Paleolithic. Although levels G and "G/H interface" contain a few Aurignacian tool types, it is difficult to determine with certainty to which Upper Paleolithic industry they belong, although it is most probable that they represent Aurignacian levels. In constrast, nosed (figure 5, nos. 7–12) and carinated endscrapers (figure 5, nos. 1–3) are frequent in the tool assemblages oflevels F, E and "E/F interface," and the industry of these levels can safely be assigned to the Aurignacian. Sidescrapers and other types were also found in those levels (see figure 5), as were bone artifacts and four perforated teeth (see Karavanić 2003, figure 9). In levels F, Eand "E/F interface," the debitage is dominated by flake technology, although blade and bladelets produced by soft hammer are also present.

Comparison of the Sites and Discussion

These four sites located in various regions of Croatia furnish important evidence for refining our understanding the Middle/Upper Paleolithic transition in central Europe and the Mediterranean. We begin by briefly highlighting several of the most conspicuous differences and similarities between the industries from the four sites discussed in this paper.

The late Mousterian level G3 of Vindija could be contemporary with some Mousterian levels from Mujina Pećina in Dalmatia. The Vindija level G3 lithic assemblage is characterized by a high frequency of retouched pieces (32% of whole assemblage) while that frequency is 26% for Mujina Pećina level B. Notches and denticulates represent the second most frequent tool category at Vindija G3 (28% of all retouched tools), whiles that category is the most frequent one in Mujina Pećina level B (38%), although the Mujina Pećina assemblage also includes becs which are absent in the Vindija one. Sidescrapers are more frequent at Vindija G3 (23% of all retouched tools) than in Mujina Pećina level B (13%). Some Levallois debitage was found in the Mujina Pećina level B assemblage, while there is no evidence that this technique was used in level G3 at Vindija. While quartz prevails in the Vindija G3 assemblage, chert is exclusively used for tool production at Mujina Pećina. However, since these represent the available kinds of lithic raw material at each site, it seems clear that Mousterian toolmakers from both sites could adapt their lithic technology to effectively exploit local raw material sources. Taking Pontinian "Micromousterian" assemblages as a reference point (e.g., Kuhn 1995) against which to compare the various Micromousterian assemblages of Croatia, it can readily be seen that notches and denticulates are significantly more frequent in the Vindija and Mujina Pećina assemblages than they are in the Pontinian, where scrapers generally dominate the tool assemblages (Kuhn 1995: table 3.3b). One possible explanation for this distinction is that, despite their shared features (i.e., the diminutive size of their blanks), assemblages in the Adriatic and Tyrrhenian areas of the Mediterranean reflect functional variants of a similar technological strategy (Rink et al. 2002).

P. T. Miracle (1998:176) has demonstrated that the lithic tool assemblages from levels G1 to Fd at Vindija differ significantly from the pattern of the typical French Aurignacian, although a few Aurignacian tools and blade technology have been identified in G/F interface and lower F complex at that site. The Neanderthal remains were found in level G1, and although some authors explain the unusual association of Neanderthal remains and Upper Paleolithic bone points in this level as the result of post-depositional mixing artifacts from different original origin (Zilhao & D'Errico 1999), this association likely represents an actual, distinct "culture assemblage" (Karavanić and Smith 1998, 2000; Straus 1999). As well, beyond the argument presented by Karavanić & Smith (1998), a number of hypotheses have been proposed about the affiliation of the Vindija level G1 artifacts. Svoboda (2001) and Miracle (1998) have suggested affinities to the Szeletian,

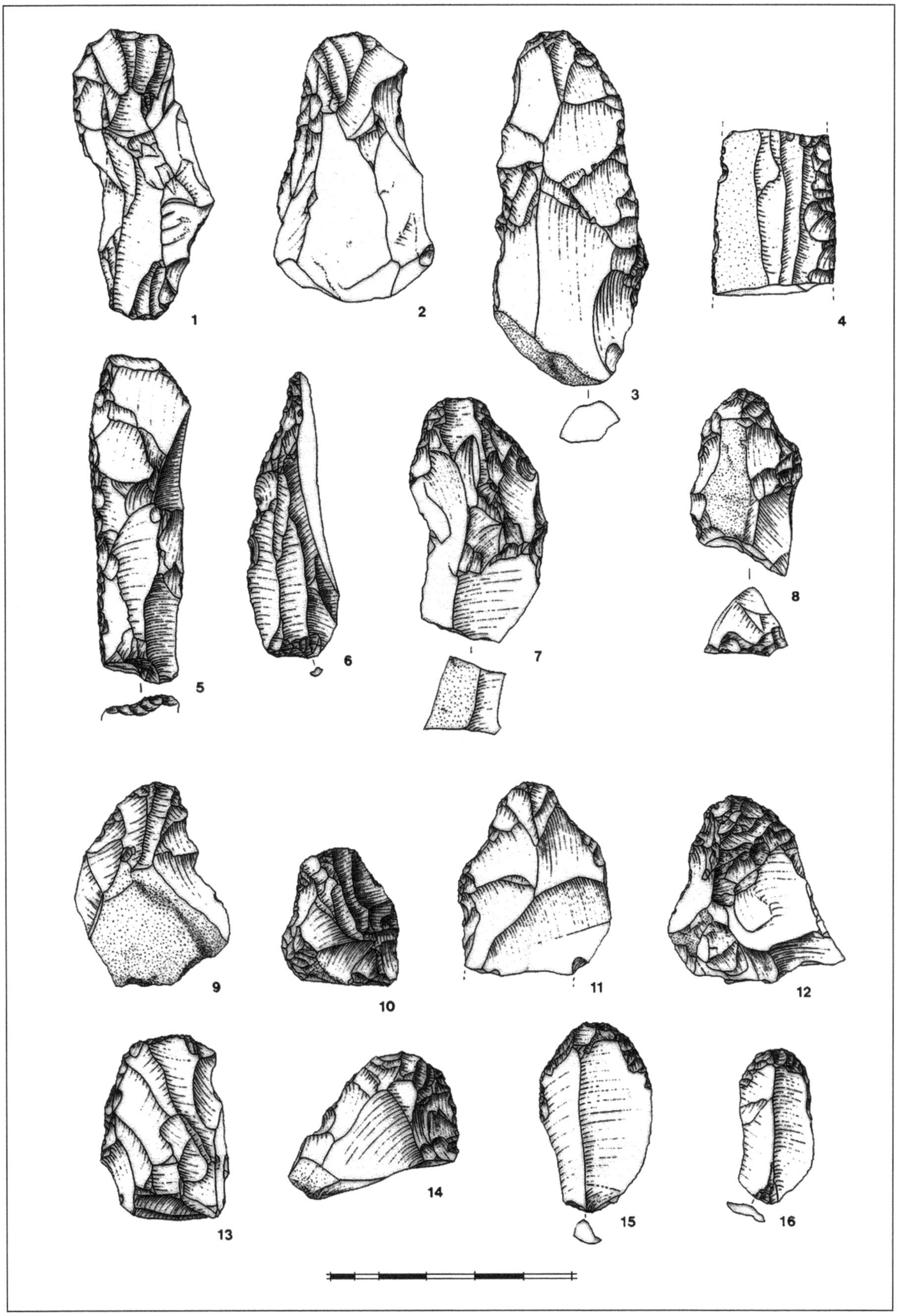

Fig. 7.5. Selected artifacts from Šandalja II.Level E: 1-3. carinated endscrapers, 4. blade with two continuously retouched edges, 5. blade with one continuously retouched edge and truncation, 6. atipical drill. Level F: 7-12. nosed endscrapers, 13. atypical carinated endscraper, 14. endscraper on flake, 15-16. simple endscrapers. Scale is in cm. Drawing: M. Bezić.

while Montet-White (1996) and Straus (1999) "see Vindija in the context of a complex pattern of cultural expressions that characterize the shift from a Mousterian-dominated to a classic Upper Paleolithic-dominated landscape in Europe" (Karavanić & Smith 2000: 839). The term "Olschewian" can be used to label one aspect of this complex pattern which existed as a regional "transitional industry" documented at the sites of Vindija (G1) and Velika Pećina in Croatia, and at some Slovenian sites which are discussed below (Montet-White 1996; Karavanić 2000b). In contrast to the characters of the lithic industry present in Vindija level G1 described above, the combination of blade technology and of a few Aurignacian tool types in the overlying "G/F interface" assemblage suggests a regional variant of what is generally called Central European Aurignacian.

However, it is unclear whether the early Upper Paleolithic industries of Vindija can be attributed to early modern humans, Neanderthals, or to both populations. It is possible that the bone tools associated with the Vindija G1 Neanderthals were the result of imitation of early modern human technologies or even trade with those people. Itseems less likely to us that Neanderthals developed boneworking technology without any external influences because, except for Vindija, no site in Croatia has yielded evidence of Upper Paleolithic bone technology associated with Neanderthal remains. Furthermore, by providing a temporal overlap of several thousand years in central Europe, the recent AMS radiocarbon dates of the Vindija G1 Neanderthals (Smith et al. 1999) lend additional support to a model of cultural and biological integration of Neanderthals and early modern people.

The combination of bone tools similar to those from Vindija and small numbers of undiagnostic (i.e., non-Aurignacian) lithic artifacts has also been identified in the early Upper Paleolithic levels of Velika Pećina, as well as the Slovenian sites of Mokriška Jama (Brodar & Osole 1979) and Divje Babe I (Turk & Kavur 1997). The small quantities of artifacts found at most of these sites suggest that they may have been occupied only for short episodes. However, another site, Potočka Zijalka (Brodar & Brodar 1983), contained an abundant stone and bone industry displaying these characteristics, although the lithic assemblage includes typical Aurignacian tools absent in other "Olschewian" assemblages (except one tool in the Vindija G1 assemblage, and a few tools in Vindija "G/F interface" and level F assemblages). Despite this minor variation, all these Croatian and Slovenian sites are distinctly similar in having yielded bone points along with their stone tool assemblages. Two point types have been identified in those assemblages (i.e., massive-base and split-base), and these have been found together at Vindija level G1 and at Mokriška Jama, which indicates that they are not mutually exclusive. At Potočka Zijalka, only one split-base point has been found among a number of massive-base points, while only massive-base points occur later both at Vindija and Mokriška Jama, suggesting that split-base points might not have been in use anymore after a certain point. The points from Mokriška Jama are distinguished by an oval-flat section similar to that of the split-base point found at Vindija and the points from Velika Pećina. On the other hand, the points from Potočka Zijalka are mainly thick and oval, like the massive-base points from Vindija.

These data suggest the distinct possibility that the archaeological assemblages from Vindija level G1, Velika Pećina level i, Divje Babe I layer 2 and Mokriška Jama represent a regional early Upper Paleolithic industry with possible roots in the local Mousterian. Overall, this industry does not to conform to the "classic" Aurignacian pattern, despite sharing with the latter industry a well-developed bone point technology that includes both split- and massive-base types. As well, while Potočka Zijalka appears closely related to this regional industry, this assemblage (as well as that from the Vindija "G/F interface") displays some affinities with the Aurignacian sensu lato, and it may thus represent a regional variant of the Central European Aurignacian.

Several years ago, Hahn (1977) suggested that small assemblages of stone tools associated with bone implements likely represented activity-specific components (specifically hunting) of the Aurignacian (see also Adams, this volume). This might be a possible explanation for "Olschewian" assemblages, and stable isotope analysis on the Vindija G1 Neanderthals has demonstrated that these people obtained the vast majority of their dietary protein from meat which must have been obtained largely by hunting (Richards et al. 2000). It should be noted, however, that there is no obvious unique hunting activity shared by all of these Croatian and Slovenian assemblages (Karavanić & Smith 1998), which casts doubt on the validity of that interpretation. Potential alternative explanations for the lack of Aurignacian stone tools in those assemblages (except at Potočka Zijalka and Vindija G/F interface) include the influence of relatively poor-quality raw materials on the lithic assemblages; the different technological skills of the occupants of these sites; and/or some type of other functional specialization (cf. Hahn 1977; Montet-White 1996; Karavanić& Smith 1998).

Šandalja II is the only Upper Paleolithic site in the Eastern Adriatic Region and, in contrast to the situation in northwestern Croatia and Slovenia, it has yielded a series of credible Aurignacian layers. Importantly, the presence of this industry elsewhere in that region has also been suggested on the basis of assemblages from several other sites. For instance, a bone point discovered in Bukovac Pećina near Lokve in the Gorski Kotar area has been attributed to the Aurignacian (Malez 1979), although it was recently noted that it might have more affinities to

Olschewian ones (Horusitzky 2004). Furthermore, Upper Paleolithic tools attributed to the Aurignacian mixed with Middle Paleolithic (Mousterian) material were discovered on the surface of two open air sites, Ražanac, near Zadar, and Panđerovica, at Dugi Otok (Batović 1988; Malez 1979). However, because both samples consist of mixed archaeological material and because both contain a very limited number of Aurignacian tool types, the attribution of these assemblages to the Aurignacian is unreliable and it would be pointless to compare the relatively rich collection of Aurignacian material from Šandalja II with the material from these two sites.

Further south, in Bosnia and Herzegovina, Z. Kujundžić (1989:12) attributed to the Upper Aurignacian (i.e., Aurignacian II) some tools discovered with a few flakes in the upper part of layer IVat Gigića Pećina, near Bosansko Grahovo. However, like at most other sites of the Eastern Adriatic Region, this assemblage contained no typical Aurignacian tools such as keeled and nosed end-scrapers. Further, it is possible that the Upper Paleolithic tools of Gigića Pećina are Epigravettian rather than Aurignacian and, in any case, the amount of tools is too small to allow any meaningful comparison with the Šandalja material. Similar problems also apply to layer X at the site Crvena Stijena, near Nikšić in Montenegro, which Đ. Basler (1979:397) classified as belonging to a "locally nuanced Aurignacian." The author also noted that the assemblage includes no typical Aurignacian tools, a feature he explained as the specific feature of the Aurignacianin the Adriatic-Mediterranean region (Basler 1979:398). A. Montet-White (1996; personal communication) doubts that this industry is Aurignacian, a view supported by other evidence. For instance, besides the lack of Aurignacian tool types, there are no dates for those layers to confirm that they were deposited during the timespan of the Aurignacian. Overall, it seems more likely that the artifacts from layer X belong to the Epigravettian.

Turning now to a discussion of the early Upper Paleolithic data from Croatia, it is clear that there are significant differences between the Aurignacian of Šandalja II and the early Upper Paleolithic assemblages of northwest Croatia. Unlike Šandalja II, both Vindija level G1 and Velika Pećina yielded no typical Aurignacian tools (except for one tool in Vindija G1) and the initial Upper Paleolithic industry from these sites therefore should not be called Aurignacian. Furthermore, the small split-base bone point found at Šandalja II (level G) is more similar to some examples from the Franco-Cantabrian Magdalenian (L. G. Straus, personal communication) than to Central European Aurignacian specimens while the bone points recovered from Vindija and Velika Pećina are very similar to examples from the contemporary Central European sites (see Albrecht et al. 1972).

It is important to note however that the sites in northwestern Croatia are situated in a different bio-province than the Adriatic sites. For instance, among the Middle Paleolithic faunal assemblages of northwestern Croatia, cave bear is the dominant species, although the concentration of these remains can be explained in large part by bear denning behavior, meaning that any human presence at many sites was likely sporadic and short (Miracle 1991). In contrast, during the late Middle Paleolithic in Dalmatia, some interesting changes in the relative representation of larger ungulates in Mujina Pećina have been identified by P. T. Miracle (personal communication). From complex D to level B the frequency of red deer drops dramatically, while that of equids and small caprids rises significantly. Carnivores are relatively rare at this site. As concerns the autochthonous faunas of the early Upper Paleolithic assemblages in Croatia, cave bear continues to dominate in northwestern Croatia (like in the Middle Paleolithic), while horse is the most frequent species in the Šandalja II assemblages, and by extension in southwestern Croatia as a whole (Miracle 1991).

The differences in contemporary industries of northwestern Croatia and the eastern Adriatic region probably reflect different forms of adaptation of local Paleolithic populations induced by different climatic and environmental conditions. Although the level of the Adriatic Sea in the late Mousterian period (i.e., Oxygen Isotope Stage 3) was lower than today, the climate in Dalmatia (Mujina Pećina) was milder than in northwestern Croatia (Vindija and Velika Pećina). The same probably held true during the early Upper Paleolithic when the Aurignacian occupants of Šandalja II enjoyed milder conditions compared those endured by the early Upper Paleolithic "Olschewian" occupants of northwest Croatia.

Conclusions

The archaeological evidence from four Croatian sites presented in this paper is important for our understanding of the Middle/Upper Paleolithic interface in Croatia and elsewhere. Vindija and Velika Pećina are located in a continental area (northwestern Croatia) distinct geographically and ecologically from the Eastern Adriatic area where the sites of Mujina Pećina and Šandalja II are found. Some of the material form Vindija Cave bears directly on the Middle/Upper Paleolithic transition, and direct radiocarbon dates of 29 and 28ka B.P. on Neanderthal bones from Vindija indicates they were late survivors in that context. These dates also provide a significant temporal overlap between these Neanderthals and early modern humans in central Europe, which would have provided considerable opportunity for interaction between these human groups. The early Upper Paleolithic industry recovered at Vindija and Velika Pećina is distinct from the typical Aurignacian documented in the early Upper Paleolithic levels of Šandalja II. Further, the Upper Paleolithic bone tools from Vindija may be have been produced by Neanderthals and are associated with Middle Paleolithic stone tools, and this association is what defines a regional "transitional" industry, the Olschewian.

In contrast to the situation in northwestern Croatia, there is no evidence of any kind of transitional industry in the Eastern Adriatic area. Although the late Mousterian of Dalmatia contains some Upper Paleolithic stone tool types, there are unambiguous chronological and technological gaps between this industry and the Aurignacian from Šandalja II. Chronologically, the youngest Mousterian levels from Mujina Pećina were dated to 39ka B.P., while one of the oldest Aurignacian level at Šandalja II yielded a date of 28ka B.P., indicating a potential separation of 11,000 years. On a technological level, blades are rare in the Mousterian of Mujina Pećina and, when present, they are the product of hard-hammer percussion and/or of the use of the Levallois method while they are very frequent and produced by soft-hammer percussion in the Aurignacian of Šandalja II. Additionally, worked bone tools and some ornaments were found in the Aurignacian layers of Šandalja II while there are no such items in the late Mousterian of Mujina Pećina. Unfortunately, neither of the Eastern Adriatic sites contains a stratigraphic sequence that includes both Middle and Upper Paleolithic layers. And while Crvena Stijena in Montenegro (Basler 1979) contains both Middle and Upper Paleolithic levels ,this site yielded no evidence of a transition—nor has any other site in the Eastern Adriatic region (see above).

Two possible reasons for the current lack of such evidence include the decidedly insufficient level of research on the Paleolithic of the Eastern Adriatic region and the rise in sea level following the end of Pleistocene which has likely left many sites dating to that period under the sea. However, not even renewed research in Istria and Dalmatia has yielded sites with "transitional" or early Upper Paleolithic archaeological material. Therefore, a third—and perhaps the main—reason for absence of such transitional industries in the Eastern Adriatic region may have been a scarce population of the area or even the actual abandonment of some parts of its at that time. Ultimately, then, the absence of transitional material in the Eastern Adriatic might well be interpretable as providing further evidence of the complexity and "mosaic"nature of the Middle/Upper Paleolithic interface in Europe.

Acknowledgments

I thank Julien Riel-Salvatore and Geoffrey A. Clark for inviting me to participate in the symposium "Upper Paleolithic "transi-tional" industries: New questions, new methods" held at the 67th Annual Meeting of the Society for American Archaeology in Denver, Colorado. I also thank Julien Riel-Salvatore for editorial assistance with the text of this paper. I also thank the American Embassy in Zagreb, Croatia, and the Ministry of Sciences and Technology of the Republic of Croatia for financial support.

References

Ahern, J., I. Karavanić, M. Paunović, I. Janković, & F. H. Smith
2004 New discoveries and interpretations of hominid fossils and artifacts from Vindija Cave, Croatia. *Journal of Human Evolution* 46:25–65.

Albrecht, G., J. Hahn, & W. G. Torke
1972 *Merkmalanalyse von Geschoßpitzen des mittleren Jungpleistozäns in Mittel- und Osteuropa.* Archaeologica Venatoria. Kohlhammer, Stuttgart.

Basler, Đ.
1979 Paleolitske i mezolitske regije i kulture u Crnoj Gori. In *Praistorija jugoslavenskih zemalja (1)* (A. Benac, ed.), pp. 387–403. Svjetlost, Sarajevo.

Batović, Š.
1988 Paleolitički i mezolitički ostaci s Dugog otoka (with German summary). *Poročilo o raziskovanju paleolita, neolita in eneolita v Sloveniji* 16:7–54.

Blaser, F., D. Kurtanjek, & M. Paunović
2002 L'industrie du site néandertalien de la grotte de Vindija (Croatie): une révision de matières premières lithiques. *L'Anthropologie* 106:387–398.

Brodar, M., & F. Osole
1979 Paleolitske i Mezolitske regije i kulture u Sloveniji. In *Praistorija jugoslavenskih zemalja (1)* (A. Benac, ed.), pp. 159–194. Svjetlost, Sarajevo.

Brodar, S., & M. Brodar
1983 *Potočka Zijalka.* Slov. akad. znan. umet., Ljubljana.

Conard, N. J.
1996 Middle Paleolithic settlement in the Rhineland. In *Middle Paleolithic and Middle Stone Age Settlement Systems* (N. J. Conard & F. Wendorf, eds.), pp. 255–68. Proceedings of the XIII U.I.S.P.P. International Congress. ABACO Edizioni, Forlì.

Crkvenčić, I., Z. Dugački, I. Jelen, P. Kurtek, A. Malić & M. Šašek
1974 Središnja Hrvatska 2. In *Geografija SR Hrvatske* (A. Cvitanović, ed), p 224. Institut za geografiju Sveučilišta u Zagrebu, Školska knjiga, Zagreb.

Hahn, J.
1977 *Aurignacien. Das Ältere Jungpaläolithikum in Mittel- und Osteuropa.* Fundamenta, Ser. A.,Vol. 9. Böhlau, Cologne.

Horusitzky, F. Z.
2004 Les artefacts en os et bois de cerf à Bukovac,Lokve (Croatie). Un seconde flûte possible? Relations entres les chasseurs de Lokve et les montagnards d'Olcheva au début du Paléolithique supérieur. *Arheološki Vestnik* 55:9–37.

Karavanić, I.
1995 Upper Paleolithic occupation levels and late-occuring Neandertal at Vindija Cave (Croatia) in the context of Central Europe and the Balkans. *Journal of Anthropological Research* 51:9–35.
1999 *Gornji paleolitik Šandalje II u okviru jadranske regije.* Unpublished dissertation. Department of Archaeology, University of Zagreb.
2000a Research on the Middle Palaeolithic in Dalmatia, Croatia. *Antiquity* 74:777–778.
2000b Olschewian and appearance of bone technologyin Croatia and Slovenia. In *Neanderthals and Modern Humans—Discussing the Transition* (J. Orschiedt & G-Ch. Weniger, eds.), pp. 159–168. Wissenschaftliche Schriften des Neanderthal Museum, Bd. 2. Neanderthal Museum, Mettmann.
2003 L'industrie aurignacienne de la grotte de Šandalja II (Istrie, Croatie) dans la région de l'Est de l'Adriatique. *L'Anthropologie* 107:577–602.

Karavanić, I., & I. Bilich-Kamenjarin
1997 Musterijensko nalazište Mujina pećina kod Trogira, rezultati trogodišnjih iskopavanja (with English summary). *Opuscula Archaeologica* 21:195–204.

Karavanić, I., & F. H. Smith
1998 The Middle/Upper Paleolithic interface and therelationship of Neanderthals and early modernhumans in the Hrvatsko Zagorje, Croatia. *Journal of Human Evolution* 34:223–248.
2000 More on the Neanderthal problem: The Vindija case. *Current Anthropology* 41:838–840.

Kozlowski, J. K.
1996 Cultural context of the last Neanderthals andearly Modern Humans in Central–EasternEurope. In *The Lower and the Middle*

Palaeolithic (O. Bar-Yosef, L. L. Cavalli-Sforza, R. J. March & M. Piperno, eds.), pp. 205–218. Proceedings of the XIII U.I.S.P.P. International Congress, vol. 5. ABACO Edizioni, Forlì.

Krings, M., C. Capelli, F. Tschentscher, H. Geisert, S. Meyer, A. von Haeseler, K. Grosschmidt, G. Possnert,M. Paunović, & S. Pääbo
2000 A view of Neandertal genetic diversity. *Nature Genetics* 26:144–146.

Kujundžić, Z.
1989 Gigića pećina - paleolitska stanica kod Resanovaca (Bosansko Grahovo). *Glasnik Zemaljskog muzeja Bosne i Hercegovine u Sarajevu* (arheologija, nova serija) 42/43:9–16.

Kurtanjek, D., & V. Marci
1990 Petrografska israživanja paleolitskih artefakata spilje Vindije (with English summary). *Rad Jugoslavenske akademije znanosti i umjetnosti* 449(24):227–38.

Malez, M.
1967 Paleolit Velike pećine na Ravnoj gori u sjeverozapadnoj Hrvatskoj (with German summary). *Arheološki radovi i rasprave Jugoslavenske akademije znanosti i umjetnosti* 4–5:7–64.
1978 Novija istraživanja paleolitika u Hrvatskom zagorju (with German summary). In *Arheološka istraživanja u sjeverozapadnoj Hrvatskoj* (Ž. Rapanić, ed.), pp. 6–69. Izdanja Hrvatskog arheološkog društva 2. Hrvatsko arheološko društvo, Zagreb.
1979 Nalazišta paleolitskog i mezolitskog doba u Hrvatskoj. *In Praistorija jugoslavenskih zemalja (1)* (A. Benac, ed.), pp. 227–76. Svjetlost, Sarajevo.
1987 Pregled paleolitičkih i mezolitičkih kultura napodručju Istre (with German summary). In *Arheološka istraživanja u Istri i Hrvatskom primorju* (V. Jurkić, ed.), pp. 3–47. Izdanja Hrvatskog arheološkog društva 11(1). Hrvatsko arheološko društvo, Pula.
1988 Prethistorijske koštane rukotvorine iz spilje Vindije (Hrvatska, Jugoslavija) (with German summary). *Radovi Zavoda za znanstveni rad Jugoslavenske akademije znanosti i umjetnosti* 2:217–252.

Malez, M., & J. C. Vogel
1970 Die Ergebnisse der Radiokarbonanalysen der quartären Schichten der Velika Pećina in Nordwest-Kroatien. *Bulletin Scientifique du Conseil des Académies des Sciences et des Artsde la RSF Yougoslavie, Section A15* (11–12):390–391.

Malez, M., & D. Rukavina
1979 Položaj naslaga spilje Vindije u sustavu članjenja kvartara šireg područja Alpi (with German summary). *Rad Jugoslavanke akademije znanosti i umjetnosti* 383(18):187–218.

Marjanac, T., Lj. Marjanac, & E. Oreški
1990 Glacijalni i periglacijalni sedimenti u Novigradskom moru. *Geološki vjesnik* 43:35–42.

Miracle, P.
1991 Carnivore dens or carnivore hunts?—A reviewof Upper Pleistocene mammalian assemblagesin Croatia and Slovenia. *Rad Hrvatske akademije znanosti i umjetnosti* 458(25):193–219.
1995 *Broad-spectrum adaptations re-examined: Hunter-gatherer responses to Late Glacialenvironmental changes in the eastern Adriatic.* Unpublished dissertation. Department of Anthropology, the University of Michigan.
1998 The spread of modernity in Europe. In *The Origins and Past of Modern Humans—Toward Reconciliation* (K. Omoto & P. Tobias, eds.), pp. 171–187. Recent Advances in Human Biology, vol. 3. World Scientific, Singapore.

Montet-White, A.
1996 *Le Paléolithique en ancienne Yougoslavie.* Jérôme Millon, Grenoble.

Obelić, B., N. Horvatinčić, D. Srdoč, I. Krajcar-Bronić, A.Sliepčvić, & S. Grgić
1994 Rudjer Bošković Institute radiocarbon measurements XIII. *Radiocarbon* 36:303–324.

Petrić, N.
1979 Mujina pećina, Trogir—paleolitičko nalazište. *Arheološki pregled* 20(1978):9.

Radovčić, J., F. H. Smith, E. Trinkaus, & M. H. Wolpoff
1988 *The Krapina Hominids: An Illustrated Catalogof Skeletal Collection.* Mladost and Croatian Natural History Museum, Zagreb.

Richards, M. P., P. B. Pettitt, E. Trinkaus, F. H. Smith, M. Paunović, & I. Karavanić
2000 Neanderthal diet at Vindija and Neanderthal predation: The evidence from stable isotopes. *Proceedings of the National Academy of Sciences* 97:7663–7666.

Rink W. J., I. Karavanić, P. B. Pettitt, J. van der Plicht, F. H. Smith, & J. Bartoll
2002 ESR and AMS based ^{14}C dating of Mousterianlevels at Mujina Pećina, Dalmatia, Croatia. Journal of Archaeological Sciences 29:943–952.

Šimek, J. F.
1991 Stone tool assemblages from Krapina (Croatia,Yugoslavia). In *Raw Material Economies among Prehistoric Hunter-Gatherers* (A. Montet-White & S. Holen, eds.), pp. 58–71. Publications in Anthropology 19. University of Kansas, Lawrence.

Šimek, J. F., & F. H. Smith
1997 Chronological changes in stone tool assemblages from Krapina (Croatia). *Journal of Human Evolution* 32:561–75.

Smith, F. H.
1976 *The Neandertal Remains from Krapina: A Descriptive and Comparative Study*. Reports of Investigation 15. University of Tennessee, Department of Anthropology, Knoxville.
1982 Upper Pleistocene hominid evolution in South-Central Europe: A review of the evidence andanalysis of trends. *Current Anthropology* 23:667–703.

Smith, F. H., & J. C. Ahern
1994 Brief communication: Additional cranial remains from Vindija cave, Croatia. *American Journal of Physical Anthropology* 93:275–280.

Smith, F. H., E. Trinkaus, P. B. Pettitt, I. Karavanić, & M. Paunović
1999 Direct radiocarbon dates for Vindija G1 and Velika Pećina Late Pleistocene hominid remains. *Proceedings of the National Academy of Sciences* 96:12281–12286.

Srdoč, D., A. Sliepčvić, J. Planinić, B. Obelić, & B. Breyer
1973 Rudjer Bošković Institute radiocarbon measurements II. *Radiocarbon* 15:435–441.

Srdoč, D., A. Sliepčvić, J. Planinić, B. Obelić, & N. Horvatinić
1979 Rudjer Bošković Institute radiocarbon measurements V. *Radiocarbon* 21:131–147.

Straus, L. G.
1999 The Neanderthal problem continued. *Current Anthropology* 40:352–355.

Svoboda, J.
2001 La question szélétienne. In *Les industries a outils bifaciaux du Paléolithique moyen d'Europeoccidentale*.(D. Cliquet, ed.), pp. 21–230. ERAUL No. 98. Liege.

Šegota, T.
1982 Razina mora i vertikalno gimanje dna Jadranskog mora od ris-virmskog interglacijala dodanas (with English summary). *Geološki vjesnik* 35:93–109.

Turk, I., & B. Kavur
1997 Survey and description of Paleolithic tools, fireplaces, and hearths. In *Mousterian "Bone Flute" and other finds from Divje Babe I Cave Site in Slovenia* (I. Turk , ed.), pp. 119–156. Opera instituti archaeologici Sloveniae (2). Znanstvenoraziskovalni center SAZU, Ljubljana.

Wolpoff, M. H.
1996 *Human Evolution*. McGraw-Hill, New York.

Zilhão, J., & F. d'Errico
1999 The Neanderthal problem continued: Reply. *Current Anthropology* 40:355–64.

8

Gulyás Archaeology: The Szeletian and the Middle to Upper Palaeolithic Transition in Hungary and Central Europe

Brian Adams
University of Illinois

Abstract

The term "Szeletian" is derived from Palaeolithic assemblages discovered at Szeleta Cave in northeast Hungary. Retouched tools from these assemblages are dominated by types referred to as "leaf points," "leaf scrapers," bifaces, bifacial foliates, etc., and are considered "type fossils" of the "transitional" Szeletian "industry." Based on the occurrence of Middle Palaeolithic assemblages in the region which also possess bifacial artifact components, two hypotheses have emerged to explain the Szeletian material: in situ *evolution, and acculturation. The former argues that Middle Palaeolithic populations which produced bifacial artifact evolved directly into Upper Palaeolithic populations, while the latter argues that indigenous Middle Palaeolithic populations interacted with incoming populations (i.e., Aurignacians) associated with Upper Palaeolithic industries. The Szeletian is thus argued to represent either an evolutionary stage or the product of acculturation. However, it can be argued that these hypotheses rely too heavily on lithic typological data and fail to incorporate other sources of information, such as lithic raw material, chronological, and cultural data. Consideration of these other data sets, together with data derived from recent excavations at open-air and cave sites in Hungary, suggest that the Szeletian material may not represent a "transitional" phenomenon, but rather an activity facies of the local early Upper Palaeolithic Aurignacian.*

> "... *gulyás* is one of the five most popular meat dishes on the American cooking scene. Of course, what is usually served under this name shouldn't happen to a Rumanian ... *Never* Frenchify it with wine. *Never* Germanize it with brown sauce ... But many variations are possible ... " (George Lang, *The Cuisine of Hungary*, 1971).

Introduction

Like the quintessential Hungarian food *gulyás*, the Szeletian means many things to many people. While Allsworth-Jones' (1986) acculturation model has gained general acceptance by western scholars, new and existing data support alternative hypotheses advocated by Hungarian and other specialists working in the area. This paper presents a brief discussion of the term "Szeletian," summarizes recent interpretations of this material, and present a new model of the Szeletian and Aurignacian in the Bükk Mountains of the north Carpathian region. This will include a discussion of new data from recent investigations conducted by the author and the University of Miskolc in Hungary at Szeleta Cave and other sites.

Szeleta Cave and the Szeletian

The Szeletian is named after Szeleta Cave in the Bükk Mountains of northeast Hungary where extensive excavations have been conducted throughout the twentieth century (Hillebrand 1913; Kadič 1916; Mottl 1945; Saád & Nemeskéri 1955; Vértes 1968) (figure 1). A sequence of five cultural levels has been recognized at the site and the Palaeolithic material has been segregated into two main groups: an "Early Szeletian" dominated by crudely fashioned foliates, retouched blades, end scrapers, retouched flakes and split-base bone points, and a "Developed Szeletian" dominated by more refined bifacial leaf points, retouched bladelets, side scrapers, and burins (Kadič 1916; Vértes 1968) (figures 2 & 3). The presence of crude bifacial artifacts in the lower levels and more finely worked

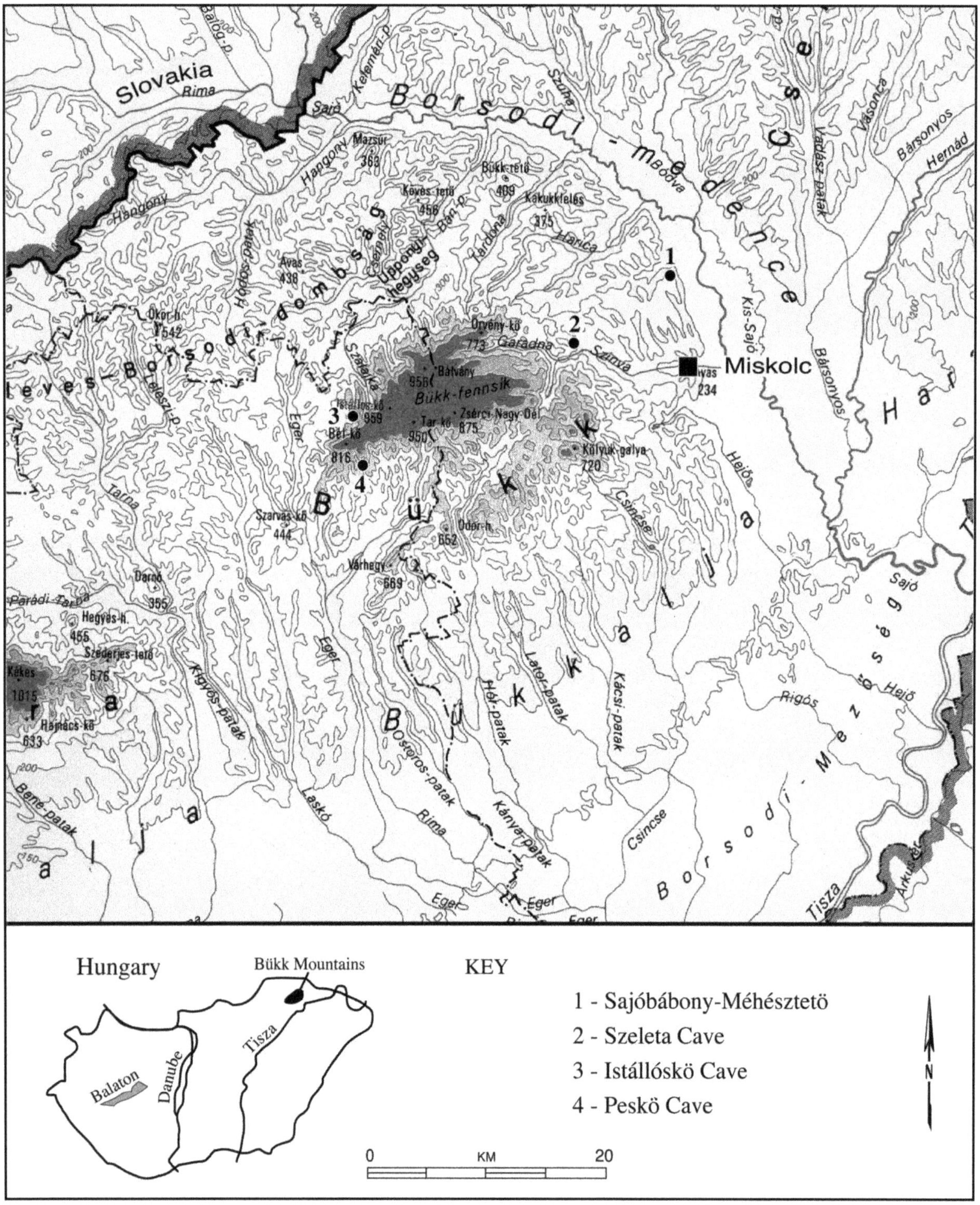

Fig. 8.1. Location of selected Bükk Mountain Palaeolithic sites.

bifaces in the upper cultural horizons has been variously attributed to *in situ* cultural evolution, cryoturbation events, and post-depositional crushing by cave bears which intensively utilized the cave as a den (Adams 1998; Allsworth-Jones 1986). No diagnostic human remains have yet been found in association with Szeletian material. Based on the analysis of the author and others, the Szeletian can be defined as a non-Levallois industry dominated by the production of bifacial tools and retouched blades/bladelets (table 1), and, as several researchers have noted, is very similar to the Aurignacian in the region (Adams 1998; Allsworth-Jones 1986; Barta 1965; Brandtner 1950; Neustupny & Neustupny 1960; Svoboda 1988). Dates of between approximately 33,000 and 43,000 years B.P.,

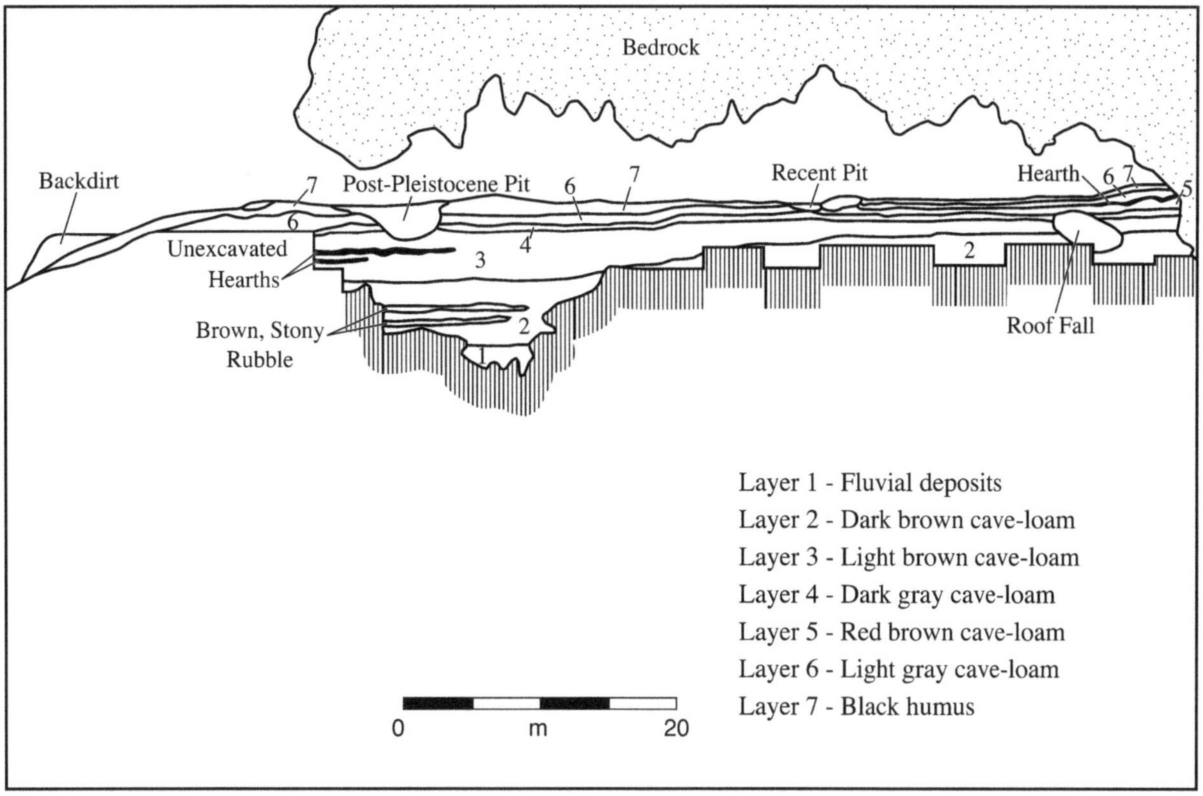

Fig. 8.2. Szeleta Cave stratigraphy (after Kadič 1916).

secured during the late 1960's by Lászlo Vértes (Vértes 1968), suggested that the Szeletian was contemporary with the early Upper Paleolithic, yet the abundance of bifacial artifacts together with some alleged "archaic" types suggested Middle Paleolithic affiliations.

Since its discovery, various hypotheses have been advanced to explain the material from Szeleta Cave, and space permits only a very brief summary and critique of the various interpretations of this material. Hungarian and western archaeologists have proposed that the material supports models of both *in situ* evolution from local Middle Palaeolithic roots and abrupt discontinuity between the Middle and early Upper Palaeolithic (Dobosi 1989; Ringer 1989, 1990; Svoboda & Simán 1989; Vértes 1959, 1968). Allsworth-Jones (1986) has presented the acculturation model, arguing that the Szeletian is the result of contact between indigenous Neanderthals and immigrating "Aurignacian" populations of modern humans. Major shortcomings of these models include (1) an over-reliance on the presence/absence of certain lithic tool types, particularly bifaces; (2) poorly defined and dated Middle Palaeolithic industries; and (3) extremely low population densities that would have made contact unlikely between populations of Neanderthals and immigrating groups (Ambrose 1998; Bar-Yosef 1988, 1992, 1995; Butzer 1982; Gamble 1983a; Hassan 1978; Klein 1989; Mellars 1996, 1998; Rollans 1990; Stringer & Gamble 1993).

Table 8.1. Lithic tool types from Szeleta Cave. Based on existing collections in Hungarian National Museum (Budapest) and Herman Ottó Museum (Miskolc) (Adams 1998).

	Lower Component		Upper Component	
Type	N	%	N	%
Biface	65	74	24	44
Retouched blade	5	6	4	7
Retouched bladelet	1	1	8	14
End scraper	5	6	3	5
Single side scraper	1	1	2	4
Conv. Side scraper	0	-	3	5
Transverse scraper	0	-	1	2
Retouched flake	4	4	0	-
Burin on truncation	1	1	0	-
Double burin	1	1	2	4
Single burin	1	1	2	4
Mult. Burin	0	-	1	2
Denticulate	2	2	1	2
Uniface	1	1	0	-
Truncation	1	1	0	-
Gravette point	0	-	1	2
Aurignacian blade	0	-	1	2
Scraper/piercer	0	-	1	2
Backed blade	0	-	1	2
Total	88	99	55	100

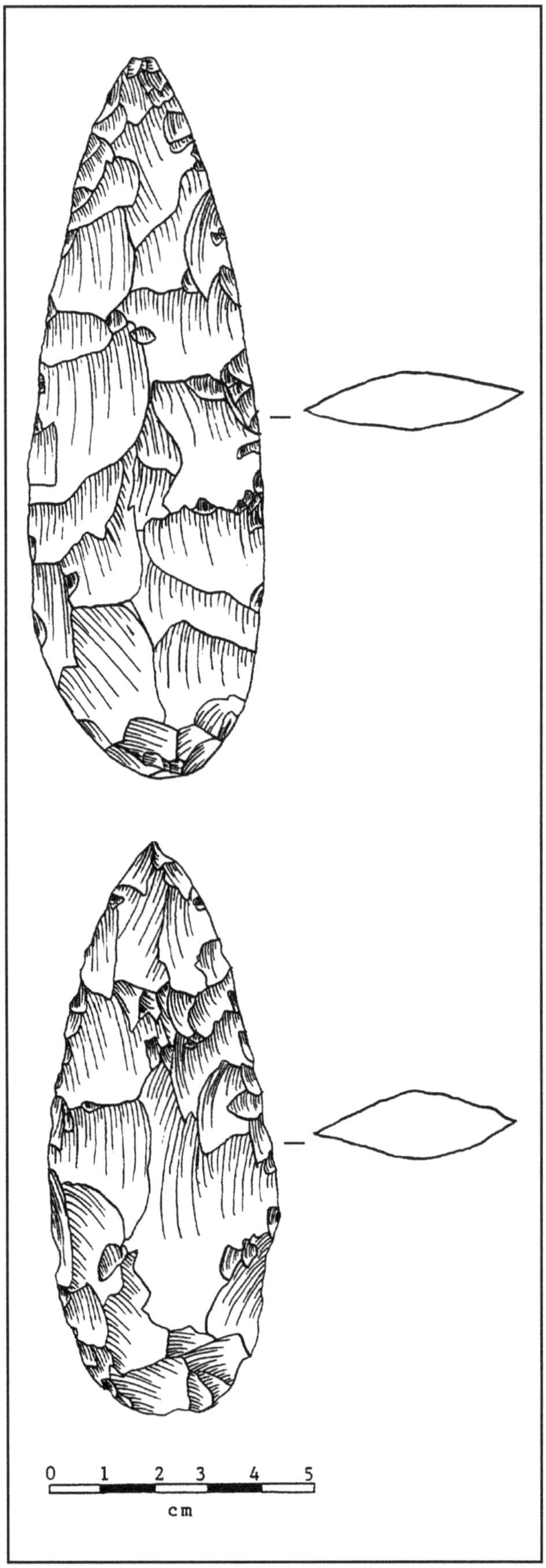

Fig. 8.3. Leaf points from Szeleta Cave ("Developed Szeletian").

It is suggested here that both the acculturation and *in situ* models of the Szeletian are based on weak and tenuous data, and proposed that the Szeletian represents a regional facies of the Aurignacian, and is therefore most likely the product of anatomically modern humans (Adams 1998). Support for the "Aurignacian facies" model of the Hungarian Szeletian as presented here is based on the following evidence: (1) Aurignacian and Szeletian material is contemporaneous in the region; (2) finely worked bifacial leaf points, made of a highly localized raw material that outcrops near Szeleta Cave, occur at sites of both types; (3) split-based bone points occur in both Aurignacian and Szeletian contexts; (4) lithic raw material movement patterns; and (5) site distribution patterns.

Bükk Mountain Aurignacian

Archaeological material classified as Aurignacian represents the initial appearance of early Upper Paleolithic cultural material in Europe. In Hungary such material has been documented at Istállóskő, Peskő, and Szeleta Caves, and open-air Aurignacian sites have been found in nearby eastern Slovakia. At Istállóskő Cave two complexes are recognized: Aurignacian II from the upper cultural levels and previously dated to about 31,000 B.P., and Aurignacian I from the lower levels with controversial dates of approximately 40,000 B.P (Svoboda & Simán 1989; Vértes 1955a) (figure 4). Peskő Cave produced a small assemblage of lithic and bone artifacts classified as Aurignacian II, radiometrically dated to about 35,000 years B.P (Gábori 1969; Svoboda & Simán 1989; Vértes 1956).

Recent Work at Szeleta and Istállóskő Caves

Recent work by the author and the University of Miskolc at Istállóskő and Szeleta caves permits a refinement of the early Upper Palaeolithic chronology in the region and demonstrates the contemporaneity of the Szeletian and Aurignacian in this area (Adams & Ringer 2004). In the summer of 2000, two stratigraphic units were exposed at Istállóskő Cave which can be equated with the previously defined Aurignacian horizons (figure 5). The upper unit consists of an extensive hearth tentatively equated with the Aurignacian II material. A bone sample from the base of this hearth produced an AMS date of approximately 28,000 B.P., very close to Vértes' dates of about 31,000 B.P. (table 2). Approximately one meter below this hearth, an irregular band of reddish brown clayey sediment was exposed that is believed to correspond to Vértes' Aurignacian I level. Bone samples from above and below this deposit produced AMS dates of approximately 33,000 B.P. Significantly, a broken leaf point made felsitic quartz porphyry, a highly localize raw material used to produce the majority of leaf points at Szeleta Cave, was found between these two stratigraphic units, and can be dated to between

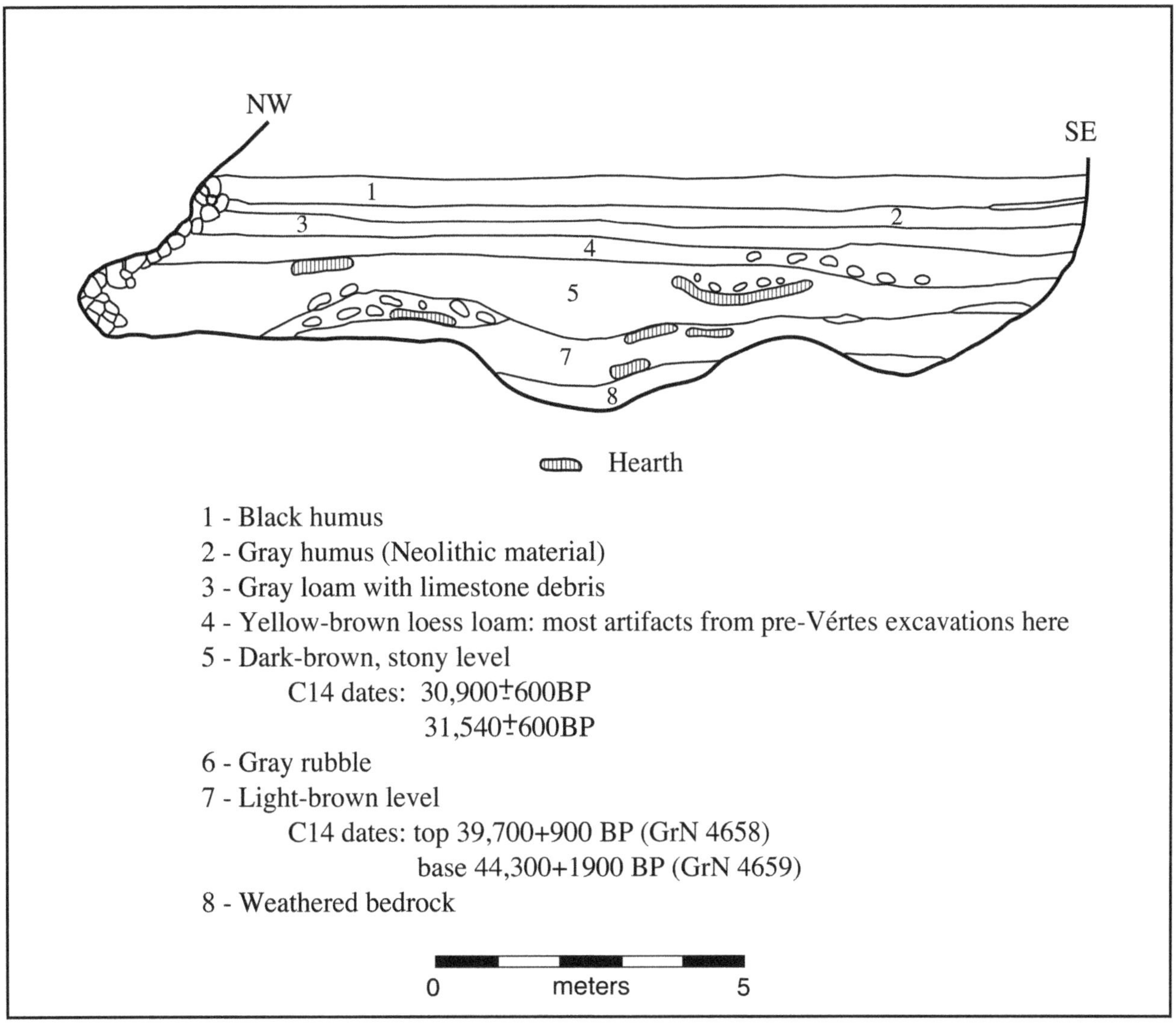

Fig. 8.4. Istállóskő Cave stratigraphy (after Vértes 1955).

about 28,000 and 33,000 year B.P., making it contemporary with the earlier Szeletian material from nearby Szeleta Cave.

In the summer of 1999, our work at Szeleta Cave exposed a stratigraphic profile at the cave entrance comparable to that reported by Ottokar Kadič (1916) in the early Twentieth Century (figure 6) (Adams & Ringer 2004). In Layer 6, at the top of the sequence, Kadič recovered the material now referred to as Developed Szeletian. A bone sample we collected from within this layer yielded an AMS date of about 22,000 B.P., and approximately 2.0 meters below this dated sample we exposed a thin hearth (figure 7). This hearth was found at the same depth as Kadič's hearth 3/b and most likely represents the southern extension of this feature. Based on data presented by Kadič (1916:241), hearth 3/b is associated with the lower part of the Early Szeletian component. A charcoal sample from immediately above this feature produced an AMS date of approximately 26,000 years B.P., while two bone fragments from the base of the hearth produced anomalous dates of approximately 12,000 and 14,000 years B.P. (table 2). As both bone samples were collected within close proximity to each other, it is likely that post-depositional contamination accounts for the anomalous dates. Approximately 20 cm above the hearth and the sample dated to 26,000 B.P., we recovered an obsidian bladelet core, which in this area is a typical Upper Palaeolithic type (figure 8). Kadič (1916:295, fig. 39) recovered an analogous obsidian core approximately 1.0 meter higher up in the sequence associated with his "Hochsolutréen" (now "Developed Szeletian") horizon in the "light yellow cave loam" (i.e., Layer 6). In the cave's main chamber, we exposed the contact between the base of Layer 3 and the top of Layer 2, approximately 5 meters below the original cave floor. The section exposed in 1999 is nearly identical to that documents by Vértes in the 1960's, and is important because it exposes the boundary between Layer 2,

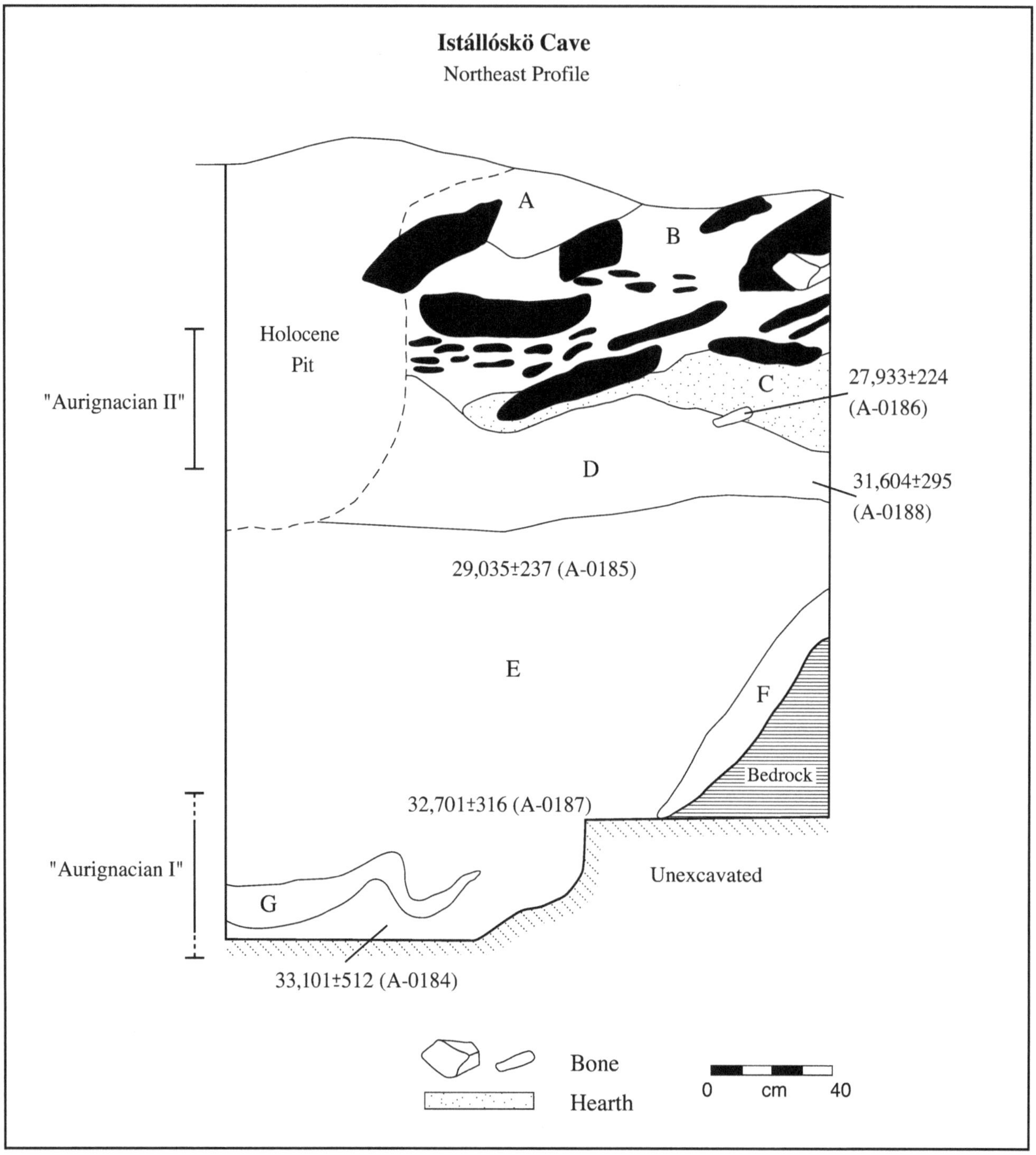

Fig. 8.5. Istállóskő Cave stratigraphy, 2000 excavations. *A:* Holocene spoil; *B:* light yellow clayey sediment with large flat limestone debris; *C:* hearth; *D:* dark brown clayey sediment with dense limestone debris ~[Vértes Layer 5/8]; *E:* light brown clayey sediment with dense limestone debris ~[Vértes Layer 7/9]; *F:* red brown clay; *G:* dark gray brown sediment/possible hearth.

which produced a very small assemblages described by some as Mousterian (Breuil 1923; Ringer 1989; Vértes 1959), and layer 3, which produced the Early Szeletian materual (figure 9). Layer 2 is a reddish brown (5YR 4/4) sediment with little limestone debris, and is analogous to that described by Kadič (1916:210). While Vértes secured a radiocarbon date of 43,000 ±1100 B.P. from this area, the precise provenance of his sample is uncertain (Allsworth-Jones 1986): Vértes claimed the date is from Layer 3, and above the Early Szeletian material, while based on the provenance data reported in the journal *Radiocarbon* (Vogel & Waterbolk 1972), this sample falls within the underlying Layer 2, *predating* the Szeletian material. During our recent excavations, several bone samples were collected from the boundary between Layers 2 and 3, approximately 1 meter *below* Kadič's (1916)

Table 8.2. New Chronometric Dates from Szeleta and Istállóskö Caves.

Site	Date	Material	Stratigraphy	Archaeology
Szeleta	22,107±130* (ISGS-A-0131)	Bone	Layer 6	"Developed Szeletian"
Szeleta	26,002 ±182* (ISGS-A-0189)	Charcoal	Layer 3	"Early Szeletian"
Szeleta	>25,200 (ISGS-4460)	Bone	Layer 3	"Early Szeletian"
Szeleta	11,761±62* (ISGS-A-0128)	Bone	Layer 3 (hearth)	"Early Szeletian"
Szeleta	13,885±71* (ISGS-A-0129)	Bone	Layer 3 (hearth)	"Early Szeletian"
Szeleta	42,960±860 (ISGS-4464)	Bone	Layer 2/3 boundary	Pre- "Early Szeletian"?
Istállóskö	27,933±224* (ISGS-A-0186)	Bone	Layer 5/8**	Aurignacian II
Istállóskö	31,608±295* (ISGS-A-0188)	Bone	Layer 5/8	Aurignacian II
Istállóskö	29,035±237* (ISGS-A-0185)	Bone	Layer 7/9**	Aurignacian I/II
Istállóskö	32,701±316* (ISGS-A-0187)	Bone	Layer 7/9	Aurignacian I level
Istállóskö	33,101±512* (ISGS-A-0184)	Bone	Layer 7/9	Aurignacian I level

* = AMS ^{14}C dates.
** Note that Vértes (1955, Figure 3 and Plate LI) presents two profiles from his excavations at Istállóskö Cave and has numbered analogous levels differently in each.

Early Szeletian component, and one of these samples has produced a radiocarbon date of approximately 43,000 years B.P., suggesting that Vértes's sample also *predates* the Early Szeletian material from Layer 3 (table 2). Based on the chronometric dates secured thus far from Szeleta, it is suggested that the Early Szeletian material from this site is no older than approximately 43,000 years, and may not be older than about 33,000 years B.P. In summary, these new chronometric data are significant because they suggest that the Szeletian and Aurignacian material is much more recent than previously thought and confirm the previously observed contemporaneity of the Early Szeletian material from Szeleta and the Aurignacian I material from Istállóskő. This is further supported by the recent discovery at Istállóskő of the fragmentary leaf point dating to between 28,000 and 33,000 years B.P. and made from the same raw material used to produce the bulk of such artifacts at Szeleta Cave.

Lithic and Bone/Antler/Ivory Points in Hungarian Szeletian and Aurignacian Contexts

Bifacial leaf points are relatively common in Central and Eastern European Aurignacian assemblages (Hahn 1977), and two leaf points were recovered from the Aurignacian II levels at Istállóskő Cave and one from the lower Aurignacian I levels (figure 10). The leaf points derived from the Aurignacian II levels are made from felsitic quartz porphyry, the same raw material used to produce 70% of the foliates from Szeleta Cave which the author has examined. As discussed above, recent work conducted at the site produced a broken leaf point made from felsitic quartz porphyry between the Aurignacian I and II layers, dating to between 28,000 and 33,000 year B.P., making it contemporary with the earlier Szeletian material in the area. Nearby east Slovakian Aurignacian sites have also yielded leaf points, some of which are made from Hungarian felsitic quartz porphyry (Adams 1998). In addition to these examples, leaf points have also been documented in Aurignacian contexts in Moravia, Slovakia, Romania, and the former Soviet Union (Hahn 1977).

Artifacts made from bone, antler and ivory represent a common component of Aurignacian assemblages, and split-base types are especially typical. In the Bükk Mountains, split-based bone points were found in the lower levels of Szeleta Cave and at the Aurignacian sites of Peskő and Istállóskő Caves (Sáad 1929; Sáad & Nemeskéri 1955; Svoboda & Simán 1989; Vértes 1956). Such points have also been found in Szeletian contexts in west Slovakia at Pállfy Cave/Dzerava skala (Hillebrand 1914). At Istállóskő Cave, a total of 30 bone points was recovered from the upper Aurignacian deposits while 114 were found in the lower culture level, 31 of which are split-based (Vértes 1955).

Lithic Raw Material Utilization

Lithic raw material data from Aurignacian and Szeletian sites also suggest similarities between the two. Using a distance of 15 km to separate "local" from "non-local" lithic raw materials, it can be demonstrated that while non-local raw materials are generally not abundant at early Upper Paleolithic sites, they are more common than at Middle Paleolithic sites in the same region (Adams 1998) (tables 3 & 4). During the early Upper Paleolithic of the Bükk Mountains, non-local lithic raw materials are derived from sources located up to 250 to 300 km distant, and consist of high quality flints from southern Poland and Ukraine. The Early Szeletian component from Szeleta Cave exhibits 23% non-local lithic raw material, consisting of Tokaj obsidian and various hydro-limnoquartzites from northeast Hungary, while the Developed Szeletian comprises 42% non-local material, some of which was

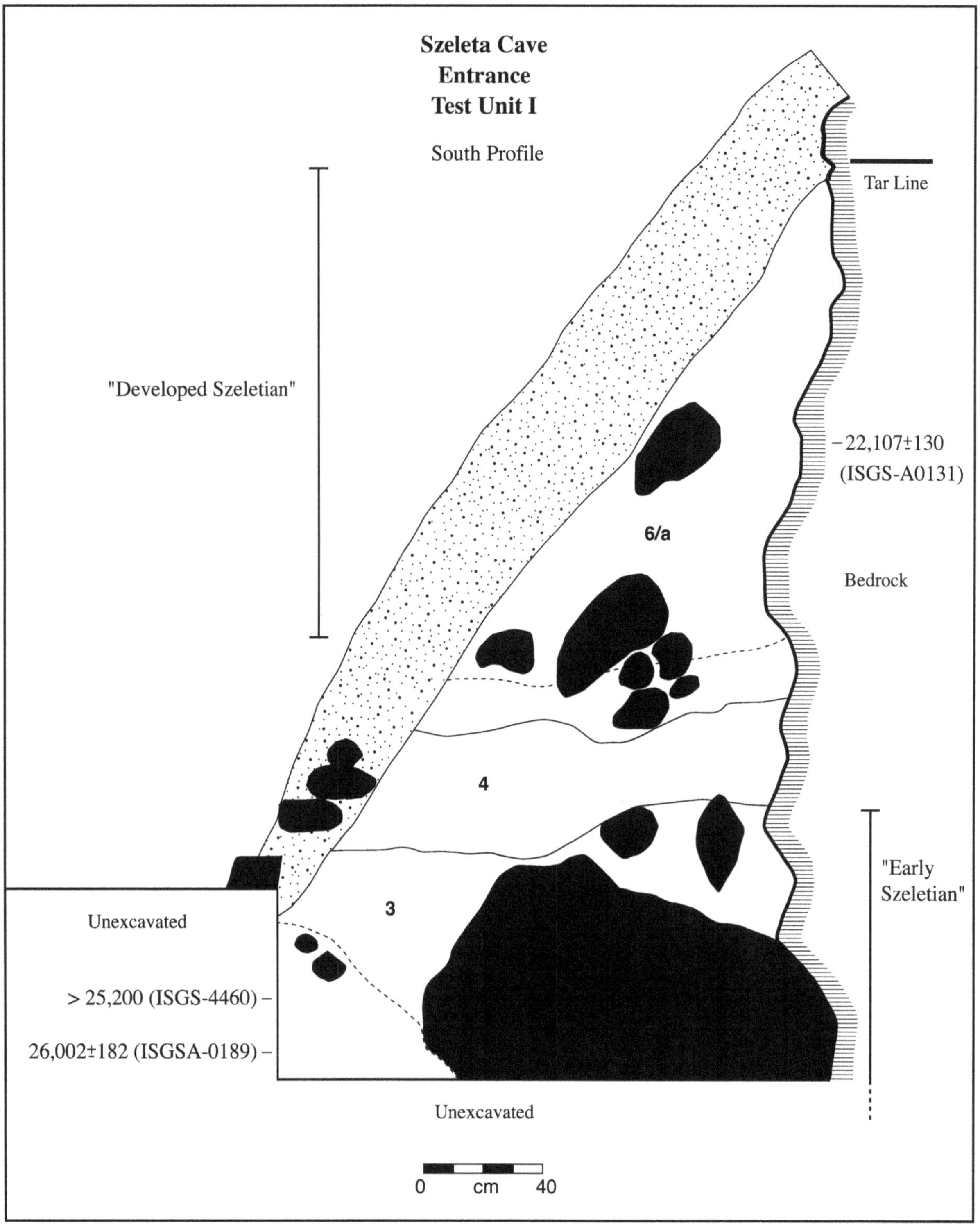

Fig. 8.6. Szeleta Cave stratigraphy, 1999 excavations.

derived from southern Poland. At Istállóskő Cave, non-local raw materials represent an even larger proportion of the assemblage, accounting for nearly 50% of the Aurignacian II collection. Most of this is derived form flint sources in southern Poland. Finally, the distribution of leaf points analogous to types recovered from Szeleta Cave and made from highly localized Hungarian felsitic quartz porphyry indicates that these tools were transported over distances of 200 to 300 km (figure 11). The data from Bükk Mountain Szeletian and Aurignacian contexts thus indicate that lithic raw material was transported over similar distances: At Istállóskő Cave, raw materials were derived from southern Poland (300 km north), while

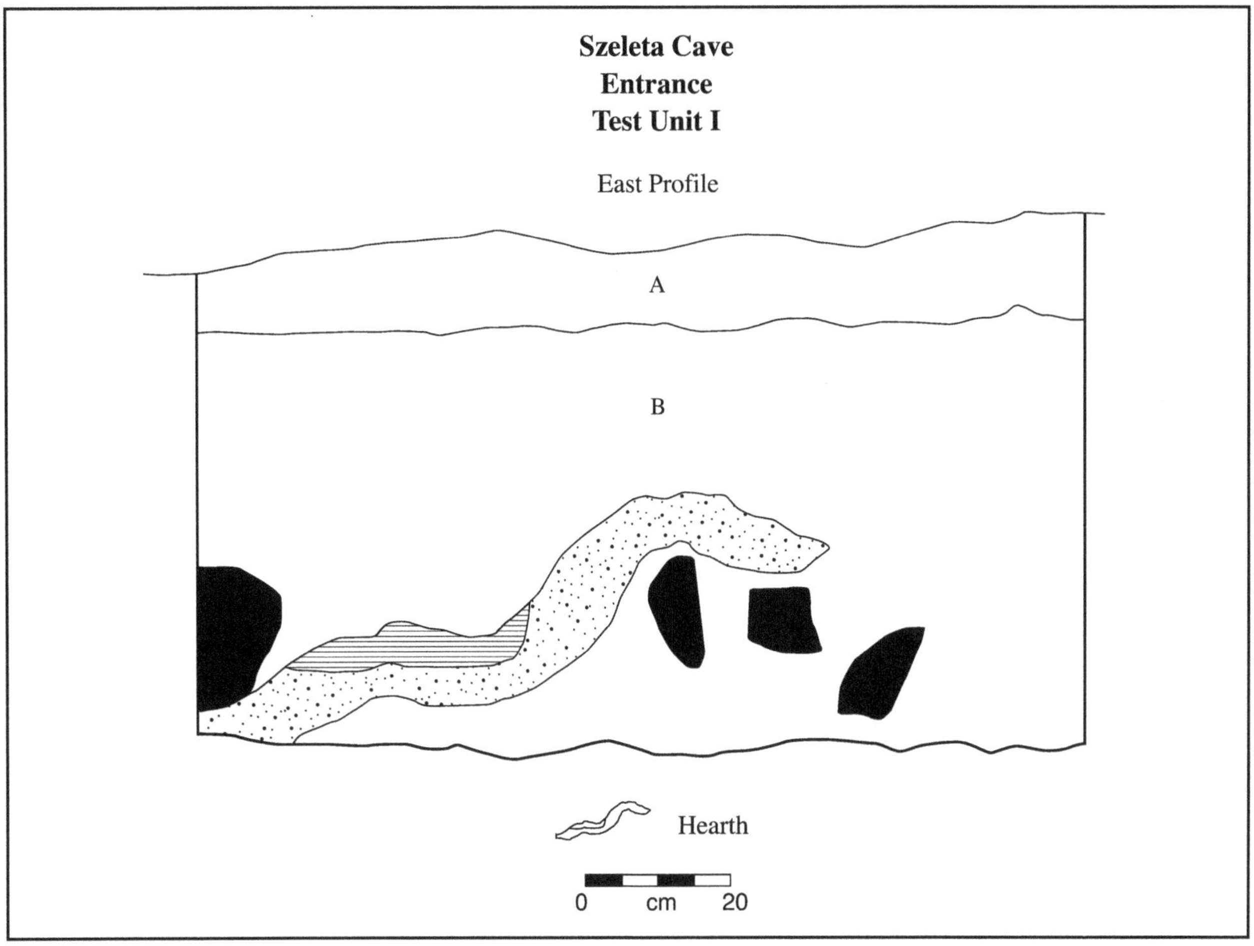

Fig. 8.7. Szeleta Cave, 1999 excavations: hearth. *A:* spoil; *B:* 10YR 4/3 brown cave sediment with dense limestone debris.

Szeletian-type leaf points made from felsitic quartz porphyry were transported up to 300 km away.

A similar pattern exists at the site of Sajóbábony-Méhésztető, a possible open-air Szeletian site rich in leaf points recently excavated by the author in cooperation with the University of Miskolc (Adams 2000; Dobosi 1987/88; Ringer 1983; Svoboda & Simán 1989; Tóth 1973). The site is located approximately 20 km northeast of Szeleta Cave on a prominent ridge overlooking the confluence of the Sajó River and the Bábony Brook (Adams 2000). Based on similarities with the nearby open-air site of Püspökhatvan (Csongrádi-Balogh & Dobosi 1995) and other sites in the region, Sajóbábony-Méhésztető can be tentatively dated to approximately 28,000 years B.P. (Adams 2000; see Ringer 2000, for an alternative dating of this site). 24% of all artifacts from Sajóbábony-Méhésztető are made from raw materials derived from 50 to 300 km from the site, and include hydro- and limnoquartzites, obsidian, opal, mountain crystal from the Zemplén Mountains in Hungary, and spotted Świeciechów flint from southern Poland (table 5) (Adams 2000; Balcer 1976). As at Szeleta Cave, bifaces were produced primarily from felsitic quartz porphyry which outcrops approximately 10 km southwest of the site.

Fig. 8.8. Szeleta Cave, 1999 excavations: obsidian bladelet core.

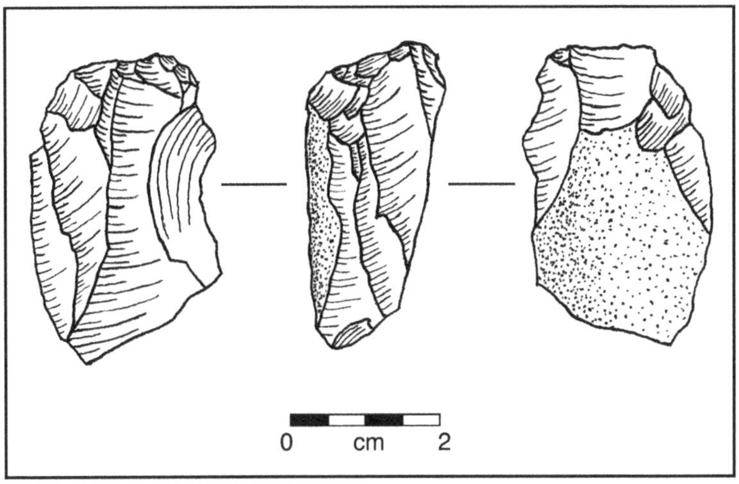

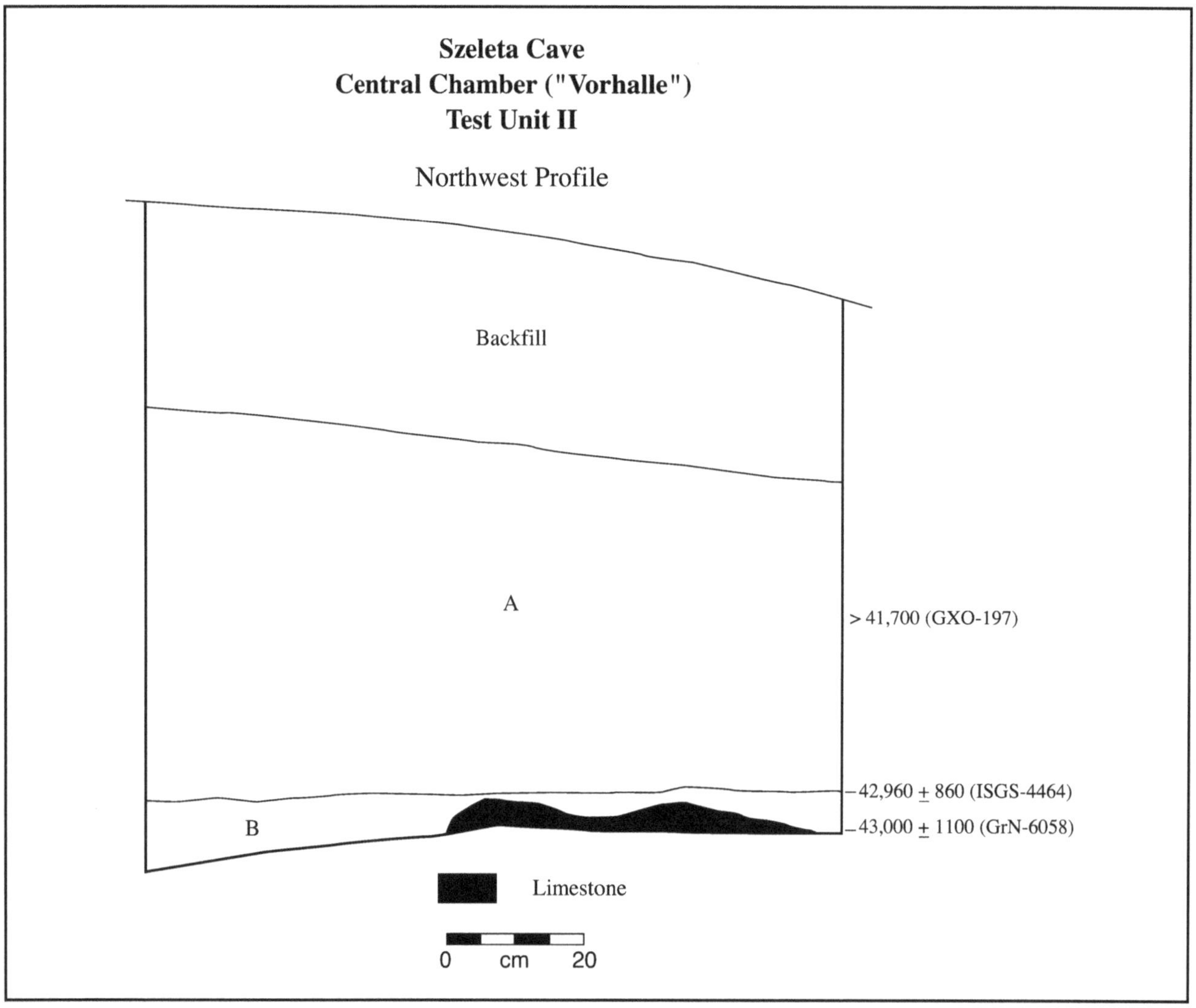

Fig. 8.9. Szeleta Cave, 1999 excavations: level 2/3 boundary. *A:* 10 YR 4/4 dark yellowish brown cave sediment with dense angular debris; *B:* 5YR 4/4 reddish brown cave sediment without limestone debris.

Fig. 8.10. Leaf points from Istállóskő cave (after Hahn 1977 and Allsworth-Jones 1986).

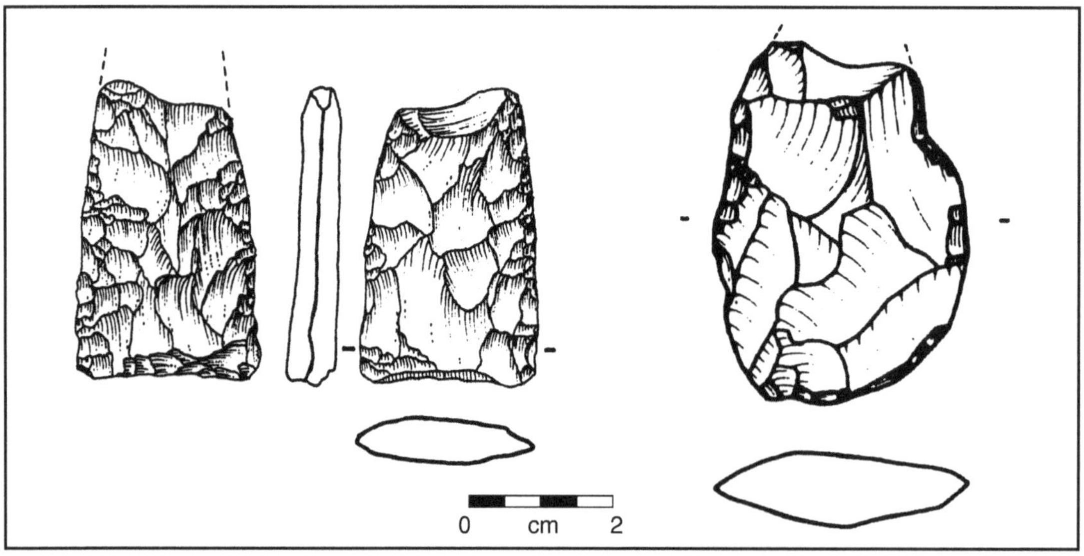

Table 8.3. Local and non-local lithic raw materials at selected Hungarian Middle and early Upper Paleolithic sites (percentages based on artifact counts).

Site/Level	Archaeology	% Local (0–15 km)	% Non-local (>15 km)
Subalyuk 3	Typical Mousterian	88	11
Subalyuk 5	Typical Mousterian	71	22
Subalyuk 11	Quina-Charentian Mousterian	76	16
Subalyuk 14	Quina-Charentian Mousterian	83	16
Büdöspest, lower	Mousterian	82	18
Büdöspest, upper	Mousterian	91	8
Szeleta Lower	"Early" Szeletian	77	23
Szeleta Upper	"Developed" Szeletian	57	42
Sajóbábony	Szeletian	76	24
Istállóskö, lower	Aurignacian I	22	33
Istállóskö, upper	Aurignacian II	25	50

Table 8.4. Detailed breakdown of proportions of non-local raw materials at selected Hungarian Middle and early Upper Paleolithic sites, presented as proportion of weight of entire assemblages).

		Distance (in km)				
Site/Level	Archaeology	15–20	50–100	100–200	200–300	300–400
Subalyuk 3	Typical Mousterian	10 %	1 %	-	-	-
Subalyuk 5	Typical Mousterian	20 %	0 .4%	-	-	-
Subalyuk 11	Quina-Charentian Mousterian	19 %	0 .03%	-	-	-
Subalyuk 14	Quina-Charentian Mousterian	15 %	-	-	-	-
Büdöspest, lower	Mousterian	5 %	0 .5%	-	0 .9%	-
Büdöspest, upper	Mousterian	16 %	7 %	-	0 .05%	-
Szeleta Lower	"Early" Szeletian	2 %	1 %	-	-	-
Szeleta Upper	"Developed" Szeletian	6 %	2 %	-	1 %	-
Istállóskö, lower	Aurignacian I	-	18 %	-	12 %	8 %
Istállóskö, upper	Aurignacian II	4 %	10 %	-	33 %	12 %

Site Distribution Patterns

The geographical distribution of early Upper Palaeolithic sites in north Central Europe reveals a tendency for Aurignacian sites to cluster at lower elevations while Szeletian sites are generally found at higher elevations, and it is suggested here that this pattern is the result of functional differences between the two types of sites. In this region, a mosaic of elevationally distinct physiographic zones, each with unique ecological characteristics, would have occurred within short distances (e.g. wide floodplains, narrow montane valleys, steep mountain slopes) (Kodym 1967; Pécsi & Sárfalvi 1977). It is suggested here that such closely packed, ecologically diverse environments would have favored economic diversification among Palaeolithic hunter-gatherer groups in the region, as it has among more recent nomadic groups in similar environments (Bánesz 1976; Butzer 1982:239; Johnson 1978:27; Peacock & Turner 2000:137; Watanabe 1972:57). Occupation of the mountainous north Carpathian region by Aurignacian groups would have necessitated adaptations to highly varied and closely juxtaposed environments resulting in the creation of varied archaeological signatures by a particular hunter-gatherer group.

Between about 30,000 and 40,000 years ago, non-mountainous areas in Hungary consisted of open, grass steppes with river valleys dominated by coniferous forests. Although faunal preservation tends to be poor at open-air sites in the region, palaeontological and limited archaeological data indicate that moose *(Alces alces),* red deer *(Cervus elaphus),* wild horse *(Equus abeli),* mammoth *(Mammuthus primigenius),* and wild pig (*Sus scrofa)* were common in grasslands and along wooded streams (Adams 1998; Járai-Komlódi 1991; Tuzson 1929; Zólyomi 1953). An especially valuable environmental data set from the period between approximately 30,000 to 40,000 years ago has been derived from the "blue clay" deposits along the Tisza River in the Great Hungarian Plain (Tuzson 1929). These deposits have produced *in-situ* tree stumps and root systems of several coniferous species (e.g. *Pinus cembra, P. montana, Larix deciduas*) and well-preserved needles and cones from *P. cembra.* Arboreal pollen from these deposits is dominated by Pinus (ca. 70%), followed by Betula (ca. 20–25%) and Salix (ca. 5–10 %). Similar environmental data has been derived from other locations in the Great Hungarian Plain and in lower elevations of Slovakia (Ložek 1967; Pécsi et al. 1979; Urban 1984).

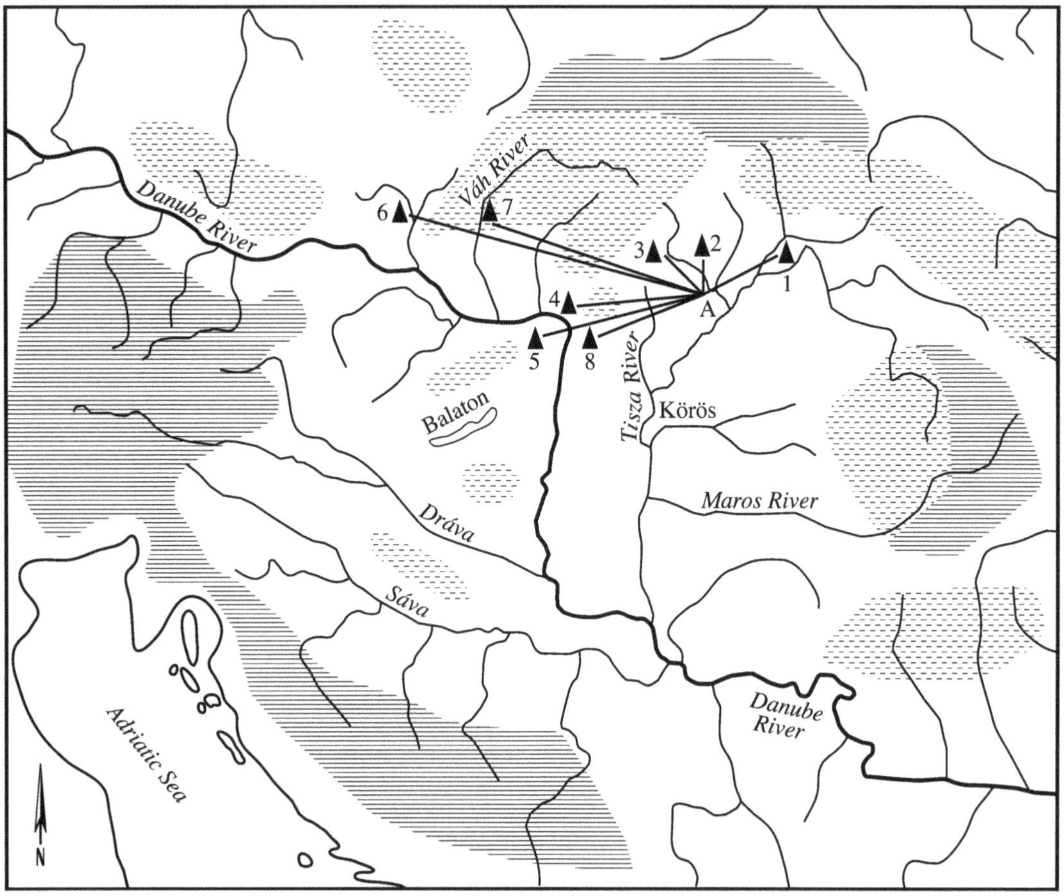

0 km 200

> 200 meters above sea level

> 500 meters above sea level

A: Felsitic quartz-porphyry source area in Bükk Mountains

1. Velky Gires (Nagygéres), Slovakia. (Isolated find) Approximately 100 km NE of source.
2. Čečejovce (Csécs), Slovakia. (Aurignacian?) Approximately 60 km N of source.
3. Domica Cave, Slovakia. (Isolated find) Approximately 70 km NW of source.
4. Szob-Öregfalűdűlo, Hungary. (Isolated find) Approximately 150 km SW of source.
5. Jankovich Cave, Hungary. (Middle Paleolithic?) Approximately 180 km SW of source.
6. Neslovice, Moravia. (Szeletian?) Approximately 340 km NW of source.
7. Moravany nad Vahom (Morávan), Slovakia. Approximately 225 km NW of source.
8. Aszod (Hungary). (Isolated find) Approximately 100 km SW of source.

Fig. 8.11. Distribution of felsitic quart porphyry bifaces in the Carpathian Basin (based on Barta 1962-63, 1990, and personal communication; Dobosi 1990; Šiška and Császta 1980; Valoch 1987).

Table 8.5. Lithic raw materials from Sajóbábony-Méhésztetö (percentages based on artifact counts).

Raw Material	% of Assemblage	Distance to Source (in km)
Felsitic quartz porphyry	76	10
Hydro- and limnoquartzites	13	50
Obsidian, opal, mountain crystal	11	50
Polish Swieciechów flint	<1	300

Compared with more open environments on the plains and in larger stream valleys, environmental conditions became increasingly varied with increased proximity to higher elevations in both northern Hungary and Slovakia. During the Middle Würm, the Bükk Mountains were dominated by pine-larch-spruce forests (Sárkány & Stieber 1955). Likewise, during the Würm interstadials of Slovakia, upland environments were characterized by coniferous forests and varied from park-taiga to steppe park-taiga to stony mountain tundra, depending on elevation (Ložek 1967). Mountain sites in this region have produced faunal inventories similar to those of open-air sites, although more types have been identified at the former due to more favorable preservation. Like many Central European cave sites, Bükk Mountain sites are usually overwhelmingly dominated by cave bear remains *(Ursus spelaeus),* followed by other carnivores such as wolf *(Canis lupus),* lion *(Felis spelaea),* and hyena *(Hyaena spelaea)* (Kadič 1916; Mottl 1938, 1941; Vörös 1984). The degree to which cave bear were actively hunted by Upper Palaeolithic groups in the region is still debated (Gargett 1996). Non-carnivore faunal material recovered from Bükk Mountain sites consists of wild ass *(Equus hemionus),* wild horse *(Equus abeli),* reindeer *(Rangifer tarandus),* red deer *(Cervus elaphus),* roe deer *(Capreolus capreolus),* moose *(Alces alces),* giant deer *(Megaceros),* woolly rhinoceros *(Coelodonta antiquitatus),* mammoth *(Mammuthus primigenius),* aurochs *(Bos primigenius),* and bison *(Bison sp.)* (Jánossy & Vörös 1979; Mottl 1938, 1941; Simán 1988). These forms are typically associated with more open environments and most likely penetrated more mountainous areas by following stream courses within montane valleys. Species unique to mountain sites are ibex *(Capra hircus)* and chamois *(Rupicapra rupicapra),* two forms that often dominate the non-carnivore portion of mountain site faunal assemblages (Vörös1984, 1987). While ibex and chamois are not as abundant in the archaeofaunas of the Bükk Mountain region as elsewhere in Central and Western Europe (Gamble 1983b), it is suggested here that the acquisition of these two species was an important objective of early Upper Palaeolithic groups in the north Carpathian area (cf. Straus 1987, 1992).

In the Bükk Mountains, Szeletian sites occur between elevations of 300 and 350 meters above sea level and are commonly found in rugged, mountainous environments, usually in caves. In nearby east Slovakia, Aurignacian sites occur between approximately 235 and 120 meters above sea level in open-air settings (Adams 1998). Exceptions in this area are Istállóskő and Peskő caves, where Aurignacian assemblages were found at high elevations in caves. However, both of these sites produced rich bone, ivory and antler point assemblages, artifacts that may have been functionally equivalent to bifacial leaf points. A similar pattern is observed in the Váh River valley of west Slovakia, and in Moravia (Ambroz et al. 1952; Svoboda 1994) (figures 12 & 13). Based on this evidence, it is proposed here that in the north Carpathian region of Central Europe, material classified as "Szeletian" is most parsimoniously interpreted as special purpose Aurignacian activity sites (cf. Ashton 1983). These data suggest that there is a correlation between Aurignacian and Szeletian assemblages and elevation, a pattern which might reflect seasonal movements of early Upper Paleolithic hunter-gatherers between lowlands and uplands (cf. Henry 1992, 1995). Binford's (1980, 1982) discussions of synchronic variability in material culture due to seasonally determined variations in site functions, the latter related to the influence of resource structure on settlement patterns, provides a useful framework to interpret the Szeletian and Aurignacian material from the north Carpathian region. As a group of hunter-gatherers moves across the landscape, individual sites are often reoccupied but utilized for different purposes. In such cases, the tool assemblages associated with the varied functions of a particular site may differ significantly even though they represent the material remains of a single group. Stated another way, a particular hunter-gatherer group will leave several archaeological signatures at the assemblage level of analysis depending on the particular activities conducted at a site or sites. It is suggested here that factors such as these account for the Szeletian/Aurignacian phenomenon in northern Central Europe. It is argued that the material culture of early Upper Palaeolithic groups exploiting resources in open environments, such as plains or wide stream valleys, will appear more "Aurignacian-like", while the archaeological remains of the same groups will appear more "Szeletian-like" at sites associated with the exploitation of montane resources, such as ibex and chamois (Adams 1998). As hunting groups moved into highland environments, their archaeological signatures would shift as they "geared up" to hunt mountain species. It is expected that evidence for the production and maintenance of certain artifacts likely associated with hunting and butchery, such

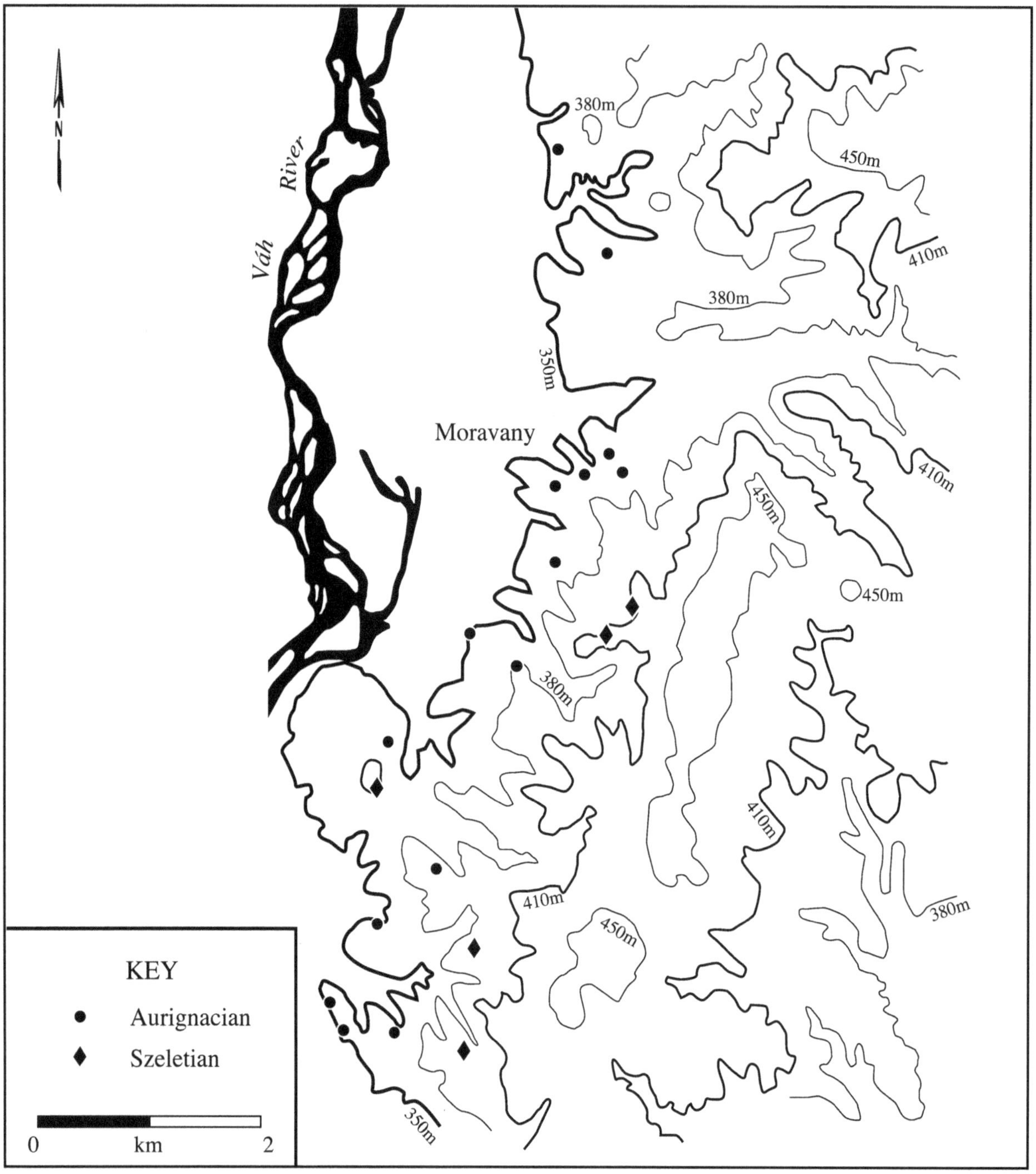

Fig. 8.12. Distribution of Aurignacian and Szeletian sites in the Váh (Vág) River valley near Moravány, west Slovakia (after Ambroz et al. 1952).

as bifacial leaf points, would become more common at upland sites focused on the exploitation of such resources.

The author has presented a hypothetical settlement system based on the early Upper Palaeolithic evidence from the North Carpathian Basin region (Adams 1998). The proposed settlement system demonstrates how the reevaluation of existing data combined with new data permit alternative interpretations of the Aurignacian and Szeletian material in the Bükk Mountain region. Utilizing varied data sets, such as assemblage variability, artifact density, lithic microwear analysis, and feature abundance and diversity, a typology of site types was proposed. As also proposed by Sklenár (1975, 1976), Slovakian open-air Aurignacian sites with pit features, such as Tibava and Barca II, are interpreted as cold season base camps occupied for several continuous months. To date, no such Aurignacian sites have been found in Hungary, although a possible candidate for a cold season base camp occupation is "Molotov utca", an open-air site along the Szinva Brook

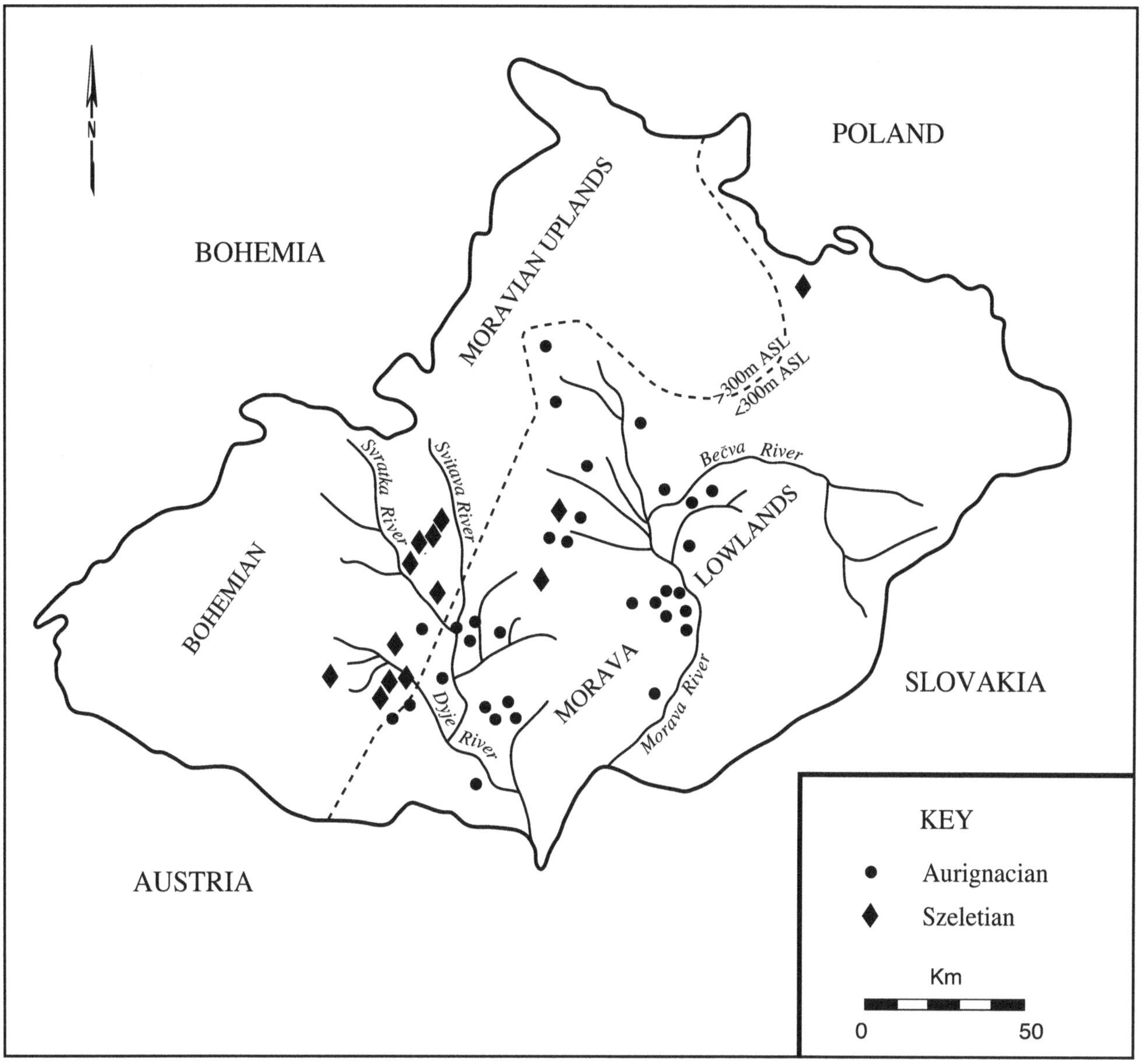

Fig. 8.13. Distribution of Aurignacian and Szeletian sites in Moravia, Czech Republic (after Svoboda 1994).

in downtown Miskolc (Ringer 1989:224, and personal communication 1993). The site was exposed in 1958 during construction activity east of the former "Magyar-Szovjet Friendship Square" and was hastily excavated, and unfortunately little contextual data exist as a result. The site produced 21 large pyramidal blade cores, 1 large leaf point, 1 Aurignacian blade, and 1 dihedral burin on a tertiary blade (Adams 1998). The abundance of blade cores suggests that tool production was an important activity at this site, and the variety of retouched artifacts suggests that a range of activities was performed. Exploitation of montane resources, such as ibex and chamois, resulted in the establishment of a variety of short-term occupations by groups utilizing streams courses in narrow ravines to access upland hunting sites. In the Bükk Mountains, the Szinva Valley represented such an environment, with steep walls lined with caves and rockshelters, including the eponymous site of Szeleta Cave. Some sites, such as the Szeletian site of Puskaporos Rockshelter (Kadič1934; Vértes 1955b), produced lithic assemblages consisting only of debitage and bifacial leaf points, and are interpreted by the author as locations where hunters "geared up" in preparation for upland hunting expeditions. Szeleta Cave, situated near the top of a steep, rocky bluff approximately 95 meters above the Szinva Brook, also produced lithic assemblages dominated by bifacial leaf points, and is interpreted as a repeatedly occupied, short-term seasonal hunting site. Functional analysis of selected Szeletian leaf points from Szeleta Cave (Adams 1998) suggests that these artifacts were used as thrusting spears. Upland species, such as ibex, would have been in prime condition for hunting in late fall/early winter, suggesting that Szeletian and Aurignacian sites in mountainous settings were occupied during these seasons (Straus 1987). As discussed above,

"Aurignacian" bone points may have been the functional equivalents of "Szeletian" bifacial leaf points. The proximity of Szeleta Cave to the preferred lithic raw material for leaf point production, felsitic quartz porphyry, likely explains its abundance at this and other sites in the Szinva Valley, while the rarity of such material in the vicinity of the contemporaneous site of Istállóskő Cave suggests that points were more commonly produced from materials such as bone derived from abundant natural accumulations typical of such sites (Gargett 1996; Simán 1986; Vértes & Tóth 1963).

Summary

In summary, recent interpretations of the Szeletian material have focused on the issues of *in situ* evolution and acculturation, models that are only weakly supported by archaeological data and based mainly on typological comparisons. As demonstrated here, such studies ignore a wide range of existing alternative data which suggest that the Szeletian represent special-purpose activity sites associated with the Aurignacian (cf. Ashton 1983). In addition, new chronometric data suggest that the Szeletian and Aurignacian material is more recent than suggested by radiocarbon dates secured from earlier excavations and confirm the previously observed contemporaneity of the material from the Early Szeletian levels at Szeleta and from the Aurignacian I levels at Istállóskő. The data presented above support recent hypotheses that the Aurignacian postdates 36,500 B.P., as well as the generally accepted idea that the Hungarian Szeletian and Aurignacian are contemporary (Adams 1998; Allsworth-Jones 1986; Vértes 1961; Zilhão & d'Errico 1999). The range of interpretations of the Szeletian material accentuates the limited explanatory power of lithic typologies alone, and demonstrates that more robust models of the Szeletian phenomenon can be developed if additional, existing data sets are utilized together with data derived from new excavations utilizing modern methods of excavation and analysis.

Acknowledgments

The author would like to thank Árpád Ringer (University of Miskolc), the National Science Foundation (Research Grant SBR 97-07008) and the Wenner-Gren Foundation (Small Grant #6351) for making this research possible.

References

Adams, B.
1998 *The Middle to Upper Paleolithic Transition in Central Europe: The Record from the Bukk Mountain Region.* BAR International Series 693. Archaeopress, Oxford.
2000 Archaeological investigations at two open-air sites in the Bükk Mountain region of north-east Hungary. In *Neanderthals and Modern Humans: Discussing the Transition: Central and Eastern Europe From 50,000–30,000 B.P.,* (Jörg Orschiedt & Gerd-Christian Weniger, eds.) pp. 169–179. Neanderthal Museum, Mettman.

Adams, B., & Á. Ringer
2004 New C14 dates for the Hungarian early Upper Palaeolithic. *Current Anthropology* 45:541–551.

Allsworth-Jones, P.
1986 *The Szeletian.* Clarendon Press, Oxford.

Ambrose, S. H.
1998 Late Pleistocene human population bottlenecks, volcanic winter, and differentiation of modern humans. *Journal of Human Evolution* 34: 623–651.

Ambroz, V., V. Ložek, & F. Prošek
1952 Mlady pleistocèn v okolí Moravan u Piešt'an nad Váhom. *Anthropozoikum* 1:53142.

Ashton, N. M.
1983 Spatial patterning in the Middle–Upper Paleolithic transition. *World Archaeology* 15:224–235.

Balcer, B.
1976 Position and stratigraphy of flint deposits, development of exploitation and importance of the Świeciechów flint in prehistory. *Acta Archaeologica Carpathica* 16:179–199.

Bánesz, L.
1976 Prírodné Prostredie, Hospodárska Základòa a Materiálna Kultúra Aurignacienu Strednej

Európy (Natürliches Milieu, wirtschaftliche Basis und materielle Kultur des Aurignacien Mitteleuropas). *Slovenská Archeológia* 24(1): 5–82.

Barta, J.
1962/63 Paläolithische Höhlenbesiedlung im Karpatischen Teil der Tschechoslowakei. Slovenska Akademija Znanosti in Umetnosti. *Archeološki Vestnik* 12–14:19–35.
1965 *Slovensko v Starsej a Strednej Dobe Kamennej*. Slovenska Academia Vied, Bratislava.
1990 Blattspitzenindustrien in der Westslowakei. In *Feuilles de pierre: Les industries à pointes foliacées du Paléolithique supérieur européen* (J. K. Kozlowski, ed.), pp. 235–238. ERAUL 42, Liège.

Bar-Yosef, O.
1988 The date of south-west Asian Neandertals. In *l'Homme de Néandertal 3: l'Anatomie* (M. Otte, ed.), pp. 31–38. ERAUL 30, Liège.
1992 The role of western Asia in modern human origins. *Philosophical Transactions of the Royal Society, Series B* 337: 193–200.
1995 The role of climate in the interpretation of human movements and cultural transformations in Western Asia. In *Paleoclimate and Evolution, with Emphasis on Human Origins* (E. S. Vrba, G. H. Denton, T. C. Partridge & L. H. Burckle, eds.), pp. 507–523. Yale University Press, New Haven.

Binford, L.R.
1980 Willow smoke and dog's tails: hunter-gatherer settlement systems and archaeological site formation. *American Antiquity* 45:1–17.
1982 The archaeology of place. *Journal of Anthropological Archaeology* 1:1–31.

Brandtner, F.
1950 Über die relative Chronologie des jüngeren Pleistozän Niederösterreich. *Archaeologica Austriaca* 5:101–113.

Breuil, H.
1923 Notes de voyage paléolithique en Europe Centrale. I. Les industries paléolithiques en Hongrie. *L'Anthropologie* 33:323–346.

Butzer, K.
1982 *Archaeology as Human Ecology*. Cambridge University Press, Cambridge.

Csongradi-Balogh, E., & V. T. Dobosi
1995 Palaeolithic settlement traces near Püspökhatvan. *Folia Archaeologica* 44:37–59.

Dobosi, V. T.
1987/88 Ásátas Sajóbábony-Méhésztetőn. *Archaeologiai Értesítő* 112–113:226–235.
1989 Data on the relationship between the Middle and Upper Palaeolithic in Hungary. *Anthropologie* 27:231–244.
1990 Leaf-shaped implements from Hungarian open-air sites. In *Feuilles de pierre: Les industries à pointes foliacées du Paléolithique supérieur européen* (J. K. Kozlowski, ed.), pp. 175–188. ERAUL 42, Liège.

Gábori, M.
1969 Regionale Verbreitung Palaolithischer Kulturen Ungams. *Acta Archaeologica Academiae Scientiarum Hungaricae* 21: 155–165.

Gamble, C.
1983a Culture and society in the Upper Palaeolithic of Europe. In *Hunter-Gatherer Economy in Prehistory* (G. Bailey, ed.), pp. 201–211. Cambridge University Press, Cambridge.
1983b Caves and faunas from last glacial Europe. In *Animals and Archaeology I: Hunters and Their Prey* (J. Clutton-Brock & C. Grigson, eds.), pp. 163–172. BAR International Series 163, Oxford.

Gargett, R. H.
1996 *Cave Bears and Modern Human Origins: The Spatial Taphonomy of Pod Hradem Cave, Czech Republic*. University Press of America, Lanham, MD.

Hahn, J.
1977 *Aurignacien, das ältere Jungpaläolithikum in Mittel- und Osteuropa*. Böhlau, Köln.

Hassan, F.
1978 Demographic archaeology. In *Advances in Archaeological Theory and Method, Volume 1* (M. B. Schiffer, ed.), pp.49–103. Academic Press, New York.

Henry, D. O.
1992 Transhumance during the late Levantine Mousterian. In The *Middle Paleolithic: Adaptation, Behavior and Variability* (H. L. Dibble & P. Mellars, eds.), pp. 143–162. University of Pennsylvania Museum Press, Philadelphia.
1995 *Prehistoric Cultural Ecology and Evolution*. Plenum Press, New York.

Hillebrand, J.
1913 A pleistocaen ősember ujabb nyomai hazánkban (Neure Spuren des diluvialen Menschen in Ungarn). *Barlangkutatás* 1:19–52.

1914 Ergebnisse meiner Höhlenforschungen im Jahre 1913. *Barlangkutatás* 2:115–124.

Jánossy, D., & I. Vörös
1979 Grossäuger-Streufunde aus dem Pleistozän Ungarns. *Fragmenta Mineralogica et Palaeontologica* 9:21–61.

Járai-Komlódi, M.
1991 Late Pleistocene vegetation history in Hungary since the last interglacial. In *Quaternary Environment in Hungary* (M. Pécsi & F. Schweitzer, eds.), pp. 35–46. Studies in Geography in Hungary 26. Akadémia Kiadó, Budapest.

Johnson, D. L.
1978 Nomadic organization of space: reflections on pattern and process. In *Dimensions of Human Geography: Essays on Some Familiar and Neglected Themes* (K. W. Butzer, ed.), pp. 25–47. University of Chicago, Department of Geography, Research Paper 186, Chicago.

Kadič, O.
1916 *Ergebnisse der Erforschung der Szeletahöhle.* Mitteilungen aus dem Jahrbuch der Könglischen Ungarischen Geologischen Reichanstalt 23(4), Budapest.
1934 A jégkor embere magyarországon (Der Mensch zur Eiszeit in Ungarn). *A Magyar Királyi Földtani Intézet Évkönyve* 30(1). Magyar Királyi Földtani Intézet Kiadása, Budapest.

Klein, R. G.
1989 *The Human Career: Human Biological and Cultural Origins.* The University of Chicago Press, Chicago.

Kodym, O. (editor).
1967 *Regional Geology of Czechoslovakia. Geological Atlas.* Geological Survey of Czechoslovakia, Prague.

Ložek, V.
1967 Climatic zones of Czechoslovakia during the Quaternary. In *Quaternary Paleoecology* (E .J. Cushing & H. E. Wright, Jr., eds.), pp. 381–392. Yale University Press, New Haven.

Mellars, P.
1996 *The Neanderthal Legacy: An Archaeological Perspective from Western Europe.* Princeton University Press, Princeton.

Mottl, M.
1938 Faunen, Flora und Kultur de ungarischen Solutréen. *Quartär* 1:36–54.
1941 Die Interglazial- und Interstadialzeiten im Lichte der ungarischen Säugetierfauna. *A Magyar Királyi Földtani Intézet Évkönyve* 35:3–33.
1945 Jelentés az 1936–38. Évi barlangkutatásról és az ősgerinces-osztály működéséről (Bericht über die Ergebnisse der Grabungen der Jahren 1936/38, sowie über die Tatigkeit der Vertebraten-Abteilung der Kgl. Ung. Geol. Anstalt) A Magyar Kir"alyi Földtani Intézet Évi Jelentési az 1936–38. *Evékrol* 4:1513–1585.

Neustupny, E., & J. Neustupny
1960 Nástin Pravekych Dejin Ceskoslovenska. *Sborník Národního Musea V Praze, Acta Musei Nationalis Pragae Series A (Historia)* 14:95–110.

Peacock, S. L., & N. J. Turner
2000 Just like a garden: traditional resource management and biodiversity conservation on the Interior Plateau of British Columbia. In *Biodiversity and Native America* (P. E. Minnis & W. J. Elisens, eds.), pp. 133–179. University of Oklahoma Press, Norman.

Pécsi, M., & B. Sárfalvi
1977 *Physical and Economic Geography of Hungary.* Corvina Press, Budapest.

Pécsi, M., E. Szebenyi, & M. A. Pevzner
1978 Upper Pleistocene litho- and chronostratigraphic type profile from the exposure at Mende. *Acta Geologica* 22:371–389.

Ringer, Á.
1983 *Bábonyien: Eine mittelpaläolithische Blattwerkzeugindustrie in Nordostungarn.* Dissertationes Archaeologicae Ser. II, No.ll, Budapest.
1989 L'origine du Szélétien du Bükk en Hongrie et son évolution vers le Paléolithique supérieur. *Anthropologie* 27:223–229.
1990 Le Szélétien dans le Bükk en Hongrie. In *Paléolithique moyen récent et Paléolithique supérieur ancien en Europe* (C. Farizy, ed.), pp. 107–109. Mémoires du Musée de Préhistoire d'Île-de-France, Nemours.
2000 Le complexe techno-typologique du Bábonyien-Szélétien de la Hongrie du Nord-Est et le Yabroudien du Levant. In *Toward Modern Humans: Yabrudian and Micoquian, 400,000–50,000 years ago* (A. Ronen & M. Weinstein-Evron, eds.) pp. 181–187. BAR International Series 850. Archaeopress, Oxford.

Rolland, N.
1990 Middle Palaeolithic socio-economic formations

in western Eurasia: an exploratory survey. In *The Emergence of Modern Humans* (P. Mellars, ed.) pp. 347–388. Edinburgh University Press, Edinburgh.

Sáad, Á.
1927 Die Ergebnisse der Ausgrabungen in der Istállósköer Höhle in Jahre 1927. *Die Eiszeit* 4: 97–98.
1929 A Bükk-hegységben végzett ùjabb kutatasók eredményei (Über die Resultate der neueren Ausgrabungen im Bükkgebirge). *Archaeologia Értesítő* 43:238–247.

Sáad, Á., & J. Nemeskéri
1955 A Szeleta barlang 1947. évi kutatásainak eredméyei. *Folia Archaeologica* 7:15–21.

Sárkány, S., & J. Stieber
1955 Anthrakotomische bearbeitung der in neuester zeit in der Höhle Istállóskő freigelegten Holzhohlenreste. *Acta Archaeologica Scientiarum Hungaricae* 5:239–258.

Simán, K.
1986 Felsitic quartz porphyry. In *International Conference on Prehistoric Flint Mining and Lithic Raw Material Identification in the Carpathian Basin* (T. Biró Katalin, ed.), pp. 271–276. Magyar Nemzeti Muzeum, Budapest.
1988 Települési formák Borsod-Abaúj-Zemplén megye területén a Paleolitikum idején. *A Miskolci Herman Ottó Múzeum Évkönyve* 25–26:55–67.

Šiška, S., & J. Czászta
1980 Prieskum mikroregión v Cecejovciach. In *Archeologické Výskumy a Nálezy Na Slovensku v Roku 1979,* pp. 297–208. Archeologický Ústav Slovenskej Akadémie v Nitre, Nitra.

Sklenár, K.
1975 Paleolithic and Mesolithic dwellings: problems of interpretation. *Památky Archeologické* 66: 266–304.
1976 Paleolitic and Mesolitic dwellings: an essay in classification. *Památky Archeologické* 67:249–340.

Straus, L. G.
1987 Upper Paleolithic ibex hunting in southwest Europe. *Journal of Archaeological Science* 14: 163–178.
1992 *Iberia Before the Iberians.* University of New Mexico Press, Albuquerque.

Stringer, C. B., & C. Gamble
1993 *In Search of the Neanderthals.* Thames & Hudson, London.

Svoboda, J.
1988 Early Upper Paleolithic industries in Moravia: A review of recent evidence. In *l'Homme de Néandertal 8: La Mutation* (J. K. Kozlowski, ed.), pp. 169–192. ERAUL 36, Liège.
1994 *Paleolit Moravany a Slezska.* Dolnovestonicke studie, svasek 1, Brno.

Svoboda, J., & K. Simán
1989 The Middle–Upper Paleolithic transition in southeast Central Europe (Czechoslovakia and Hungary). *Journal of World Prehistory* 3: 283–322.

Tóth, L.
1973 Mezolit-telep Sajóbábonyban. *A Miskolci Herman Ottó Múzeum Közleményei* 12:17–25.

Tuzson, J.
1929 Adatok a Magyar Alföld őskori növényzetének ismeretéhez. *MTA Matematika és Természettudományi Értesítő* 46:442–457.

Urban, B.
1984 Palynology of Central European loess-soil sequences. In *Lithology and Stratigraphy of Loess-and Paleosols,* (M. Pécsi, ed.), pp. 229–248. Geographical Research Institute, Hungarian Academy of Sciences, Budapest.

Valoch, K.
1987 Raw materials used in the Moravian Middle and Upper Paleolithic. In *International Conference on Prehistoric Flint Mining and Lithic Raw Material Identification in the Carpathian Basin, Budapest-Siimeg, 20–22 May, 1986, Vol. 2,* (K. T. Biró), pp. 263–268. KMI Rota Press, Budapest.

Vértes, L.
1955a Neure Ausgrabungen und Palätolithische Funde in der Höhle von Istállóskő. *Acta Archaeologica Academiae Scientiarum Hungaricae* 5(3–4):111–131.
1955b Paläolithische Kulturen des Würm I/II-Interstadials in Ungarn. *Acta Archaeologica Academiae Scientiarum Hungaricae* 5(3–4):261–277.
1956 Gruppen des Aurignacian in Ungarn. *Archaeologia Austriaca* 19/20:15–27.
1959 Das Mousterian in Ungarn. *Eiszeitalter und Gegenwart* 10:21–40.
1968 Szeleta-Symposium in Ungarn. *Quartär* 19: 381–390.

Vértes, L., & L.Tóth
1963 Der Gebrauch des Glasigen Quartzporphyrs im

Paläolithikum des Bükk-Gebirges. *Acta Archaeologica Academiae Scientiarum Hungaricae* 15:3–10.

Vogel, J. C., & H. T. Waterbolk
1972 Gronigen Radio-carbon Dates X. *Radiocarbon* 14:6–110.

Vörös, I.
1984 Hunted mammals from the Aurignacian cave bear hunters' site in the Istállóskő Cave. *Folia Archaeologica* 35:7–31.
1987 Large mammalian faunal changes during the late Upper Pleistocene and early Holocene times in the Carpathian Basin. In *Pleistocene Environment in Hungary* (M. Pécsi, ed.), pp. 81–101. Hungarian Academy of Sciences, Geographical Research Institute, Budapest.

Watanabe, H.
1972 *The Ainu Ecosystem: Environment and Group Structure.* The American Ethnological Society, Monograph 54. University of Washington Press, Seattle.

Zilhão, J., & F. d'Errico
1999 The chronology and taphonomy of the earliest Aurignacian and its implications for the understanding of Neanderthal extinction. *Journal of World Prehistory* 13:1–68.

Zólyomi, B.
1953 Die Entwicklungsgeschichte der Vegetation Ungarns seit dem letzen Intergläzial. *Acta Biologica Academia Scientorum Hungaricae* 4: 367–430.

9

Hafting and Technological Innovation in the Western Eurasian Middle Paleolithic

Michael S. Bisson
McGill University

Abstract

Evidence for the chronology, nature, and distribution of the hafting of stone tools during the Middle Paleolithic, and the efficiency of various hafting techniques is summarized. Using attribute analysis, hafting modifications from layer B1 at Skhūl cave (Israel) are contrasted to the hand-held tools characterizing the rest of the Mount Carmel sequence. The overall record for Middle Paleolithic hafting indicates episodic rather than incremental technological change, and limited transmission of useful innovations across time and space. Hafting of tools by the earliest anatomically modern Homo sapiens *may have compensated for lesser muscularity rather than imparting an adaptive advantage over contemporary Neanderthals.*

Although the theme of this volume is transitional industries, this paper investigates technological change from a different perspective. Instead of discussing a single industry, or the Middle to Upper Paleolithic transition in a specified region, I will trace the history of a single useful innovation, the hafting of stone tools on handles or shafts. The extensive use of hafting, particularly of tools other than spear points, has often been included in the inventory of traits distinguishing Upper from Middle Paleolithic industries (Mellars 1996), and the increase in tool efficiency gained by hafting has been implicitly linked to claims for greater cognitive complexity and adaptive superiority of anatomically modern *Homo sapiens* (AMHs) over the Neanderthals (Mellars 1989a). Yet evidence for at least some forms of hafting extends far back into the Middle Paleolithic, and the dichotomy between a predominantly hand-held Middle Paleolithic technology and a predominantly hafted Upper Paleolithic technology may not be as clear cut as once seemed. In order to follow the origins of hafting, I will present a brief discussion of the various techniques and the evidence used to identify it on archaeological specimens. This will be followed by a summary of the chronology and distribution of Eurasian sites where hafting of Middle Paleolithic tools has been identified, including a detailed study of the data from Skhūl and Tabūn caves in Israel, where components associated with early AMHs and Neanderthals respectively show distinct differences in the frequency and types of hafting modifications employed. The identification of these hafting modifications was facilitated by the use of attribute analysis as opposed to traditional typological classification. These data will serve as the starting point for a discussion of the nature and pace of technological innovation in the Western Eurasian Middle Paleolithic.

Methods and Functions of Hafting

Although hafting is considered to have been one of the most important innovations in the history of technology, the search for its origins—and thus for that of composite tools[1]—has been surprisingly unsystematic. The most recent comprehensive treatment of this topic was a C.N.R.S. symposium held in Lyon, France, in 1984, which covered the Middle Paleolithic through the Neolithic (Stordeur 1987). Many of the papers featured the results of micro-wear analysis (i.e., Anderson-Gerfaud & Helmer 1987; Beyries 1987b; Plisson 1987; Moss 1987; Vaughan 1987; Casper & Cahen 1987; Unger-Hamilton et al. 1987), often combined with experimental replication. In introducing the topic, Stordeur (1987: 12–30) presented a useful summary and classification of hafting techniques based on the position and orientation of the tool in its shaft or handle. This classification proceeds in a hierarchy from a basic division of "families" of simple composite tools, those with only one element fixed to the handle/shaft (as in a stone point on a spear), and complex composite tools, in which multiple elements are employed (as in multiple backed blades set in a sickle handle). The type of articulation of the tool to the handle can be either "male,"[2] in which a projection on the tool fits into a socket or slot on the handle; "female", in which the tool is perforated to fit it to a handle; or "juxtaposed", in which the tool is placed parallel to a tongue in the handle. This latter technique requires particularly tight and durable binding to be effective in most circumstances. All three of these may be either "direct", in which the tool is attached directly to the support, or "indirect", in which the tool is mounted in a foreshaft, which is in turn attached to a shaft/handle. The "placement of the tool" refers to its overall position on the shaft/handle, "terminal" at the end; "latero-distal" on the side of the shaft near the end; and "lateral", on the side of the shaft near the middle. The "direction of the tool" refers to the relationship of the long axis of the tool to the shaft/handle. In "axial" the long axis is oriented in the same direction as the shaft (the tip of an arrow, spear, or knife, where the blade protrudes directly from the end of the shaft). In "transverse" directions, the axis of the tool is either perpendicular to the shaft/handle (as in a short adze blade or endscraper), or "oblique", in which the tool is mounted at a diagonal to the shaft/handle (as in microlithic arrow barbs). The "orientation of the active part" refers to the orientation of the working edge of the tool to the long axis of the shaft/handle. This includes "parallel" (as in a knife blade mounted in either the end or the side of the handle), "perpendicular", and "oblique."

The following discussion will use the term 'mode of hafting' to refer to the four variables that best describe a haft. In the order given by Stordeur, these are the type of articulation, the placement of the tool, the direction of the tool, and the orientation of the active part. "Direct" or "indirect" are omitted since indirect hafting has not been identified on Middle Paleolithic tools. Thus a knife blade hafted into a slot at the end of a handle would be "male-terminal-axial-parallel" (figure 1*A*), and an adze blade hafted onto a tongue on an angled handle would be "juxtaposed-terminal-transverse-perpendicular" (figure 1*B*). This combination of direction of the tool and orientation of the active part (i.e. transverse-perpendicular) is of particular interest here as it requires more secure binding and possibly more elaborate modification of the tool to fit the handle than does the axial-parallel direction and orientation.

Hafted elements are attached to a support through one or a combination of friction/pressure, bindings, and adhesives. Friction/pressure includes socketed or perforated specimens in which the tool is attached to the support often with the aid of a wedge or jamb to keep it in place. Common bindings include plant fibers derived from bark and roots, animal sinews and strips of hide. Hide and sinew are particularly useful in that they are pliable and easy to manipulate when wet, and shrink when they dry, thus tightening the binding. Adhesives include a variety of plant saps, starch pastes, bee's wax, and animal products including blood glue (Mills & White 1994).

Hafting serves three basic, often overlapping functions. One is to make the tool easier and more comfortable to manipulate. For example, wrapping one side of a stone tool blunts sharp edges, and prevents the users hand from being cut. In the case of very small tools such as thumbnail scrapers and microlithic blades, hafting is the only practical way to employ them. A second function of hafting is to extend the reach of the tool. A spear or arrow shaft allows a relatively small piercing and cutting blade to be projected considerable distances, and in the case of hunting implements, allows the hunter to minimize close and potentially dangerous contact with the prey. The third and perhaps the most significant function of hafting is to increase the efficiency of the tool by taking advantage of the leverage added by the handle (Clark 1983; Stordeur 1987). Discovering the origin of this last function is critical because the recognition of mechanical advantage is conceptually different from the direct use of a tool.

Evidence for Hafting

Direct evidence for hafting in the form of specimens actually mounted on handles or shafts has not yet been found in the Middle Paleolithic. Indeed, lithics retaining their hafts are exceedingly rare, even from later periods when hafting is thought to have been common. Perishable materials (wood for handles, plant fibers, hide and sinew, plant and animal adhesives for mastic) seem to have been employed in hafting until the Epipaleolithic. Bone or antler handles, which would survive in limestone cave deposits, are not only lacking in Middle Paleolithic contexts, but are also absent in the succeeding early Upper Paleolithic of Europe and the Levant. Thus, evidence for hafting in the Middle Paleolithic is derived from interpretations of forensic and morphological data.

It is now generally accepted (Shea 1988; Mellars 1996) that the earliest hafted Middle Paleolithic tools in Eurasia were spear points (a male or juxtaposed-terminal-axial-parallel mode). However, the lack of obvious hafting modifications on most Middle Paleolithic points is a striking contrast between them and their Upper Paleolithic successors (Otte & Caspar 1987). In the majority of cases, a Mousterian point tip is opposite the striking platform and bulb of force of the flake blank. This is true of most Mousterian points, and all Levallois points. Since platform and bulbar thinning are rarely present, and the exterior platform angle of most Middle Paleolithic flake blanks is 70° or greater, then the part of the point actually in contact with the shaft is inevitably blunt. This situation differs from almost all known projectile points from later periods, where the butt of a point is prepared for mounting by one or a combination of thinning to create a symmetrical wedge, a stem, and/or narrowing or notching of the base to attach the binding. Lacking these modifications, the only logical means of affixing a Mousterian or Levallois point to a shaft is to prepare an L-shaped notch

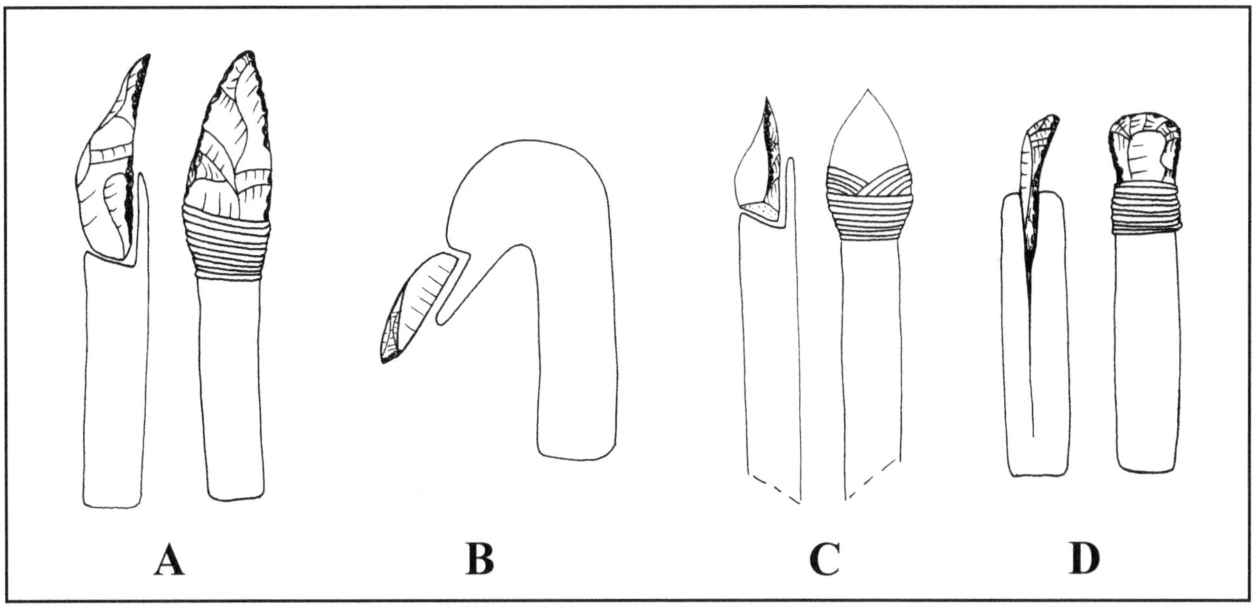

Fig. 9.1. Common hafting modes attributed to Middle Paleolithic tools.
A: Convergent scraper or knife, juxtaposed-terminal-axial-parallel; *B:* Scraper mounted as an adze, juxtaposed-terminal-transverse-perpendicular; *C:* Mousterian point, juxtaposed-terminal-axial-parallel; *D:* Endscraper, male (split shaft) -terminal-axial-perpendicular.

(figure 1*C*) cutting across the end of the shaft, setting the point base in this notch, and fixing the point in place with binding and mastic in a juxtaposed articulation (Beyries 1987a, b). Alternatively, Henry (1995) believes that the convex base shape on Levallois points was fit into a slot with a shallow socket in the center, a design that would stabilize the point during impact. How such a narrow slot could be created with Mousterian tools is unclear, since thin burins or burin-like chisels are rare or absent in Mousterian assemblages. In either case, it is absolutely essential that the center of the base of a point be oriented at precisely a right angle to the long axis. Without this feature, the point will inevitably tilt in its haft on impact, significantly diminishing its penetrating ability. Juxtaposed hafting of points is relatively easy, but is only minimally efficient. In experiments conducted by the author employing replicas of Mousterian and Levallois points hafted in this manner and thrown or stabbed into damp clay soil, the point stayed in place when impacts were perpendicular to the shaft, but was easily dislodged when a slashing or transverse impact occurred. Thus this type of hafting mode is inherently unstable when lateral forces are applied to the blade (figure 1*D*). This could possibly be counteracted by the use of extremely strong binding and mastic, but as yet very little evidence of mastic has survived, and the well-dated examples (Königsaue, Germany) come from late in the Middle Paleolithic sequence (Grünberg 2002). The variety of potentially usable mastic in Pleistocene Levantine and European environments was certainly more limited than that available to recent tropical hunter-foragers (Mills & White 1994). Any mastic and binding applied to points must have relatively high efficiency per unit volume for the simple reason that they must not form a bulge at the base of the point that is so large that it inhibits deep penetration of the weapon into the prey. This would have been particularly important for both Mousterian and Levallois points, which tend to be small and have short blades when compared to ethnographically documented spear heads made of more modern materials. A stone tip is also not essential for a spear to function effectively. As the Schöningen (Thieme 1997) and Lehringen (Movius 1950) specimens demonstrate, sharpened wooden spears could effectively kill large mammals. The increase in sharpness imparted by a stone point would have to improve the penetrating ability of the spear sufficiently to overcome the added resistance created by the binding. Perhaps the greatest advantage of stone points to Middle Paleolithic hunters was the ability to replace a dulled or broken tip more rapidly than a wooden tip could be re-carved.

Micro-wear traces attributed to hafting of Middle Paleolithic tools other than points were first reported by Anderson-Gerfaud (1990) and Beyries (1987a, b). These diagnostic traces include striae on the faces of tools (often oriented perpendicular to the edges), surface polishing located away from edges, and crushing localized on exterior ridges. Tool types bearing evidence of having been hafted include convergent, transverse and end scrapers identified as having been used to scrape or strike wood. The inferred hafting had these tools bound in leather strips or partially covered with a leather sleeve, and then attached either to a split shaft or to an L-shaped notch in the end of the handle using a male (or juxtaposed)-terminal-axial-parallel and male (or juxtaposed)-terminal-transverse-perpendicular modes (Beyries 1987a, b;

Anderson-Gerfaud 1990). Binding without mastic was employed, since mastic would prevent tool movement and thus inhibit micro-wear formation. Small opposed lateral notches, interpreted to facilitate binding, were observed in some cases. Bulbar thinning and other types of modification to the platform end of the blank were not found to be associated with micro-wear hafting traces.[3]

Some of the most interesting evidence of Middle Paleolithic hafting in Eurasia is the recent discovery of traces of mastic in four sites. The only examples from Western Europe are two lumps of birch-bark pitch found during rescue excavations at Königsaue, Germany, a Central European Micoquian site of the *Keilmessergruppen*. These specimens, originally identified as tree resin (Mania & Toepfer 1973), have since been determined to have a high betulin content, proof that they are derived from birch-bark (Grünberg 2002). One of the pieces was undoubtedly used in hafting as it has the imprint of a bifacially flaked artifact on one side, and the imprint of a wooden haft on the other. Unfortunately, that specimen is not complete enough to determine the mode of hafting, although a lateral orientation is more likely than terminal. This mastic was found in horizons A and B of the site, with the one bearing the hafting traces coming from the earlier horizon. Radiocarbon dates for these layers are 43,800±2,100 B.P. (OxA 7124) and 48,400±3,700 B.P. (OxA 7125) respectively (Hedges et al. 1998:229). It is particularly noteworthy that birch-bark pitch is difficult to obtain and must be distilled in an oxygen-free environment (Pollard & Herron 1996). Lacking ceramic or other fire-proof containers, how Neanderthals accomplished this is a mystery.

In the Levant, dark material and black stains representing "bitumen" have been found on tools from the Final Mousterian sites of Hummal and Umm el Tlel in the Palmyra basin of central Syria (Boëda et al. 1998). The chemical composition of these residues is uncertain, but some of the stains are manganese and iron, which, as coprecipitates, are known to replace bitumen. Of the 16 stained artifacts from Umm el Tlel, one double scraper (described as a "convergent scraper" [*ibid:* 186]) showed coverage of most of the tool in bitumen except for the presumed cutting edge. The inferred handle morphology of this tool is uncertain, and the bitumen may have served the same purpose as "wrapping" to allow the tool to be comfortably used without a handle. All the illustrated specimens suggest juxtaposed-terminal-axial-parallel and juxtaposed-latero-distal-axial-parallel hafting modes, with one blade probably having been mounted in an oblique direction. There is no evidence of modification of the pieces themselves for hafting.

At another Levantine Mousterian site, Quneitra, located on the Golan Heights, Israel, Friedman et al. (1994) attribute differential patination on the platform end and approximately half of an adjacent lateral margin of one (of over 12,000 artifacts) atypical convergent scraper to that part of the specimen having been imbedded in some form of adhesive. The asymmetrical distribution of the hafting traces was attributed to a juxtaposed-latero-distal-axial-parallel mode of hafting. In other words, the handle included a tongue at the end with a slot in it into which the proximal part of the lateral side of the scraper was fit. No specific function was attributed to this scraper, and any traces of micro-wear were obliterated by the patination. The reconstructed mode of hafting would only be consistent with its use as a knife. No platform modification or other retouch possibly attributable to hafting was reported for this or any other specimen at Quneitra.

Both micro-wear and residues have been linked to hafting on a sample of 31 tools from the Middle Paleolithic site of Starosele in the Crimea (Hardy & Kay 1999). In this analysis, micro-wear traces were given priority in the identification of hafting, and the criteria employed were the presence of striae limited to one part of the tool. Sixteen specimens were found to have been hafted. All five typological points in the study had traces of hafting, as did two pointed typological scrapers that were interpreted as functional points. Eight other scrapers and one denticulate were also identified as having been hafted. In many cases, the striae were associated with starch grains, plant fibers, and an unidentified black residue, all of which were interpreted as the remains of binding and mastic. Modes of hafting were probably juxtaposed-terminal-axial-parallel, but may include some that were mounted with transverse-perpendicular directions and orientations. The residue analysis indicates that these tools were used in meat procurement, animal product processing, wood working, and in the processing of starchy plant storage organs. This latter function has rarely been recorded in previous analyses.

Hafting at Mugharet es-Skhūl

The Levant has provided some of the best evidence of a temporal overlap between Neanderthals and anatomically modern *Homo sapiens,* with the Mount Carmel caves of Mugharet es-Skhūl and Mugharet et-Tabūn being among the most important sites (Valladas et al. 1987, 1988; Schwartz et al. 1988; Stringer et al. 1989; Grün et al. 1991). At present it is not certain whether these hominids occupied the area simultaneously, or in alternation due to Pleistocene climactic fluctuations, during which Neanderthals found refuge in the area during the cold period of Oxygen Isotope Stage 4, to be replaced by AMHs during warmer episodes (Tchernov 1981, 1998; Bar-Yosef 1987, 1988). The question of the precise chronological relationships of key fossils and assemblages is the subject of ongoing debate which will only be resolved with further refinements in dating techniques and additional field data (Jelinek 1991). Notwithstanding this issue, there is general agreement that the Levantine archaeological record includes two hominid populations with significant anatomical differences (Howell 1958; Stringer 1978; Trinkaus 1984, 1991, 1996; Vandermeersch 1981). What has always puzzled archaeologists is that these biological differences do not coincide with clear-cut differences in their associated archaeological assemblages (Howell 1957; Brace 1967; Jelinek 1982; Clark & Lindly 1989). This phenomenon was recognized as soon as variability in the Levantine fossil record became evident (McCown & Keith 1939). Since morphology was assumed to be related to behavior (Stringer 1990), some evidence of behavioral difference was expected in the archaeological residues (Shea 1989). This has not been forthcoming (Churchill 1997; Straus 1997). Prior to the past two decades, virtually all investigations have been undertaken employing technological and typological perspectives that focus primarily on end products, either expended cores (and their intended products, large whole flakes or blades), or retouched "formal" tools. From this traditional techno-typological perspective, the statement by Bate (Garrod & Bate 1937:110) that there was "no essential difference" between assemblages associated with each type of hominid has yet to be contradicted (Bar-Yosef 1989). Even forensic analysis of the functions of artifacts failed to find statistically significant differences in the activities represented by the artifacts associated with the Kebara Neanderthal and the Qafzeh early AMHs fossils (Shea 1989, 1998), although the higher frequency of points in Levantine sites attributed to Neanderthals led Lieberman and Shea to hypothesize that although Neanderthals and early AMHs employed almost identical technologies, they may have focused on different environments and employed "different mobility and resource exploitation strategies" (1994:300).

The following is an attribute rather than a typological analysis of samples of tools from Skhūl and Tabūn. It shows that the original assessment of the similarities of blank production technology and tool typology made by Garrod and Bate (1937) for those sites was essentially correct, but demonstrates that significant albeit more subtle differences do occur between tools from one layer at Skhūl (B1) and the other Mt. Carmel collections in the deliberate modification of formal tools for hafting, as well as in the placement of retouch on the flake blanks. Modifications of the Skhūl B1 tools attributed to hafting are the primary focus here. Differences in the placement of retouch will be briefly discussed, and are dealt with at length elsewhere (Bisson 2001).

Mugharet es-Skhūl is located in the Wadi el-Mughara on the slopes of Mt. Carmel, south of Haifa, Israel. It was the smallest of the caves excavated as part of a project led by Dorothy Garrod that took place from 1929 to 1934 (Garrod & Bate 1937). The site itself is a small cave and rockshelter with a terrace at its mouth. The terrace was the source of most of the archaeological materials as well as of all the burials. The excavations directed by Theodore McCown (1937) occurred in 1931–32 and stripped the terrace of its archaeological contents, which were contained in brecciated sediments. Because excavation by controlled layers was considered impossible, heavy picks and hammers were employed. These extremely rough methods prevented the excavators from directly observing any grave pits in which the hominids were buried, and make attributions of artifacts to specific layers problematic.

McCown recognized three major strata at Skhūl, of which Layer B is relevant here. It began c. 50 cm below surface, and was described as hardened earth and concrete-like breccia with a few horizontally limited thin lenses of sterile stalagmite. These deposits became increasingly brecciated with depth. This layer was at the maximum slightly more than two m. thick, with no visible strata or features other than the burials. However, the condition of artifacts changed at a depth of "-2.00 m" (McCown 1937:96), although whether this referred to depth below surface or depth below the datum point is not clear. The changes were primarily manifested in patination. Lithics from the upper part of this layer were heavily patinated white or gray. Those from the lowest, most solidly brecciated gray-colored sediments were rarely patinated. This distinction was used to subdivide Layer B into an upper horizon, B1, above the ambiguous "-2.00 m." boundary, and B2, from below that point to the bottom of the layer. It was probably B1 that was the source of most if not all of the early anatomically modern *Homo sapiens* fossils.

Although Garrod recognized some variation in the collections from Layer B, she concluded that: "There is no essential difference between the industries of B1 and B2" (Garrod & Bate 1937:110). Her conclusion was that Skhūl Layer B was most similar to the lower part of Layer C at Tabūn, which she characterized as Lower Levalloiso-Mousterian. Modern evaluation of this assertion have been hampered by her artifact descriptions, which were made before adoption of the standardized Bordesian typology (Bordes 1961). In addition, her apparent discard of notches and denticulates (A. Jelinek, personal communication) renders numerical comparisons with better documented collections problematic. Table 1 lists the inventories of formal tools from Skhūl Layer B as published by Garrod, but with her type names converted to their closest Bordesian equivalents.[4]

The age and chronological relationships of Skhūl Layer B have undergone a number of revisions since Garrod's original assessment. A more detailed metrical analysis by

Table 9.1. Inventory of formal flake tools in Skhul Layer B (Garrod and Bate 1937) compared to the sample from that collection in the Redpath Museum, McGill University, and the sample excavated by Jelinek from the dump of the original Skhul excavations.

	Garrod Collection				McGill Sample				Dump Sample			
	B1		B2		B1		B2		B1		B2	
Type of tool	N	%	N	%	N	%	N	%	N	%	N	%
Points	177	18.6	88	17.1	7	12.5	2	9.1	2	6.3	2	6.5
Single Side Scrapers	561	59	323	62.7	20	35.7	11	50.0	7	21.9	10	32.3
Double Side Scrapers	116	12.2	69	13.4	11	19.6	2	9.1	3	9.4	3	9.7
Transverse Scrapers	15	1.6	12	2.3	3	5.4	4	18.2	0		1	3.2
Endscrapers	6	0.6	0		1	1.8	0		1	3.1	0	
Burins	50	5.3	13	2.5	1	1.8	0		0		1	3.2
Notches	13	1.4	9	1.7	3	5.4	2	9.1	12	37.5	7	22.6
Pièce Esquillée	5	0.5	0		0		0		0		0	
Denticulates	8	0.8	0		10	17.9	2	9.1	7	21.9	7	22.6
Total	951		515		56		22		32		31	

Note: Attributions to layers for the Dump Sample are based on patination (patinated = B1, unpatinated = B2). Garrod's type names have been modified to correspond more closely to modern descriptive terminology. Because she used few type designations compared to Bordes (1961), only composite categories can be used in these comparisons. Her "squamous flakes" are here listed as denticulates, and "chisels" as *pièce esquillée*. Based on her brief descriptions, Garrod's "variously retouched flakes and fragments" included both intentionally retouched pieces, and pieces with pronounced utilization damage and/or post-depositional surface modification. They are therefore excluded from these comparisons.

Jelinek (personal communication) placed it significantly later in the Mt. Carmel sequence, perhaps contemporary with the Tabūn chimney deposits (Layer B). This interpretation has been strengthened by the recent application of absolute dating techniques. Burned flints from Skhūl Layer B yielded TL dates of approximately 119±18 kyr (Mercier et. al. 1993), and ESR dates average 81±15 kyr (early uptake) to 101±12 kyr (linear uptake) (summarized in Bar-Yosef 1994). Unfortunately, the published reports do not indicate from which subdivision these samples were derived.

Two samples are analyzed here. The first consists of a total of 77 retouched tools plus 12 Levallois flakes and points derived from McCown's original Skhūl collection and donated by Garrod to the Redpath Museum, McGill University. Attributions of artifacts to Layers B1 and B2 were made by her at the time of donation. The question of whether the McGill sample is representative of her collection as a whole can be judged by classifying it using her very broad artifact types (table 1). If notches and denticulates, which were rarely recognized by Garrod, are deleted from the McGill sample, the type frequencies are broadly similar. The largest discrepancies are that proportionately more double scrapers from B1 were donated to McGill, and that for B1 and B2, points are much less frequent in the McGill collection. This last feature results from the probability that convergent scrapers were included in her point category, as they often were in early typologies (Debénath & Dibble 1994), combined with the fact that this study employed much more stringent criteria of symmetry and sharpness to define points (Bisson 2000). Overall, the McGill sample appears to have been selected by Garrod to resemble her complete collection, and she seems to have taken pains to have included at least one example of the rare forms such as endscrapers and burins. Since Garrod makes no mention of platform modifications in her descriptions of the Skhūl artifacts beyond noting the percentage of specimens with platform faceting, we must assume that those attributes were unrecognized and played no part in her selection of specimens for donation.

The second sample includes 62 retouched flake tools and 77 unetouched flakes with maximum dimensions greater than 5 cm collected from McCown's waste dumps at Skhūl by Jelinek in 1972. The overall composition of this sample is significantly different from the original collection in that notches and denticulates are much more common. This sample cannot be used to characterize the original composition of the Skhūl assemblage since larger artifacts and more obvious point and scraper types were winnowed out by the original excavators.[5] Nevertheless, excluding the notches and denticulates, and the points which, because of easy recognition, were not discarded as frequently, proportions of formal tools are similar to those in the Garrod collection. Single scrapers outnumber double forms, and other tool categories are rare. Moreover, because Jelinek did not discard any specimens, there is no selection either for or against unusual platform modifications. If evidence of hafting modifications also occurs in this sample, it strengthens the interpretation that Garrod's collection was not subject to either conscious or unconscious bias in favor of those characteristics.

The hafting modifications in the Skhūl collection were discovered during an attribute analysis rather than a traditional typological study. The initial focus of that analysis was not evidence for hafting modifications, but was instead to investigate the strategies employed by Middle Paleolithic hominids in the production of scrapers, and the nature of "imposed form" (Mellars 1989) on those simple tools. The goal was to identify what, if any, behavioral patterns governed the choice of where retouch was placed on scrapers. Two primary rules, each strongly influenced by the initial form of the flake being retouched, could be seen operating in a great majority of specimens:

1. The platform (and any naturally blunt edge) was left unretouched to facilitate manual prehension.

2. The longest edge of the blank with the most acute spine-plane angle was retouched first (Bisson 2001).

These rules were seen to operate in over 90% of scrapers in samples from four French and Levantine sites, including Tabūn layers D and E, and Skhūl B2 totaling 311 specimens. Both Dibble (1995) and Mellars (1996) had previously noted these as common characteristics of Middle Paleolithic scrapers from Europe, but did not specify how frequently they occurred. I have argued elsewhere that in the case of scrapers imposed form can only be identified as statistically significant violations of these rules, since that would indicate deliberate shaping of the tool that was not dictated by the original form of the flake-blank and the morphological requirements of a hand-held tool (Bisson 2001).

The Skhūl Layer B1 scrapers were found to violate these rules in nearly 55% of the 31 cases in the McGill sample, with deliberate modification of the platform end of the tool independent of retouch on the effective (working) edge occurring on 10 specimens (32.3%). Since platform modification and bulbar thinning have often been linked to hafting (Bordes 1961), all of the Skhūl formal tools were subject to careful scrutiny to see if these modifications were widespread in this collection. In addition to the identification of patterns of platform modification, these tools were also compared to the bases of Mousterian points, since there is general agreement that many points were indeed hafted (Stordeur 1987; Mellars 1996). Each tool was also inspected under both a binocular dissecting microscope at magnifications of 10–50X to identify any micro-flaking related to hafting, and an incident-light microscope at magnifications of 100–200X to identify hafting polishes (Anderson-Gerfaud 1990). Unfortunately, the heavy patination that characterizes the B1 specimens rendered the latter technique ineffective, and it was not employed on the Dump sample. Platform modification was not automatically assumed to represent hafting. The shape created by the modification had to be plausibly suitable for use in that manner, and not all specimens with modified platforms are included in the descriptions below.

Four patterns of platform modification interpreted to facilitate hafting were observed on the McGill sample (table 2):

1. The platform was either unifacially or bifacially removed to create an asymmetrical wedge at the butt-end of the tool (figure 2*A–C*). This was found on four specimens including a single convex scraper, a double convex scraper, a Mousterian point and a naturally backed knife. The most efficient haft for this modification would be a split shaft.

2. The entire platform was removed to create either a short pointed or blunt stem that could fit in a socketed mount (figure 2*E, F*). This was found on four specimens including two convex side scrapers, a double convex-concave scraper, and a transverse raclette on the interior surface of a flake. The platform modifications on the two side scrapers produce very short points and may simply be truncations. Both have very flat interior surfaces and thin or acute edge angles opposite the retouched edge, which may indicate that these tools were hafted with transverse-perpendicular directions and orientations. The lengths of the retouched (working) edges are 39.5 and 40.6 mm. Some of the Skhūl B1 stems produce blunt convex butts that resemble a pronounced *châpeau de gendarme* platform, and this may have been the reason that Garrod and McCown failed to recognize them as discrete retouch episodes made after the blank was produced.

3. Half of the platform was removed to create a shoulder (figure 2*G*). This occurs on two tools, a convergent scraper and a second scraper, probably convergent, which is missing the distal end due to heat spalling. This modification would be most effective in a juxtaposed articulation.

4. The entire platform was removed unifacially, truncating the blank,[6] a modification that would be best employed in a juxtaposed articulation (figure 2*H, I*). This was found on two specimens, one single slightly convex scraper in which the bulb was thinned to flatten the interior surface of the tool, and a second single straight scraper with a very flat interior surface. The cortical edge opposite the retouched edge in both specimens is convex and acute, and the location of small areas of retouch of accommodation suggest that although both are technically side scrapers, they were attached to a handle with the working edge perpendicular to the handle. These are very similar in size to the two scrapers described in point 2 above, with working edges measuring 38.6 and 41.8 mm. Although the sample size is small, the similarity in size of these four tools suggests a standardized handle.

Specimens where the longest and thinnest edge was not retouched, a violation of the second rule, may also be interesting from the perspective of hafting. Failure to place retouch on an otherwise ideal edge may indicate the use of that edge to fit into a split or slotted haft, particularly if

Table 9.2. Hafted tool types from Skhul, McGill Collection.

Acc. #	Tool Type	Mode of Hafting: Type	Placement	Orientation	Direction	Utilization Scars
Bifacial Platform Thinning						
131	straight lateral scr.	direct	terminal	axial	parallel	bi-directional with normal terminations
161	double straight- convex scr.	direct	terminal	axial	parallel	bi-directional with normal terminations
139.1	Mousterian point	direct	terminal	axial	parallel	none
184	naturally backed knife	direct	terminal	axial	parallel	bi-directional with normal terminations
Platform Removed Creating Short Stem or Point						
124	convex lateral scr.	direct	terminal	axial	parallel	bi-directional, normal uni-directional on unret. distal end
132	transverse raclette	direct	terminal	transverse	perpendicular	uni-directional, normal terminations
137	convex lateral scr.	direct	terminal	axial	parallel	none
139.6	double convex- concave scr.	direct	terminal	axial	parallel	none
Half of Platform Removed Creating a Shoulder						
123	convergent scraper? (tip broken)	direct	terminal	axial	parallel	none
138	convergent scraper	direct	terminal	axial	parallel	none
Platform Entirely Removed Creating a Slightly Convex Truncation						
106	straight lateral scraper	direct	terminal	transverse	perpendicular	uni-directional, nml. & step term.
170	slightly convex lateral scraper	direct	terminal	transverse	perpendicular	uni-directional, normal
Tools With Evidence of Hafting But Lacking Platform Modification						
80	Mousterian point	direct	terminal	axial	parallel	none
86	concave lateral scraper	direct	terminal	transverse	perpendicular	none

Note: The mode of hafting refers to the classification of the articulation of the tool and handle proposed by Stordeur (1987). Levallois points (4) and Mousterian points (1) lacking obvious hafting modifications are not listed here.

that edge is opposite the retouched edge. This characteristic is found on three single straight lateral scrapers, one concave lateral scraper, one straight transverse scraper, and one Mousterian point. Of these, however, only the concave lateral scraper, and the Mousterian point exhibit damage to the thin edges that might correspond to hafting wear (table 2).

The Dump sample was mixed, and attributions to layers B1 or B2 were made based on patination (table 3). Because McCown was somewhat more careful to retain artifacts than was Garrod at Tabūn, there were proportionally fewer formal tools in this sample, and easily recognizable points were particularly rare. Of the 32 retouched tools attributed to B1, 10 are scrapers, of which 5 violate the scraper production rules, almost exactly the proportion found in the McGill sample, suggesting that the unusual production rules for scrapers including significant numbers of hafting modifications are not idiosyncratic to the McGill sample. Only one of the 14 scrapers attributed to B2 violates these rules. The B1 scrapers with apparent hafting modifications include two single convex scrapers in which the platform has been removed by bifacial flaking to create a truncated butt. An atypical endscraper has bifacial thinning of the proximal end. One single convex scraper has thinning of the distal end which could facilitate hafting. Two end-notched flakes have bifacially thinned platforms consistent with hafting. In both cases, the notch is precisely opposite the platform end, strongly suggesting transverse-perpendicular direction and orientation. One denticulate has a thinned back that could have possibly been embedded in a handle. One Soyons point (figure 2*D*) has the bulb thinned to straighten what was undoubtedly a pronounced longitudinal curve, an unambiguous hafting modification. The three unretouched and one retouched second order Levallois points had no evidence of hafting modifications beyond their *châpeau de gendarme* platforms.

The platform modifications in Skhūl B1 reported here not only differ from the B2 sample, but also from the artifacts in the entire sequence at Tabūn. Analysis of a sample of scrapers from a stratigraphic column at Tabūn excavated by Jelinek (1982) in 1967–1972 revealed nothing corresponding to the stemmed or wedge-shaped butts of Skhūl B1 scrapers, although hafting modifications on points appear as early as bed 75S1[7] (table 4, figure 2*J*). In the Acheuleo-Yabrudian levels below the Amudian (Garrod's Layers Eb through Ed), not one of the 139 retouched tools had a modified platform. Likewise, in the Amudian levels (Garrod's lower Ea) with their high frequency of elongated blanks and backed knives, platform modification was entirely lacking on tools other than points, and backing

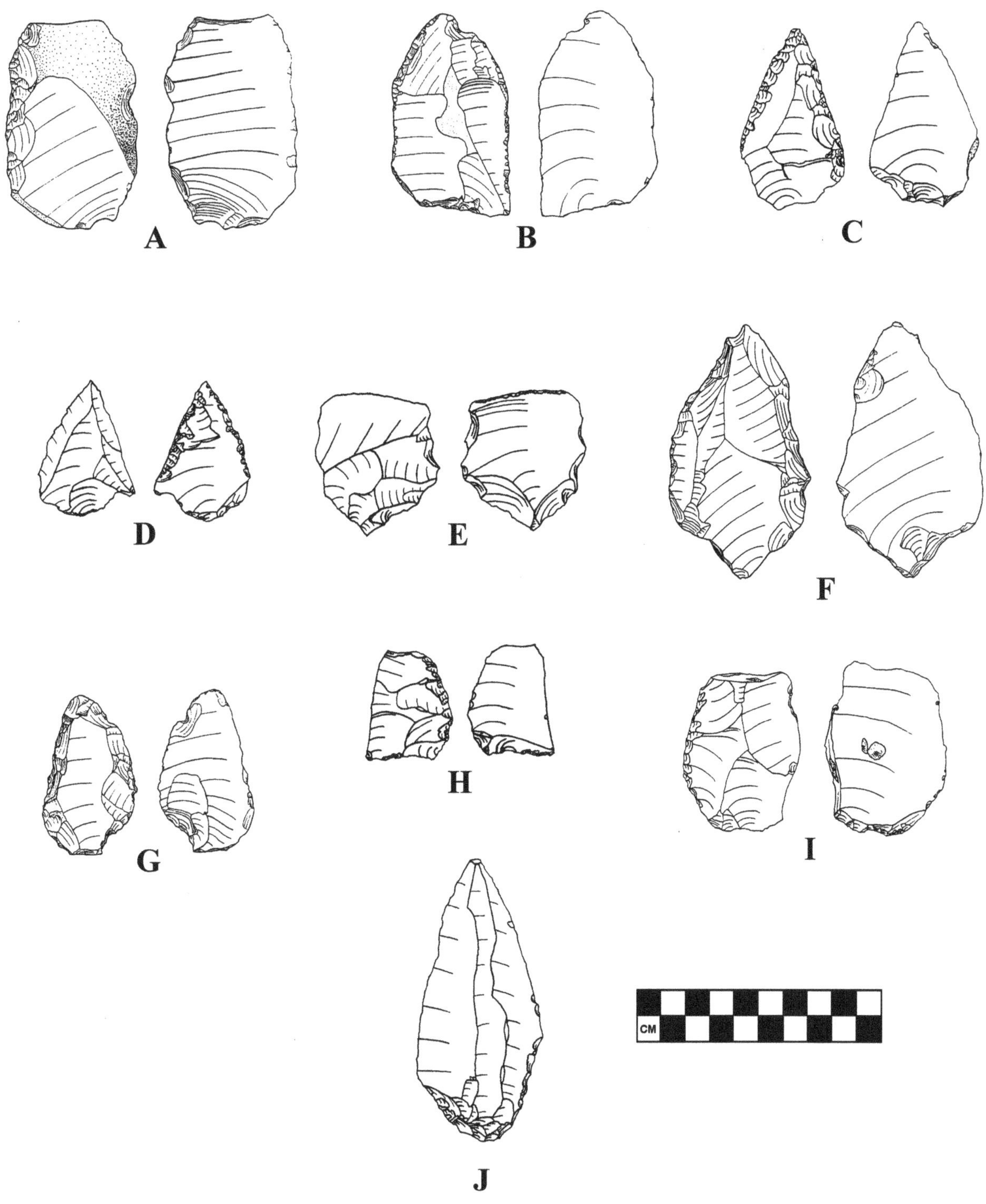

Fig. 9.2. Hafting modifications at Skhūl.
A: Single scraper with unifacial basal thinning; *B:* Minimally retouched double scraper with unifacial basal thinning; *C:* Mousterian point with unifacial basal thinning; *D:* Retouched levallois point ("Soyons" style) with unifacial basal thinning ("Emireh" style); *E:* Raclette with bifacial basal retouch creating a short stem; *F:* Double convex-concave scraper with unifacial basal retouch creating a short stem; *G:* Double scraper with unifacial basal notch creating a shoulder; *H, I:* Single scrapers with platforms removed by abrupt retouch (truncations); *J:* Point with extensive basal hafting retouch from Bed 75S1, Tabūn. Skhūl specimens *A–C* and *E–G*, McGill sample; *D, H, I,* Dump sample.

Table 9.3. Hafted tool types from Skhul, Dump Collection.

Acc. #	Tool Type	Mode of Hafting: Type	Placement	Orientation	Direction	Utilization Scars
Layer B1						
Platform Removed Creating a Straight Slightly Convex Truncation						
24	convex lateral scr.	direct	terminal	axial	parallel	none
25	convex lateral scr.	direct	terminal	axial	parallel	bi-directional with normal terminations
Bifacial Platform Thinning						
54	atypical endscraper	direct	terminal	transverse	perpendicular	none
7	end notched flake	direct	terminal	transverse	perpendicular	none
20	end notched flake	direct	terminal	transverse	perpendicular	none
Bulbar Thinning						
52	Mousterian point	direct	terminal	axial	parallel	none
Thinned Back						
14	denticulate	direct	latero-distal	axial	parallel	bi-directional with normal terminations
Thinned Distal End						
62	convex lateral scr.	direct	terminal	axial	parallel	none
Layer B2						
Double Notches on Interior Surface at Either Side of Platform						
564	double convex scr.	direct	terminal	axial	parallel	none

Note: Levallois points (3) and Mousterian points (1) lacking obvious hafting modifications are not listed here.

mirrored the outline of the blank rather than creating a straight or uniformly curved edge suitable to fit in a slotted handle. The Mugharan levels above the Amudian (Garrod's upper Layer Ea) had platform modifications in extremely low frequencies (0.8%) and no evidence of other hafting modifications, and the few specimens with thinned bulbs and thinned backs were too irregular to easily fit into handles. Indeed, bulbar thinning is not necessarily a hafting modification, as was noted by Bordes (1961). In experimenting with the use of scrapers as push-planes on wood, it was found that a pronounced bulb sometimes prevents the tool from being employed effectively, and might need to be removed. This is one of the many problems of equifinality plaguing the archaeological record. The most common type of secondary retouch in the Tabūn sequence was blunting of small, sharp projections opposite the main retouched edge. Following Bordes *(ibid),* this "retouch of accommodation" probably facilitated more comfortable manual prehension.

Similar platform modifications are generally rare or absent in other Levantine Mousterian sites. Of note, however, is a recently published drawing of a Levallois point from Qafzeh layer XV in which the base is carefully trimmed to a straight, wedge-shaped profile, and lateral retouch on both sides of the tool near the base produced small opposed notches suitable to assist binding (Shea 1998:52, figure 2e). The frequency with which this attribute occurs in the Qafzeh assemblage is not reported.

The final issue to be confronted is whether or not the Skhūl B1 platform modifications were actually for hafting. Heavy patination ruled out the identification of micro-wear polishes, but the unifacially and bifacially retouched butts were inspected for micro-flaking, and no patterns attributable to unidirectional use such as scraping were found. It is therefore unlikely that these modifications were done to create an additional working edge and indeed many of these retouched bases were so blunt that they could not have served as cutters or scrapers. The removal of these platforms was deliberate, but only in a couple of cases was the resulting form as symmetrical as some of the stems and other hafting modifications found in the Upper Paleolithic. This limited forensic evidence does not provide proof of hafting, but it does not contradict that interpretation, and no support is present for other options.

Discussion

There is currently no consensus on the function of the stems and other basal retouch that have occasionally been found on Eurasian Middle Paleolithic tools. Debénath and Dibble (1994) note that conclusive proof of stems as hafting modifications has not yet been found. The thinned bases characterizing bifacial foliates, Kostienki knives, scrapers, as well as some stemmed pieces found in the transitional Middle to Upper Paleolithic industries of Northwest Europe (Otte 1990), are generally attributed to

Table 9.4. Secondary Retouch on Tabun scrapers (Layers are from the Jelinek excavations followed by Garrod's designations in parentheses; counts for each attribute include percentages in parentheses.).

		Secondary Retouch									
Layers		Backing	Thinned Bulb	Thinned Back	Stem	Thinned Platform	Blunting	Distal Thinning	Point	Trunc.	N
5–32	(C)	1(1.5)				3 (4.5)	4 (6.0)				67
35–60	(C–D)					6 (5.4)	3 (2.7)		5 (4.8)	2(1.9)	105
61–67	(D)	3(3.1)		1(1.0)		0	8 (8.3)	3(3.1)	1 (1.0)		96
68–72s	(D–E)		14 (9.5)	3(2.1)		1 (0.7)	3 (2.1)	1(0.7)	4 (2.8)		145
73S–75S2	(Ea)		21 (3.2)	9(1.4)		5 (0.8)	30 (4.6)	1(0.2)	17 (2.6)		649
75S–75I2	(Ea–b)		3 (1.1)	1(0.4)		0	7 (2.6)		8 (3.0)		270
76S2–82BS	(Eb–d)		2 (1.4)	1(0.7)		0	3 (2.2)		5 (3.6)		139

Note: The scarcity or complete absence of thinned or otherwise modified platforms throughout the entire sequence of the Mugharan Tradition and the Mousterian. A slight increase in platform modification does occur above Layer 60, but unlike Skhûl B1, not a single specimen was found to have either a stem or a platform modified into a wedge shape suitable for hafting.

hafting, but only recently has forensic evidence supported this interpretation, and then, with the exception of Königsaue (Grünberg 2002), only for the Upper Paleolithic (Otte & Caspar 1987). In contrast, Anderson-Gerfaud argues that there is "no requisite morphology for hafting, nor that tool morphology is a clear-cut guide to whether the tool was hafted or not (1990:407)." As evidence, she cites Swiss Neolithic tools with preserved hafts in which the stone inserts have no proximal end modifications. In her view, split shafts or handles can be hollowed to accommodate individual tools. While true, this would seem to negate one of the advantages of a wooden handle, which is that it can be re-used after each lithic insert is discarded. Since repeated hollowing is a subtractive process, it would increasingly restrict the shapes of the inserts. It is easier to use retouch to accommodate the insert to the handle than to re-work the handle for each insert. Hafted Neolithic and Upper Paleolithic stone tools may not have required discrete hafting modifications because the blanks on which they were made were produced by standardized techniques that tended to produce uniform and symmetrical tool morphologies (i.e. blades and short flakes) much more conducive to hafting than the less regular forms produced by Middle Paleolithic techniques. Indeed, only the multifaceted convex *châpeau de gendarme* platforms produced by some Levallois strategies produces a base which may be specifically designed to permit hafting without further modification (Henry 1995).

Excluding points, deliberate modification of tool blanks to facilitate hafting has been very rarely reported in Middle Paleolithic assemblages (see Mellars 1996). Bordes (1961) noted that platform and bulbar thinning, as well as thinning the backs of tools, might be for hafting, but since no widespread pattern of these modifications similar to those known to be for hafting in the Upper Paleolithic (i.e. opposed notches, elongated stems, uniform and systematic backing) was evident in the Mousterian, no special category of hafted tool types was included in the standard typology, and possible methods of hafting were rarely considered. Indeed, the platform and bulbar thinning and backing categories in Bordes' typology are technological observations of the position of retouch relative to flake landmarks rather than descriptions of the forms created by that retouch. This was not an oversight on the part of Bordes and his predecessors. As has been frequently noted, Middle Paleolithic technology is primarily designed to be hand held (see Trinkaus 1983, 1987; Mellars 1996), and the demands of manual prehension play a large role in determining the formal characteristics of retouched tools (Bisson 2001).

The majority of scrapers identified as having been hafted in earlier studies are convergent forms, often with pointed ends and relatively acute angles of their effective edges at the point that they intersect. In other words, they have a sharp tip. As noted above, some of these probably were the functional equivalents of points or knives. In virtually all cases recorded thus far, these were mounted in a male or juxtaposed-terminal-axial-parallel mode (figure 1*A*). Points are perhaps the easiest to haft because the pressures applied to them during use are perpendicular to the mount, and force the tool further into the haft. Even with juxtaposed articulations like the L-shaped notch (Beyries 1988; Anderson-Gerfaud 1990), a point can achieve its functional requirements. Knives present different hafting problems, but don't require a binding that keeps them immobile. Specimens identified as knives tend to be elongated (often they are elongated convergent scrapers with acute edge angles), and a significant portion of the length can thus be encased in the haft to hold it in place if it is mounted like a point. The handle increases the leverage that can be employed when cutting with a reciprocal motion parallel to the long axis of the haft, and the blade can move slightly in the haft as long as enough of it is encased that a small amount of wobble does not dislodge it. This is because the forces placed on the haft by parallel motions do not tend to stretch the haft open. This method is less effective with shorter specimens such as Mousterian points, since achieving a stable mount would require the bindings to cover so much of the length of the tool that its use as a knife would be inhibited.

Functional scrapers present different problems for hafting. If a tool is clamped in a split or notched haft, and is then pulled across the worked material in the classic unidirectional scraping motion with the interior surface forming the leading edge, then strong perpendicular forces are placed on the binding and/or mastic. If the worked material is soft, such as fresh hide, these forces might not be great enough to dislodge the scraper. However, the forensic evidence seems to show wood to be the most common worked material (Anderson-Gerfaud & Helmer 1987; Beyries 1987a, b, 1988), and this could place much higher strain on hafts. It is unlikely that brittle adhesives could withstand these stresses, and as noted above, the adhesives available to Middle Paleolithic peoples in Europe and the Levant were limited. Bee's wax is too soft to function effectively, particularly when warm. Blood glue can hold objects in place, but experiments show it to be brittle and mainly suitable if the tool fits tightly in the haft. The unmodified butts that characterize Middle Paleolithic scrapers would be too irregular for use with blood glue. Bitumen distilled from birch-bark or soft wood was commonly used in later times, but this can only be produced by heating in a low oxygen environment, and with the exception of Königsaue (Grünberg 2002), is not found before the Neolithic. It also has problems with brittleness. At Starosele in the Crimea, paste made from unidentified plant starch and possibly mixed with feathers or other binding agents, suggested as a mastic by Hardy (1999), may have been used as a filler in hafting, but trials with modern starch paste (flour mixed with various combinations of water and egg-whites) show it to be brittle when dry and subject to rapid degradation when wet (Friedman et al. 1994). Since the precise plants represented by the Starosele residues have not yet been identified, the adhesive properties of substances derived from them cannot be evaluated. Pine or spruce pitch is too flexible when warm, and brittle when dried or cold. Petroleum bitumen, a natural asphalt that can seep from oil-bearing geological deposits, makes an excellent mastic, but has very limited distribution and appears to have been employed only in limited areas of the Middle East (Boëda et al. 1998).

In the absence of elongated stems or other unambiguous hafting modifications, bindings, perhaps combined with adhesives, would need to enclose a large portion of the tool to keep it stable. The pliability that allows binding materials to be wrapped and tied is also a disadvantage in that almost all of them stretch when wet. This could be prevented to a degree by the use of waterproofing agents such as fat or conifer pitch. Repeated stretching of bindings by perpendicular motions is the most serious problem confronting hafted scrapers, and is one of the reasons why Cauvin and Stordeur (1987) reserved judgment on the presence of transverse-perpendicular directions and orientations in the Middle Paleolithic. Clearly, trimming the tool to fit more tightly into the handle would assist the creation of durable hafted scrapers.

Chronology and Distribution of Hafting

The chronology and distribution of sites where hafting of tools other than points and knives has been identified provides important clues as to the nature of technological innovation in the Eurasian Middle Paleolithic. The earliest site, Biache-Saint-Vaast in northern France, dates to Oxygen Isotope Stage 7, sometime between 200 kyr and 150 kyr B.P. (Tuffreau & Sommé 1988). A convex scraper on a blade from the Würm I levels at Combe Grenal, southwestern France, was interpreted as being hafted in a juxtaposed-latero-distal-axial or oblique-parallel or oblique mode and used to cut plants, possibly grasses. Because the specific layer in which this specimen was found was not identified, its date can only be bracketed between c.128 kyr and 75 kyr B.P. (Anderson-Gerfaud & Helmer 1987:38). At Mount Carmel, TL and ESR dates for the Skhūl B1 specimens range between 119±18 kyr (TL) and 81±15 kyr (ESR-EU), with a consensus estimate of c. 100 kyr B.P. (Bar-Yosef 1994). The specimens from Starosele in the Crimea range in age from 80 kyr to 40 kyr, with the majority coming from Level 3, which dates c. 50 kyr B.P. (Hardy 1999). The specimens with traces of bitumen from Umm el Tlel, Syria, could not be precisely dated. They are described as Final Mousterian, and an overlying Early Upper Paleolithic layer was dated to 36±2.5 kyr B.P.. Given this attribution, these specimens should be no earlier than 70–80 kyr B.P. (Boëda et al. 1998). The Quneitra, Israel, scraper is dated to 54,000 kyr B.P. (Friedman et al. 1994). The specimens from Pech de l'Azé and Corbiac, southwest France, are Mousterian of Acheulian Tradition and dated to the Würm II. Laville et al. (1980) place Pech de l'Azé between 50–60 kyr B.P., whereas Mellars (1988) believes that site to be 45–50 kyr B.P.. As noted above, the birch-bark pitch from Königsaue dates to between 49,000 and 43,000 B.P. (Hedges et al. 1998).

Outside of Eurasia and the Mediterranean basin, there is a growing body of data suggesting an even earlier development of hafting, as well as many other elements indicating behavioral complexity, in Sub-Saharan Africa (McBrearty & Brooks 2000). The Lupemban of central Africa, which new absolute dates indicate may have begun before 400 kyr, has a variety of tool types including notched bifacial points, tranchets, and large crescents that may have been hafted in male or juxtaposed-terminal-axial-parallel or perpendicular modes and served as knives, the tips of spears and possibly adzes or scrapers (Barham 2001; Clark & Brown 2001). Importantly, these hafting modifications are part of a distinct regional technological tradition with a long time duration that precedes the evolution of AMHs and also includes other traits, such as widespread pigment use, that has often been cited as evidence of behavioral modernity (Barham 2002). In fact, more complex hafting modifications occur throughout Africa in a number of different Middle Stone Age regional traditions (Clark 1982). In most cases, the modifications appear on points and knives, but the well known Howieson's Poort industry of South Africa, which begins c. 100 kyr, also includes geometrics, diagonally truncated blades and endscrapers on blades (Singer & Wymer 1982) which may indicate the use of male-terminal (or lateral)-axial-perpendicular hafting modes for scrapers and some types of cutting tools. In North Africa, the Aterian has abundant stemmed tools, including some scrapers, believed to have been hafted (Ferring 1975; Clark 1983), but dates for the beginning of the Aterian remain subject to debate, with Hublin (1993) arguing that it is a development at the end of the North African Mousterian and thus not much older than 40 kyr B.P., contrary to what has been argued by Wendorf et al. (1990). In fact, at Bir Tarfawi, Egypt, the earliest stemmed tools appear in Gray Phase Middle Paleolithic sites dated by a variety of techniques to c. 135–120 kyr B.P. (Wendorf et al.1993).

These dates indicate that hafting was practiced, and may have originated, in the late Lower or early Middle Paleolithic of Africa, but was initially limited to male or juxtaposed-terminal-axial-parallel modes on tools serving as spear points, and knives used to cut resistant materials such as wood. Among the earliest hafted tools, only some points bore modifications specifically designed to facilitate hafting. In all areas, including Africa, evidence of hafting on scraping tools is much less common. In Eurasia and the Mediterranean basin, modifications for hafting first appear in North Africa in the form of the stemmed pieces dating to perhaps as early as 135 kyr B.P. in the Middle Paleolithic technologies leading to the Aterian, where they become particularly common after 70 kyr B.P. (Hublin 1993), and in the two Middle Paleolithic complexes of the Nile corridor (summarized in Van Peer 1998). Although these industries are closer by land to the Levant than they are to Europe, and share some technological similarities in the form of the high Levallois index, the hafting modifications found in Skhūl B1 do not resemble those in the Aterian and the Nile Valley complex, although they do resemble some modifications on points in the Nubian complex. In contrast, basal thinning dominates the Skhūl B1 material described here, and the few stems that are present are generally shorter and broader than Aterian and Nubian complex examples. The wide variety of hafted tools in these various traditions and at Skhūl suggests multiple hafting modes including transverse directions of the tool and perpendicular orientations of the active part, both of which require more secure attachment to the handle to be effective. Although forensic evidence for male or juxtaposed-terminal-axial-parallel modes of hafting appears as early or earlier in Western Europe than they do in North Africa and the Nile Corridor, the Levant and Eastern Europe, in Western Europe, actual modifications of tools for hafting is not found until late in the Middle Paleolithic and the transitional industries to the Upper Paleolithic in northwest Europe (Otte 1990), as well as in the Châelperronian and the Uluzzian of southwestern Europe (Mellars 1996).

Outside of Africa, none of the Middle Paleolithic hafting modifications on scrapers are particularly well developed. This is probably one of the reasons why those at Skhūl B1 went unrecognized. The temporal and spatial data show no single area of origin for these modifications, followed by a spread of a particular mode of hafting from that source. Instead, different strategies were employed at various times and places. In some cases, such as in Africa and the Middle Paleolithic of Northwestern and Eastern Europe, hafting modifications became part of local technological traditions with relatively long durations. In the Levant, it appears that deliberate modification for hafting was employed by some of the earliest examples of anatomically modern *Homo sapiens* at Skhūl, but that these modifications did not become perpetuated in the same way that stems became commonplace elsewhere. Similar modifications have not been reported in the Qafzeh assemblage (Shea 1989), although both the chronology and hominids from Skhūl and Qafzeh are similar, and they are also almost entirely absent from Tabūn. Distinct hafting modifications do not re-appear in the Levant until the late Middle Paleolithic Emireh points, with their thinned, wedge-shaped bases (Mellars 1989).

Taken in the long perspective, these data might be interpreted to represent incremental technological change as discussed by Kuhn and Stiner (1998), since evidence of hafting does appear to become more common, widespread and technologically sophisticated near the Middle–Upper Paleolithic transition. On closer inspection, however, the actual situation seems to be occasional experimentation with different hafting techniques in a model of technological change that could be better described as episodic rather than incremental (Straus 1997). If what we are seeing is not simply a consequence of which collections have been subject to careful scrutiny for evidence of hafting, then hafting modifications were probably developed and abandoned at a number of different places and times in Eurasia. A similar circumstance is already well documented in the Levant, in which the blade dominated early Middle Paleolithic Amudian facies both follows, and is succeeded by, flake-based Acheulian and Jabrudian facies in a number of sites (Jelinek 1990).

Continuous variability characterizes most Middle Paleolithic tool forms (Dibble 1995; Mellars 1996), and I have argued elsewhere (Bisson 2000, 2001) that during the Middle Paleolithic it is only the effective edge, rather than the entire shape of a tool, that is subject to the degree of imposed form sufficient to create an edge or point suitable for the desired use. Given both the variability of edge forms in Middle Paleolithic tools, and the equally informal nature of the hafting modifications identified at Skhūl B1, a notion of what might be called "necessary form" to achieve the desired functional characteristics appears to have been somewhat less developed in terms of the shapes of retouched tools than it was in some blank production strategies (Levallois and blade), which were as complex as those in the Upper Paleolithic (Simek 1992), and required a high degree of control of form to execute (Dibble & Bar-Yosef 1995). As has frequently been noted, the variability in Middle Paleolithic tool forms in part reflects a desire to create expedient tools capable of multiple uses (Mellars 1996). Overall, it would appear that Middle Paleolithic tools are generally less standardized than those in the Upper Paleolithic (but for an opposing view, see Simek 1992), although the degree to which the transition between the two is gradual or abrupt remains an extremely active debate (d'Errico et al. 1998; Mellars 1999). The same can be said for hafting modifications.

Finally, in the case of Mt. Carmel, do the data from Skhūl B1 indicate the technological superiority of early anatomically modern *Homo sapiens* over contemporary Neanderthals? Although the technology employed at Skhūl has subtle but significant differences with that of the rest of the Mount Carmel Middle Paleolithic sequence, the question of whether these technological differences imparted a significant adaptive advantage to that population is not clear. Recent comparisons of technological and economic patterns for Archaic and early AMHs in the Levant (Shea 1989, 1998; Lieberman & Shea 1994) have found that the greatest difference was not technological, but rather in the degree of mobility, strategies of resource exploitation, and in the ecological niches occupied. Studies of faunal evidence in other areas show there to have been little difference between Middle Paleolithic and early Upper Paleolithic subsistence, and that European Middle Paleolithic hominids, i.e. the Neanderthals, were effective hunters (Chase 1989).

Thus, although the modifications described here may have created somewhat more effective hafted tools, the question of whether this imparted a significant adaptive advantage to the Skhūl hominids remains unresolved. That said, an important characteristic of Levantine early AMHs is that their upper limbs were less robust than those of the Neanderthals (Trinkaus et al. 1998), whose hands were designed for a powerful fingertip grip which was used to hold scrapers and other simple tools (Trinkaus 1987). More frequent and effective hafting of tools might have given the Skhūl hominids an advantage, but might also only serve to compensate for less strength in their arms and hands.[8] In other words, their adaptation may have been different but not inherently better. Although, unlike Straus (1997), I think differences can be found in the technologies and adaptations of early AMHs in the Levant when compared to those of the Neanderthals, I agree that these differences do not provide sufficient evidence of superior adaptation. Interpretations of behavior and adaptation based on biological and artifact typologies are particularly susceptible to paradigmatic bias (Churchill 1997). The assumption that Levantine early AMHs must have had an advantage over Neanderthals is contradicted by the long temporal overlap of these two forms. Modern morphology and Upper Paleolithic technology eventually prevailed, but in the Levant it took more than 60 kyr to do so. Although this study has identified subtle technological differences between the two hominids at c. 100 kyr B.P., these differences are still small compared with their similarities. The Skhūl hominids appear to have been doing more experimenting with methods of tool use, and were showing the first glimmerings of imposed form on their finished products, but they were still fundamentally Middle Paleolithic in their behavior, as were their descendants for a very long time. If anything, these data provide support for a gradualist rather than a punctuationalist interpretation of early cultural evolution (McBrearty & Brooks 2000). The initial development of more sophisticated, efficient, and varied hafted tools does not appear to have triggered sudden or dramatic overall changes in lifestyle, and potentially useful information about hafting innovations does not seem to have been easily transmitted between regions and across time. In this respect, Eurasia differs from Sub-Saharan Africa, where these modifications appear to be much more widespread and long-lasting (Barham & Robson-Brown 2001). The reason for this difference is uncertain, but I think that the most likely explanation is demographic rather than differences in cognitive and/or communicative potential. Population densities in the temperate latitudes of late Middle Pleistocene Eurasia were probably significantly lower (Mussi & Roebroeks 1996) compared to the warmer climates of Africa, and thus there may simply have been less opportunities for contact and cultural transmission of technological innovations in colder environments. Whatever the reason, one thing is becoming increasingly clear. Experimentation and technological innovation was being carried out by Archaic *Homo sapiens* in many parts of the world, and the development of so-called behaviorally modern characteristics is not restricted to the increasingly nebulous Middle to Upper Paleolithic transition. In any case, further study of the incidence of hafting and other innovations in the

Middle Paleolithic will require systematic inspection of assemblages for either forensic evidence or the kinds of attributes described here, rather than the typological analyses that have prevailed for much of the past.

Endnotes

1. The term "composite tool" is used here to refer to any tool that combines more than one element. Thus a wooden spear shaft with a single stone tip would qualify. Stordeur (1987) uses the term to refer only to implements with multiple inserts such as an arrow with multiple microlithic barbs. This paper will use "simple composite tool" for the former, and "complex composite tool" for the latter.

2. "Wrapping" is a special category of "male" in which the "part of prehension" (Bordes 1961) is enclosed in some pliable material (leaves, bark, hide, or plastic substances such as bitumen) to make it more comfortable to hold. Stordeur describes this as the most primitive form of hafting and questions whether it should be considered equivalent in complexity to the other techniques described here.

3. These interpretations of micro-wear are not without problems. Both Mansur-Franchomme (1987) and Unger-Hamilton (1987) have questioned the attribution of unusual wear patterns to hafting. Planing wood, for example, creates wear away from the contact edge that mimics some of these purported hafting-wear patterns.

4. The author has elsewhere advocated the use of an attribute-based descriptive system in comparing collections (Bisson 2000), but Garrod's published data are inadequate to allow that system to be employed here, and Bordes' typology is the only viable alternative.

5. Because the removal of artifacts from the Skhūl breccia was a slower process, McCown's workers appear to have discarded fewer specimens than did Garrod's workers at Tabūn (Jelinek, pers. comm.).

6. These differ from "truncated-faceted" pieces, which in the Levant was a technique to transform flakes into cores. Truncated-faceted pieces occur in small numbers in Tabūn, but none occur in the Skhūl samples described here.

7. A point with extensive basal hafting modification was found in this bed, dating to c. 300 kyr. Levallois points first appear at almost the same date (A. Jelinek, personal communication).

8. Brace (2000) argues that the long term evolutionary consequences of these types of technological differences may account for the differences in robusticity between Neanderthals and African and Levantine early AMHs.

References

Anderson-Gerfaud, P.

1990 Aspects of behavior in the Middle Paleolithic: Functional analysis of stone tools from southwest France. In *The Emergence of Modern Humans: An Archaeological Perspective* (P. Mellars ed.), pp. 389–418. Edinburgh University Press.

Anderson-Gerfaud P., & D. Helmer

1987 L'Emmanchement au Moustérien. In *La Main et L'Outil: Manches et Emmanchements Préhistoriques* (D. Stordeur, ed.), pp.37–54. Maison de L'Orient, Lyon.

Barham, L. S.

1901 The emergence of regional identity in the Middle Pleistocene of central Africa. In *Human Roots: Africa and Asia in the Middle Pleistocene* (L. Barham & K. Robson-Brown, eds.), pp. 65–80. Western Academic and Specialist Press, Bristol.

2002 Systematic Pigment Use in the Middle Pleistocene of South-Central Africa. *Current Anthropology* 43:181–190.

Barham, L. S., & K. Robson-Brown (editors)

1901 *Human Roots: Africa and Asia in the Middle Pleistocene.* Western Academic and Specialist Press, Bristol.

Bar-Yosef, O.

1987 Pleistocene connections between Africa and Southwest Asia: An archaeological perspective. *African Archaeological Review* 5:29–38.

1988 The date of the Southwest Asian Neanderthals." In *l'Homme de Néandertal 3* (M. Otte, ed.), pp. 31–38. ERAUL, Liège.

1989 Geochronology of the Levantine Middle Paleolithic. In *The Human Revolution: Behavioral and Biological Perspectives on the Origins of Modern Humans* (P. Mellars & C. B. Stringer, eds.), pp. 589–610. Edinburgh University Press, Edinburgh.

1994 The contributions of Southwest Asia to the study of the origins of modern humans. In *Origins of Anatomically Modern Humans* (M. H. Nitecki & D. V. Nitecki, eds.), pp. 23–66. Plenum Press, New York.

Baumler, M., & J. Speth
1993 A Middle Paleolithic assemblage from Kunji cave, Iran. In *The Paleolithic Prehistory of the Zagros-Taurus* (D. I. Olszewski & H. L. Dibble, eds.), pp. 1–74. University Museum, University of Pennsylvania, Philadelphia.

Beyries, S.
1987a *Variabilité de l'Industrie Lithique au Moustérien Approche Fonctionnelle sur Quelques Gisements Français.* BAR International Series No. 328, Oxford.
1987b Quelques exemples de stigmates d'emmanchements observés sur des outils du Paléolithique Moyen. In *La Main et L'Outil: Manches et Emmanchements Préhistoriques* (D. Stordeur, ed.), pp. 55–62. Maison de L'Orient, Lyon.
1988 Functional Variability of Lithic Sets in the Middle Paleolithic. In *Upper Pleistocene Prehistory of Western Eurasia* (H. L. Dibble & A. Montet-White, eds.), pp. 213–221. University Museum, University of Pennsylvania, Philadelphia.

Bisson, M. S.
2000 Nineteenth Century Tools for Twenty-First Century Archaeology? Why the Middle Paleolithic Typology of François Bordes Must Be Replaced. *Journal of Archaeological Method and Theory* 7:1–48.
2001 Interview with a Neanderthal: An experimental approach for reconstructing scraper production rules, and their implications for imposed form in Middle Paleolithic tools. *Cambridge Journal of Archaeology* 11:165–184.

Boëda, E., J. Connan, & S. Muhesen
1998 Bitumen as Hafting Material on Middle Paleolithic Artifacts from the El Kowm Basin, Syria. In *Neanderthals and Modern Humans in Western Asia* (T. Akazawa, K. Aoki & O. Bar-Yosef, eds.), pp. 181–204. Plenum Press, New York.

Bordes, F.
1961 *Typologie du Paléolithique Ancien et Moyen.* Delmas, Bordeaux.

Brace, C. L.
1967 *The Stages of Human Evolution.* Prentice Hall, Englewood Cliffs.
2000 *Evolution in an Anthropological View.* Rowman and Littlefield, New York.

Caspar, J. P., & D. Cahen
1987 Emmanchement des outils Danubiens de Belgique données techniques et tracéologiques. In *La Main et L'Outil: Manches et Emmanchements Préhistoriques* (D. Stordeur, ed.), pp. 185–195. Maison de L'Orient, Lyon.

Cauvin, J., & D. Stordeur
1987 Quelques réflexions sur l'évolution préhistorique des emmanchements. In *La Main et L'Outil: Manches et Emmanchements Préhistoriques* (D. Stordeur, ed.), pp. 331–336. Maison de L'Orient, Lyon.

Chase, P. G.
1989 How different was Middle Paleolithic subsistence? A zooarchareological perspective on the Middle to Upper Paleolithic transition. In *The Human Revolution: Behavioral and Biological Perspectives on the Origins of Modern Humans* (P. Mellars & C. B. Stringer,eds.), pp. 321–337. Edinburgh University Press, Edinburgh.

Churchill, S. E.
1997 Morphological evolution, behavioral change, and the origins of modern humans. In *Conceptual Issues in Modern Human Origins Research* (G. A. Clark & C. M. Willermet, eds.), pp. 202–219. Aldine de Gruyter, New York.

Clark, G. A., & J. Lindly
1989 Modern human origins in the Levant and Western Asia: the fossil and archaeological evidence. *American Anthropologist* 91:962–985.

Clark, J. D.
1982 The Cultures of the Middle Paleolithic/Middle Stone Age. In *The Cambridge History of Africa Vol. 1, From the Earliest Times to c. 500 BC* (J. D. Clark, ed.), pp.148–341. Cambridge University Press, Cambridge.
1983 The significance of culture change in the early Later Pleistocene in Northern and Southern Africa. In *The Mousterian Legacy: Human Biocultural Change in the Upper Pleistocene* (E. Trinkaus, ed.), pp. 1–12. BAR International Series S164, Oxford.

Clark, J. D., & K. Brown
1901 The Twin Rivers Kopje, Zambia: Stratigraphy, fauna and artifact assemblages from the 1954 and 1956 excavations. *Journal of Archaeological Science* 28:305–330.

Debénath A., & H. L. Dibble
1994 *Handbook of Paleolithic Typology Vol. 1.* University Museum, University of Pennsylvania, Philadelphia.

D'Errico, F., J. Zilhão, M. Julien, D. Baffier, & J. Pelegrin
1998 Neanderthal acculturation in western Europe? A critical review if the evidence and its interpretation. *Current Anthropology* 39:S1–S44.

Dibble, H. L.
1987 The interpretation of Middle Paleolithic scraper morphology. *American Antiquity* 52:109–117.
1991 Mousterian assemblage variability on an interregional scale. *Journal of Anthropological Research* 47:239–257.
1995 Middle Paleolithic scraper reduction: Background, clarification, and a review of the evidence to date. *Journal of Archaeological Method and Theory* 2:299–368.

Dibble, H., & O. Bar-Yosef (editors)
1995 *The Definition and Interpretation of Levallois Technology.* Prehistory Press, Madison, WI.

Dibble, H. L., & S. Holdaway
1993 The Middle Paleolithic Industries of Warwasi. In *The Paleolithic Prehistory of the Zagros-Taurus* (D. I. Olszewski & H. L. Dibble, eds.), pp. 75–100. University Museum, University of Pennsylvania, Philadelphia.

Ferring, C. R.
1975 The Aterian in North African Prehistory. In *Problems in Prehistory: North Africa and the Levant* (F. Wendorf & A. E. Marks, eds.), pp. 113–126. Southern Methodist University Press, Dallas.

Friedman, E., N. Goren-Inbar, A. Rosenfeld, O. Marder, & F. Burian
1994 Hafting during Mousterian times—further indication. *Journal of the Israel Prehistoric Society* 26:18–31.

Garrod, D. A. E., & D. M. Bate
1937 *The Stone Age of Mt. Carmel Vol. 1.* Calrendon Press, Oxford.

Grün, R., C. B. Stringer, & H. P. Schwartz
1991 ESR dating of teeth from Garrod's Tabūn Cave collection. *Journal of Human Evolution* 20: 231–248.

Grünberg, J. M.
1902 Middle Paleolithic birch-bark pitch. *Antiquity* 76:15–16.

Hardy, B. L.
1999 Microscopic residue analysis of stone tools from the Middle Paleolithic site of Starosele. In *The Middle Paleolithic of Western Crimea Vol. 2.* (V. P. Chabai & K. Monigal, eds.), pp. 179–196. ERAUL, Liège.

Hardy, B. L., & M. Kay
1999 Stone tool function at Starosele: Combining residue and use-wear evidence. In *The Middle Paleolithic of Western Crimea Vol. 2* (V. P. Chabai & K. Monigal, eds.), pp. 197–209. ERAUL, Liège.

Hedges, R. E. M., P. B. Pettit, C. Bronk Ramsey, & G. J van Klinken
1998 Radiocarbon dates from the Oxford AMS system: Archaeometry datelist 25. *Archaeometry* 40:227–39.

Henry, D. O.
1995 *Prehistoric Cultural Ecology and Evolution Insights from Southern Jordan.* Plenum Press, New York.

Howell, F. C.
1957 The Evolutionary Significance of Variation and Varieties of "Neanderthal" Man. *The Quarterly Review of Biology* 32:300–347.
1958 Upper Pleistocene man in the southwest Asian Mousterian. In *Hundert Jahre Neanderthaler* (G. H. R. von Koeningswald, ed.), pp. 185–198. Kemink en Zoon, Utrecht.

Hublin, J-J.
1993 Recent human evolution in northwestern Africa. In *The Origin of Modern Humans and the Impact of Chronometric Dating* (M. Aitken, C. B. Stringer & P. Mellars, eds.), pp. 118–131. Princeton University Press, Princeton.

Jelinek, A. J.
1975 A preliminary report on some Lower and Middle Paleolithic industries from the Tabūn cave. Mount Carmel (Israel). In *Problems in Prehistory: North Africa and the Levant* (F. Wendorf & A. Marks, eds.), pp. 297–316. Southern Methodist University Press, Dallas:
1982 The Tabūn Cave and Paleolithic man in the Levant. *Science* 216:1369–1375.
1990 The Amudian in the context of the Mugharan tradition at the Tabūn Cave (Mount Carmel), Israel. In *The Emergence of Modern Humans: An Archaeological Perspective* (P. Mellars, ed.), pp. 81–90. Edinburgh Unversity Press, Edinburgh.

Kozlowski, J. K.
1998 The Middle and Early Upper Paleolithic around

the Black Sea. In *Neanderthals and Modern Humans in Western Asia* (T. Akazawa, K. Aoki & O. Bar-Yosef, eds.), pp. 461–482. Plenum Press, New York.

Kuhn, S. L., & M. C. Stiner
1998 Middle Paleolithic 'creativity': reflections on an oxymoron? In *Creativity in Human Evolution and Prehistory* (S. Mithen, ed.), pp. 143–164. Routledge, New York.

Laville, H., J-Ph. Rigaud, & J. R. Sackett
1980 *Rock Shelters of the Périgord.* Academic Press, New York.

Lieberman, D. E., & J. J. Shea
1994 Behavioral differences between archaic and modern humans in the Levantine Mousterian. *American Anthropologist* 96:300–332.

Mania, D., & V. Toepfer
1973 *Königsaue, Ökologie, und mittelpaläolithische Funde der letzten Eiszeit.* Berlin: Veröffentlichungen des Landsmuseums für Vorgeschichte Halle 26.

McBrearty, S., & A. S. Brooks
2000 The revolution that wasn't: a new interpretation of the origin of modern human behavior. *Journal of Human Evolution* 39:453–563.

McCown, T. D.
1937 Mugharet es-Skhūl. Description and Excavations. In *The Stone Age of Mt. Carmel Vol. 1* (D. A. E. Garrod & D. M. Bate, eds.), pp. 91–107. Clarendon Press, Oxford.

McCown, T. D., & A. Keith
1939 *The Stone Age of Mount Carmel, Vol. 2, The Fossil Human Remains from the Levalloiso-Mousterian.* Clarendon Press, Oxford.

Mellars, P.
1988 The chronology of the southwest French Mousterian: a review of the current debate. In *l'Homme de Néandertal 4: La Technique* (M. Otte, ed.), pp. 97–120. ERAUL, Liège.
1989a Technological changes in the Middle–Upper Paleolithic Transition: economic, social and cognitive perspectives. In *The Human Revolution Behavioral and Biological Perspectives on the Origins of Modern Humans* (P. Mellars & C. B. Stringer, eds.), pp. 338–365. Edinburgh University Press, Edinburgh.
1989b Major issues in the emergence of modern humans. *Current Anthropology* 30:349–385.
1996 *The Neanderthal Legacy.* Princeton University Press, Princeton.
1999 The Neanderthal Problem Continued. *Current Anthropology* 40:341–50.

Mercier, N., H. Valladas, O. Bar-Yosef, B. Vandermeersch, C. B. Stringer, & J. L. Joron
1993 Thermoluminescence date for the Mousterian burial site of es-Skhūl, Mt. Carmel. *Journal of Archaeological Science* 20:169–74.

Mills, J. S., & R. White (editors)
1994 *The Organic Chemistry of Museum Objects.* Butterworth Heinemann, Oxford.

Moss, E.
1987 Polish G and the question of hafting. In *La Main et L'Outil: manches et emmanchements préhistoriques* (D. Stordeur, ed.), pp. 97–102. Maison de L'Orient, Lyon.

Movius, H. L.
1950 A wooden spear of Third Interglacial Age from Lower Saxony. *Southwestern Journal of Anthropology* 6:139–142.

Mussi, M., & W. Roebroeks
1996 The Big Mosaic. *Current Anthropology* 37: 697–699.

Otte, M.
1990 From the Middle to the Upper Paleolithic: the nature of the transition. In *The Emergence of Modern Humans: An Archaeological Perspective* (P. Mellars, ed.), pp. 438–456. Edinburgh University Press, Edinburgh.

Otte, M., & J-P. Caspar
1987 Les pointes de la Font Robert: outils emmanchés? In *La Main et L'Outil: Manches et Emmanchements Préhistoriques* (D. Stordeur, ed.), pp. 65–74. Maison de L'Orient, Lyon.

Pollard, A. M., & C. Herron
1996 *Archaeological Chemistry.* The Royal Society of Chemistry, Cambridge.

Plisson, H.
1987 L'emmanchement dans l'habitation No1 de Pincevent. In *La Main et L'Outil: manches et emmanchements préhistoriques* (D. Stordeur, ed.) pp. 75–88. Maison de L'Orient, Lyon.

Shea, J. J.
1988 Spear points from the Middle Paleolithic of the Levant. *Journal of Field Archaeology* 15:

441–450.
1989 A functional study of the lithic industries associated with hominid fossils in the Kebara and Qafzeh caves, Israel. In *The Human Revolution: Behavioral and Biological Perspectives on the Origins of Modern Humans* (P. Mellars & C. B. Stringer eds.), pp. 611–625. Edinburgh University Press, Edinburgh.
1998 Neanderthal and early modern human behavioral variability: A regional-scale approach to lithic evidence for hunting in the Levantine Mousterian. *Current Anthropology* 39:S45–S78.

Simek, J. F.
1992 Neanderthal cognition and the Middle to Upper Paleolithic Transition. In *Continuity or Replacement: Controversies in* Homo sapiens *Evolution* (G. Bräuer & F. H. Smith, eds.) pp. 231–245. A.A. Balkema, Rotterdam.

Singer, R., & J. Wymer
1982 *The Middle Stone Age at Klasies River Mouth in South Africa.* University of Chicago Press, Chicago.

Stordeur, D. (editor).
1987 *La Main et L'Outil: manches et emmanchements préhistoriques.* Maison de L'Orient, Lyon.

Straus, L. G.
1997 The Iberian situation between 40,000 and 30,000 B. P. in light of European models of migration and convergence. In *Conceptual Issues in Modern Human Origins Research* (G. A. Clark & C. M. Willermet, eds.), pp. 235–252. Aldine De Gruyter, New York.

Stringer, C. B.
1978 Some problems in Middle and Upper Pleistocene hominid relationships. In *Recent Advances in Primatology* (D. J. Chivers & Joysey, eds.), pp. 395–418. Academic Press, Ondon.
1990 Comment on 'Symbolism and modern human origins' by J. M. Lindly and G. A. Clark. *Current Anthropology* 31:248–249.

Stringer, C. B., R. Grün, H. P. Schwartz, & P. Goldberg
1989 ESR dates for the hominid burial site of Es Skhūl in Israel. *Nature* 338:756–758.

Tchernov, E.
1981 The biostratigraphy of the Middle East. In *Préhistoire du Levant* (J. Cauvin & P. Sanlaville, eds.), pp. 67–97. Éditions du CNRS, Paris.
1998 The faunal sequence of the Southwest Asian Middle Paleolithic in relation to hominid dispersal events. In *Neanderthals and Modern Humans in Western Asia* (T. Akazawa, K. Aoki & O. Bar-Yosef, eds.), pp. 77–90. Plenum, New York.

Thieme, H.
1997 Lower Paleolithic hunting spears from Germany. *Nature* 385:807–810.

Trinkaus, E.
1983 Neanderthal postcrania and the adaptive shift to modern humans. In *The Mousterian Legacy* (E. Trinkaus, ed.), pp. 165–200. BAR International Series, Oxford.
1984 Western Asia. In *The Origin of Modern Humans: a World Survey of the Fossil Evidence* (F. H. Smith & F. Spencer, eds.) pp. 251–293. Alan R. Liss, Inc., New York.
1987 Bodies, brawn, brains and noses: Human ancestors and human predation. In T*he Evolution of Human Hunting* (M. H. Nitecki & D. V. Nitecki, eds.), pp.107–145. Plenum Press, New York.
1991 Morphological Contrasts Between the Near Eastern Qafzeh-Skhül and Late Archaic Human Samples: Grounds for a Behavioral Difference? In *The Evolution and Dispersal of Modern Humans in Asia* (T. Akazawa, K. Aoki & T. Kimura, eds.), pp. 277–294. Hokusen-sha, Tokyo.
1996. Near Eastern late archaic humans. *Paléorient.* 21:9–23.

Trinkaus, E., C. B. Ruff, & S. E. Churchill
1998 Upper limb versus lower limb loading patterns among Near Eastern Middle Paleolithic hominids. In *Neanderthals and Modern Humans in Western Asia* (T. Akazawa, K. Aoki & O. Bar-Yosef, eds.), pp. 391–404. Plenum Press, New York.

Tuffreau, A., & J. Sommé
1988 *Le Gisement Paléolithique Moyen de Biache-Saint-Vaast (Pas-de-Calais): Stratigraphie, Environnement, Études Archéologiques.* Mémoires de la Société Préhistorique Française, Tome 21.

Unger-Hamilton, R., R. Grace, R. Miller, & C. Bergman
1987 Drill bits from Abu Salabikh, Iraq. In *La Main et L'Outil: Manches et Emmanchements Préhistoriques* (D. Stordeur, ed.), pp. 269–285. Maison de L'Orient, Lyon.

Valladas, H., J. L. Joron, G. Valladas, B. Arensburg, O. Bar-Yosef, A. Belfer-Cohen, P. Goldberg, H. Laville, L Meignen, Y. Rak, E. Tchernov, A.-M. Tillier, & B. Vandermeersch
1987 Thermoluminescence dates for the Neanderthal burial site at Kebara, Israel. *Nature* 330:159–160.

Valladas, H., J. L. Reyes, J. L. Joron, G Valladas, O. Bar-Yosef, & B. Vandermeersch
1988 Thermoluminescence dating of Mousterian 'Proto-Cro-Magnon' remains from Israel and the origin of modern man. *Nature* 331:614–616.

Vandermeersch, B.
1981 *Les Hommes Fossiles de Qafzeh (Israel).* Éditions du CNRS, Paris.

Van Peer, P.
1998 The Nile Corridor and the Out-of-Africa Model. *Current Anthropology* 39:115–140.

Vaughan, P.
1987 Positive and negative evidence for hafting on flint tools from various periods (Magdalenian through Bronze Age). In *La Main et L'Outil: Manches et Emmanchements Préhistoriques* (D. Stordeur, ed.) pp.135–144. Maison de L'Orient, Lyon.

Wendorf, F., A. Close, R. Schild, A. Gautier, H. P. Schwarcz, G. Miller, K. Kowalski, H. Krolik, A. Bluszcz, D. Robins, & R. Grün
1990 Le dernier interglaciaire dans le Sahara oriental. *L'Anthropologie* 94:361–391.

Wendorf, F., R. Schild, & A. Close (editors)
1993 *Egypt during the Last Interglacial: The Middle Paleolithic of Bir Tarfawi and Bir Sahara East.* Plenum Press, New York.

10

Issues in the Development of the Early Upper Paleolithic, and a "Transitional" Industry from the Zagros Region

Deborah I. Olszewski
University of Pennsylvania Museum

Abstract

The plethora of recognized Middle to Upper Paleolithic "transitional" industries across much of Eurasia speaks to a broad scale geographical development of lithic technologies that are, for all intents and purposes, archaeologically contemporary. Some of these industries are believed to be associated with the last remnants of Neandertal adaptations, while others are thought to be the work of anatomically modern humans. It is somewhat surprising from a scientific point of view that there is a startling lack of corroborating hominin finds to support such assertions. The fact that these hominin attributions to "transitional" industries are made and persist to the degree that they do, has meant that less attention has been focused on the meaning of "transitional" industries from the standpoint of the adaptations that these industries document across time and space. Using the earliest of the Upper Paleolithic occupations (Levels AA–LL) from the site of Warwasi in Iran, I examine the technological and typological parameters of this "transitional" industry from the Zagros region. Both technology and typology here show an interesting meld of what are considered by many to be Middle and Upper Paleolithic elements, for example, prismatic blade(let) technology, small, radially prepared flake cores, and tools such as carinated endscrapers and sidescrapers of various types. These elements are appraised in the context of the criteria necessary and/or sufficient to document a "transition," as well as in the milieu of the development of Aurignacian lithic industries. Issues of typology comparability between the Middle and Upper Paleolithic and how typologies (even typologies of technology) are used by different researchers, while not new to most scholars, are at the heart of why most Middle and Upper Paleolithic industries have been seen as so characteristically dissimilar. It is these differences, in part, that fuel the notion that different hominins must have been responsible for at least some of the "transitional" industries across Eurasia.

The consideration of the development of Early Upper Paleolithic (EUP) lithic industries in Western Eurasia involves several key aspects. Among these have been issues related to *in situ* technological and typological transitions, migrations of human groups across the Pleistocene landscape, associations of hominin species with particular chipped stone traditions, the advent of modern human behaviors, and the characterization of "transitional" lithic industries. The adaptations that these EUP industries may reflect, however, have constituted less concern than perhaps they otherwise deserve. This, in turn, has resulted in an overemphasis on hypotheses and models that presently are difficult to substantiate on the basis of available archaeological data. To some extent, then, depictions of the transition(s) to the EUP fall far short of scientific reliability, and, as a consequence, researchers generate interpretations that often stand at either end of a behavioral continuum.

Recognition that EUP lithic industries in Western Eurasia develop out of a Middle Paleolithic foundation is widely shared by most researchers. It is most readily conceded in cases, for example, such as the local development of the Châtelperronian (Harrold 1989, 2000; Pelegrin 1990), Bohunician (Allsworth-Jones 1986; Svoboda et al. 1996: 107–110), Szeletian (Allsworth-Jones 1986), Streletskayan (Monigal 2001:61), and Uluzzian (Mussi 1990) industries. Complete agreement is not evident for some specific instances, such as the Boker Tachtit case (Marks 1983: 66–84, 1993:9–12; Bar-Yosef 1998:154), while other sequences are not often extensively commented on in contexts other than a transition of some sort, such as for the Ksar 'Akil sequence (Azoury 1986; Copeland 1986; Bergman 1987; Ohnuma 1988; but see Iovita 2002 and Marks 1983:65–66 for other perspectives).

Of particular note is the fact that the origins of the Aurignacian lithic industry, which must have come from a Middle Paleolithic base in Eurasia, is almost never discussed in this fashion. One gains the impression, rather, that the Aurignacian sprang indiscriminately "from the head of Zeus" (or some appropriate Paleolithic deity) in some as yet undefined region(s). Yet, provided that the archaeological methods are available to recognize it, there must, at some level, also be transitional lithic industries leading to the Aurignacian. To a great degree, it would appear that we do not "see" the origins of the Aurignacian lithic industry because current paradigms obscure it in the morass of attempts to attribute this industry to anatomically modern humans, to a migration of these people who then encounter remnant Neandertal groups, to the notion that lithic traditions can have only one origin area (and thereby are representative of specific archaeo-cultures), and to the paradox of the Out of Africa 2 model (anatomical modern human wave) in which migrating humans lack lithic assemblages that are recognizably Aurignacian in the areas from which they originate.

Issues of the Early Upper Paleolithic

It would appear that for many researchers, despite the lack of corroborating evidence, the coupling of the appearance and spread of the Aurignacian with an alleged spread of anatomically modern humans (AMHs) is too tempting to avoid (for example, Bar-Yosef 1998:156; d'Errico et al. 1998:S22; Harrold & Otte 2001:8; Rigaud 2001:117; among many others). The assertion that AMHs were responsible for manufacture of the Early Aurignacian is, however, unsupported, as all such interpretation is based on the association of AMH fossil remains with temporally much later examples of Aurignacian industries (Mann 1995:39; Straus 1999:354). The period of time representative of the Early Aurignacian is lacking in fossils except for the attribution of Neandertals to the Aurignacian at Vindija in Croatia (Smith & Ahern 1994; Karavanič 1995; but see d'Errico et al. 1998:S2, for an opposing view). It is premature, therefore, for researchers to hazard an association between AMH and the Early Aurignacian, and such an association detracts substantially from the reliability of behavioral reconstructions of EUP in Eurasia because there is no definitive evidence to support it.

If EUP lithic assemblages are decoupled from specific hominin groups, however defined, this does not obviate the issue of linking lithic assemblages to specific groups of early people, as is the practice by many researchers. These are attempts to reconstruct a social history of sorts, clearly articulated, for example, by Bar-Yosef (1991:375–376) for Upper Paleolithic, Epipaleolithic, and Neolithic periods of the Near East. The foundation of this approach is the notion that such groups can be recognized in the archaeological record for much later periods of time, and as such, their histories of development can be traced farther back in time. One of the main difficulties with such tracing, however, is that the further one recedes into the past, the less available are multiple lines of archaeological evidence that potentially could serve as indicators of specific groups of people. We lose the written record, elaborate architecture, densely packed settlements, specialized labor groups (including much in the way of identifiable crafts), metal, ceramics, and distinctions between groups who practice different subsistence economies (farmers, herders, and so forth). By the time our reach extends back to the period of the EUP, the archaeological record is largely mute on all but lithic assemblages, site placement in the landscape, hunter-gatherer economies (when faunal remains, or more rarely, plant remains, are preserved), exotic materials in the form of long-distance transport of shells or certain lithic raw materials, and, depending on conditions of preservation, indications of the use of organic technologies (bone, wood, antler, and ivory) and/or art.

The most widely preserved materials are, of course, lithic assemblages, and these therefore form the basis of much of what is reconstructed in terms of linkages to specific groups, forming what could be described as archaeo-cultures. Both typological and technological studies are used to describe EUP lithic traditions such as the Aurignacian, the Châtelperronian, or the Early Ahmarian (Levantine Near East). There is no doubt that there are often typological differences, as well as a certain degree of technological distinction, between what are recognized as various EUP lithic industries or traditions. More importantly, though, are considerations of which explanatory models best account for both the similarities within lithic traditions and the differences between those that appear to be distinct. It is now commonplace to remark upon the fact that lithic assemblages are the products of reductive processes (e.g., Whittaker 1994:199–206; Dibble 1995; Andrefsky 1998:136–188; Clark 2001:43). As such, there is a far greater likelihood that convergence in lithic technology and in some aspects of lithic typology will be widespread because of the limited range of options available to EUP flintknappers (e.g., choices between bifacial reduction or variations on prismatic blade(let) techniques, or radial, subradial, and multiple platform core for flake manufacture). The reconstruction of core reduction sequences as part of the larger *chaîne opératoire* for various sites has shown remarkable heterogeneity from core to core within assemblages, as each core requires individual decisions as to how to best reduce it, as well as the fact that the reduction life of many cores demonstrates an ever-changing process of technological strategy, e.g., from single platform blade to opposing platform blade to radial flake (Baumler 1988; Iovita 2002:32). Lithic assemblages thus do not reflect single archaeo-cultures that manage to "invent" a particular technology or tool at solely one point in time and in one geographical area, so much as they show similar choices, from a limited array of options, made by many archaeo-contemporary prehistoric groups.

Using lithic assemblages to help document "migrations," for instance, of the Early Aurignacian across the European landscape, is more of a chimera than a reality (Clark 1994:

331–334). There is no evidence to date that the Aurignacian lithic industry originated in one place at one time and then spread out from there, either as accompanying the movement of people or as an idea.[1] Nor is there any basis to assume that different types of stone tools or differences in technology necessarily signal different groups of people. There are simply too many variables (raw material abundance, size, and quality; relative mobility of groups; curation; function; reuse; resharpening; and so forth), especially across a large geographic region such as Eurasia, that interact to affect the final form of artifacts recovered from the archaeological record (Bamforth 1991:217–218, 231). While in some cases progress has been made in delineating how some of these variables pertain to specific assemblages, these are still far distant from pinpointing specific groups of people.

Although the focus of this paper is on lithic assemblages of the EUP, it is worth mentioning briefly the additional, important aspect of organic technologies, especially where they interface with what has been defined as the Aurignacian EUP. In many instances, the European archaeological record (and to a much lesser extent, the Levantine context) yields examples of points—split-base bone points in the Early Aurignacian—made of organic materials such as antler and bone (Hahn 1988; Leroy-Prost 1975, 1979). If organic technology such as this is absent from an archaeological assemblage, then the question is whether this absence is due to preservation constraints or to lack of manufacture or use of such points in that site/area, with implications for whether or not the assemblage should be classified as Aurignacian.

Recently, this aspect has also been addressed from the perspective of availability of organic materials to make "bone" points. In fact, such "bone" points tend not to be made as often from bone as they are from antler, which is far better suited to point manufacture (Knecht 1993:139; Bernaldo de Quirós et al. 2001:27). As Bernaldo de Quirós et al. (2001 :27) assert, then, the availability of antler is thus a critical factor in the appearance and abundance of Aurignacian points in an assemblage. This difference is seen, for example, in the contrast between French Aurignacian assemblages which tend to be rich in this organic aspect and the Spanish Aurignacian which has far fewer examples of "bone" points. The major difference is the presence of numerous reindeer in France, in which antler is shed year-round when both female and male reindeer are considered. In red deer in Spain, on the other hand, only males have antlers—antler is therefore less abundant as well available only during a portion of the year. Obviously, this observation has important implications as well for other regions of Eurasia in which reindeer were not present, such as the Levant or the Zagros regions of the Middle East. One would expect Aurignacian "bone" points to be rare, at best, in these other locales.

Last but not least among this list of concerns, is the issue of how to define or recognize a "transitional" industry. This, of course, begs the question of whether we should expect "transitional" industries to exist or not—that is (borrowing evolutionary terminology), are lithic transitions punctuated or gradual? If punctuated, then the origin lithic industries will share less in common with the "descendant" lithic industries. If gradual, then perhaps there will be industries that exhibit characteristics that are common to both the origin lithic industry and the succeeding lithic industries. Other confounding factors include whether or not the archaeological record provides instances in which "pure" moments of lithic behavior can be discerned, that is, how often are archaeological "Pompeiis" actually found? It is much more likely that the majority of the behavioral sets that populate the archaeological record are palimpsests to one degree or another, so that much of what is interpreted is a mixture of behaviors across an unknown degree of time (Olszewski 2001:84–85). The current paradigm for this issue, however, is to examine the archaeological record for evidence of "transitional" lithic industries, and in keeping with that perspective, the following sections discuss these endeavors first in the Levantine region, and then for the Zagros EUP (or Early Zagros Aurignacian facies).

The Levantine Region

One of the early attempts to discuss transitions from the Middle to Upper Paleolithic can be found in the work of Garrod (1951, 1955), who isolated the Emiran as a "transitional" industry. She described it as based on the production of flakes that were identical to the preceding Levantine Mousterian, but also containing true blade technology, blade cores, blade tools such as endscrapers, burins, and backed knives. A particularly important tool was the Emireh point, a Levallois point on which the proximal end has been thinned. Subsequent questions about this industry's integrity, however, were raised in Bar-Yosef and Vandermeersch (1972), who discussed aspects of the stratigraphy of the caves in which Garrod had identified the Emiran. The conclusion of most researchers was that the Emiran represented a mixed industry rather than an actual example of a "transitional" industry (Marks 1983:57–58). The Emiran point, however, is still used as a tool type pertinent to some Levantine "transitional" assemblages (see below).

Work in Lebanon, particularly at the site of Ksar 'Akil, has provided one of the best known examples of attempts to document the transition from the Middle to Upper Paleolithic in the Levant (Azoury 1986; Copeland 1975; Iovita 2002:23–39; Marks 1983:60–66; Ohnuma & Bergman 1990). The Ksar Akil Phase A (Levels XXV – XXI) assemblages are the most pertinent. They were initially thought to represent a variation on Levallois technology (or "evolved" Levallois) with an Upper Paleolithic typology in which EUP tools, such as endscrapers and burins, were made on Levallois blanks (Marks 1983:58). Of special note is the presence of chamfered pieces, which appear to be primarily a Lebanese phenomenon. The picture of the Ksar Akil Phase A assemblages that has since emerged suggests rather that blade technology is already in place by the time of Ksar Akil Phase A. Technologically the earlier levels (XXV–XXIV) are not oriented toward point production and thus do not contain morphological Levallois points; such points do become common in Levels XXIII–XXI. They are not, however, manufactured using Levallois technology (Marks 1983:63; but see Iovita 2002:24 for a different perspective). The sequence at Ksar 'Akil is described as one that early on shows a technological emphasis on elongated blanks and is comparable to the Boker Tachtit Level 4 as a technological transition into the EUP (Ohnuma & Bergman 1990:133–134). Its tool component is essentially EUP in character throughout Ksar Akil Phase A (Marks 1983:65).

The technology of the "transitional" assemblages from Boker Tachtit in the Negev, on the other hand, were largely documented through the process of refitting debitage and tools to cores (Marks 1983:66–84, 1990:67–71, 1993:6–12; Volkman 1983). This technique showed a shift in core reduction strategies from the earliest assemblage (Level 1) to the latest, full-blown EUP industry (Level 4)—a sequence in which Levallois points are characteristic of the earlier phases, while the last phase of occupation has morphological Levallois points, i.e., points that were not made using Levallois technology. As is the case for Ksar 'Akil, the Boker Tachtit assemblages are mainly characterized by EUP tools, such as endscrapers and burins, although there are also retouched and non-retouched Levallois points in Levels 1, 2, and 3 (Marks & Kaufman 1983). Of particular note is the presence of Emireh points in Levels 1, 2, and 3, although these are most frequent in the Level 2 assemblage (Marks & Kaufman 1983:99). For reasons having primarily to do with how Levallois is defined and identified in assemblages, not all researchers accept that the Boker Tachtit sequence documents a transition from the Middle to Upper Paleolithic. This alternative perspective describes the technological shifts as occurring wholly within the EUP (Bar-Yosef 1998:154), much as Marks (1983:64–65) and Ohnuma and Bergman (1990:133–134) claim for the Phase A levels at Ksar 'Akil.

Umm el-Tlel, in Syria, has also been recently proposed as a site with a sequence documenting an industry intermediate between the Middle and Upper Paleolithic (Boëda & Muhesen 1993:54–56; Bourguignon 1998). In Levels IIbase and III2a, a volumetric-based Levallois point technology (recurrent, unidirectional) is described. This technology is paired with a tool assemblage that primarily contains EUP tools, such as burins and endscrapers, although there are two sidescrapers and three Nahr Ibrahim tools present as well (Bourguignon 1998:711–712). There are no Emireh points or chamfered pieces; instead, the Umm el-Tlel point, which is defined in part as having a few small removals from the proximal end, along the same exterior ridges that guide the point to be detached, is present. Perhaps the most puzzling aspect of this intermediate Umm el-Tlel industry is that it is stratigraphically sandwiched between two Upper Paleolithic occupations—what is described as a "poor" (low quantity) Ahmarian occupation (Level III2b) and the Levantine Aurignacian (Levels II4b and II4c). In order to explain this disjunction, an argument is made for a group with a late-surviving Middle Paleolithic technology occupying the site in this interval (Bourguignon 1998:716–717). In other words, in a fashion reminiscent of the Bordian interpretation of ethnic groups, people with EUP technology and tools co-exist in the Levantine region with groups that manufacture this intermediate industry, and they live at Umm el-Tlel at different points in time.

Recently, the site of Tor Sadaf in the Wadi al-Hasa region of Jordan, was described as containing a "transitional" Middle to Upper Paleolithic industry (in two phases), overlain by an Early Ahmarian Upper Paleolithic (Coinman & Fox 2000; Fox 2000). There are several technological trends that have been noted in this sequence. First, the "transitional" units contain Levallois and Levallois-like points, but few recognizable Levallois cores (Coinman & Fox 2000:129; Fox 2000:77). Second, faceting of platforms is quite common in the earliest "transitional" phase, and is slightly less frequent in the second "transitional" phase (Coinman & Fox 2000:129–130). Finally, the tools are essentially EUP, with endscrapers and burins, but also Levallois blade-points and Levallois flake-points (Fox 2000:72). There are no Emireh points or chamfered pieces. Continuing work on the "Transitional A and B" phases has now concluded that this sequence is already within the purview of the EUP, in fact being comparable to Level 4 at Boker Tachtit and Levels XXIII–XXI at Ksar 'Akil (Fox & Coinman 2004). Accordingly, the phases are now classified by these researchers as Tor Sadaf A and Tor Sadaf B, with the terminology of "transitional" eliminated.

Although there are some specific tool type differences, especially in the differential presence of Emireh points and chamfered pieces, the Levantine examples generally accepted as "transitional" assemblages (at least by some researchers), appear to document Levallois technology oriented toward point (or blade?) production which gradually changes to an EUP prismatic technology based on elongated blade blanks. Associated tool types are those widely recognized as Upper Paleolithic, with Middle

Paleolithic types such as sidescrapers either rare or absent. There is no overlay of EUP tools made on Middle Paleolithic Levallois blanks (Marks 1983); in fact, the change in tool types would appear to be "punctuated," while technological change is gradual.

Warwasi (Levels AA–LL), Iran: An Example of a "Transitional" Industry

Upper Paleolithic lithic assemblages from the rockshelter at Warwasi in the Zagros region of the Middle East have recently been published in some detail (Olszewski 1993, 1999, 2001; Olszewski & Dibble 1994). Of relevance to the notion of "transitional" industries is the lithic assemblage from Levels AA–LL at this site. This has been classified as an early phase of the Zagros Aurignacian, one in which a developmental sequence from the underlying Middle Paleolithic into a full-blown version of the Zagros Aurignacian (Levels P–Z) can be seen (see discussions of the Zagros Aurignacian classification in Olszewski & Dibble 1994; Olszewski 1999, 2001). Particularly striking is the combination of tool types characteristic of the Middle Paleolithic (sidescrapers and truncated-faceted pieces) and the Upper Paleolithic (endscrapers on blades, burins, and some microliths—Dufour bladelets, Font Yves points, and finely retouched bladelets). In this regard, the Warwasi assemblage is clearly distinct from those of the Levantine region. Additionally, there is no indication of Levallois technology, nor of an intentional production of points, either Levallois or morphologically Levallois, nor are there Emireh points or chamfered pieces.

There is a marked dichotomy in the Levels AA–LL sequence in the final morphology of the cores. This can be seen in the contrast between the cores of Levels AA–EE and those from Levels FF–LL (table 1). In the lower levels (FF–LL), there are virtually no cores with a final morphology related to blades or bladelets. This lower part of the sequence is instead dominated by flake cores, particularly of the single platform, subdiscoidal, and multiple platform types. In the upper part of the sequence, cores with final removals corresponding to bladelets become part of the reduction sequence, although it is important to note that single platform and subdiscoidal flake cores remain important in the technological process of blank manufacture. One should also bear in mind that the final morphologies of cores do not necessarily reflect the complete process of core reduction strategies, which can often change as an individual core is knapped, for example, from unidirectional blade removals to subdiscoidal/discoidal (subradial/radial) flake removals. There is, in fact, some preliminary evidence from the debitage and tool blanks to suggest that this is the case for core reduction in the Levels AA–LL sequence at Warwasi.

Distinctions between Levels AA–EE and Levels FF–LL are not as clear-cut in the debitage (table 2). Blades, for example, are about equally represented in both portions of the EUP sequence. While bladelets are present in Levels FF–LL, they are more than double in frequency in Levels AA–EE. The fact that blade and bladelet cores are largely absent from Levels FF–LL, yet this debitage is present in some quantity argues against post-depositional mixture of assemblages from different occupations,[2] and most likely is explained by shifts in core reduction strategies throughout the EUP occupations representative of the Levels AA–LL sequence. Burin spalls, which indirectly represent EUP tool types, are also present in about equal quantities in Levels AA–EE and Levels FF–LL.

Tool blanks from Levels AA–LL mimic the pattern seen in the debitage (table 3). That is, tools on flakes are the

Table 10.1. Final Core Morphology in Levels AA–LL at Warwasi.

	Warwasi AA–EE		Warwasi FF–LL	
Cores	N	%	N	%
Bladelet:				
single platform	37	11.2	1	0.3
opposed platforms	20	6.1	-	-
ninety-degree platforms	6	1.8	-	-
subdiscoidal	1	0.3	-	-
subpyramidal	2	0.6	-	-
multiple platforms	1	0.3	-	-
Blade:				
single platform	1	0.3	-	-
opposed platforms	-	-	1	0.3
Flake:				
single platform	56	16.9	37	12.4
opposed platforms	13	3.9	23	7.7
ninetydegree platforms	7	2.1	15	5
subdiscoidal	38	11.5	32	10.7
discoidal	-	-	12	4
multiple platforms	19	5.8	51	17
Mixed:				
single platform	5	1.5	1	0.3
opposed platforms	4	1.2	-	-
ninetydegree platforms	7	2.1	2	0.7
subdiscoidal	2	0.6	-	-
subpyramidal	1	0.3	1	0.3
multiple platforms	7	2.1	6	2
Fragments	103	31.2	117	39.1
Total	330		299	

Table 10.2. Debitage from Levels AA–LL at Warwasi.

	Warwasi AA–EE		Warwasi FF–LL	
Debitage	N	%	N	%
Bladelet	1165	29.3	239	12.0
Blade	328	8.3	181	9.1
Flake	1650	41.6	1099	55.0
Small flake (< 2.5 mm)	684	17.3	343	17.2
Burin spall	103	2.6	59	2.9
Shatter	35	0.9	76	3.8
Total	3965		1997	

Table 10.3. Tool Blanks from Levels AA–LL at Warwasi

	Warwasi AA–EE		Warwasi FF–LL	
Tool blanks	N	%	N	%
Bladelet	82	15.8	23	4.8
Blade	85	16.4	81	17
Flake	333	64.3	406	85.5
Core	9	1.7	6	1.3
Other	9	1.7	5	1.1
Total	518		475	

most common blank type in both Levels AA–EE and Levels FF–LL and tools on blades are nearly equal in both portions of this EUP sequence. The greatest distinction between the upper and lower levels is the much larger quantity of tools on bladelets in Levels AA–EE. Again, the fact that blades and bladelets are available as blanks for tools, especially in the near absence of blade and bladelet cores in Levels FF–LL, suggests that the final morphology of the cores does not adequately represent the reduction processes that each core underwent during flintknapping.

Finally, the tool typology of Levels AA–LL supports the characterization of an industry with both Middle and Upper Paleolithic types (table 4). It is quite interesting that sidescrapers are a consistently abundant type throughout this sequence;[3] the majority in both Levels AA–EE and Levels FF–LL are single sidescrapers.[4] Somewhat paradoxically, about 29% of the sidescrapers in Levels FF–LL are made on blade blanks, while only ca. 15% are manufactured on blade blanks in Levels AA–EE. Notch/denticulates and retouched pieces are also consistently present, although these tend to be relatively ubiquitous to lithic assemblages of many time periods and regions, and therefore do not have diagnostic value. The EUP tool component has a slightly higher representation in the Levels AA–EE portion of the sequence, but the entire EUP lithic assemblage here contains examples of carinated endscrapers, carinated burins, and Dufour bladelets, which are characteristically Aurignacian for many regions of Europe, as well as in the Levantine Aurignacian from Umm el-Tlel (Soriano 1998).

Although the Warwasi Levels AA–LL sequence has been labeled EUP (early Zagros Aurignacian) in this and other published works, this designation should not necessarily detract from a consideration of this lithic assemblage as an example of a "transition." Unlike the Levantine cases, which are clearly Upper Paleolithic in tool typologies, and perhaps debatably "transitional" in technologies, Warwasi Levels AA–LL contains what are undeniably recognizable as both Middle and Upper Paleolithic tool types. These are present in a sequence which appears to show a shift toward more bladelet production through time, and perhaps a trend toward less alteration in core reduction strategies for each specific core. In other words, by the later part of the sequence (Levels AA–EE), rather than each core beginning life as a single or bidirectional blade core and later being reduced in a radial fashion to become a subdiscoidal flake core, there may be a shift toward many cores ending their use life as they began, as single platform, bidirectional, or subdiscoidal. This preliminary observation requires that further technological study of the cores and debitage be undertaken at some future date.

Conclusion

One of the most pertinent questions to be resolved is what the expectations of a "transitional" lithic assemblage encompass. These include:

- whether a transition is "punctuated" or "gradual"—it is the latter approach that currently frames most discourse in archaeology;
- the combination of characteristics that constitute a "transition"—gradual shifts in technology from the Middle to Upper Paleolithic and implications for tool typologies;
- whether there are valid examples of "transitional" industries in the Middle East; and
- if a "transition" must be single origin, single place, or if it can consist of many different trajectories which are each situational with respect to local ecologies and conditions.

The "punctuated change" idea is not generally considered during the course of most research on the shift from Middle to Upper Paleolithic lithic assemblages, likely due to preconceptions about how culture and cultural change should operate over time. The closest one comes to a notion of "punctuated"—albeit applied incorrectly in evolutionary

Table 10.4. Tools from Levels AA–LL at Warwasi.

Tools	Warwasi AA–EE N	Warwasi AA–EE %	Warwasi FF–LL N	Warwasi FF–LL %
Scrapers:				
on blade	9	1.7	5	1.1
carinated	11	2.1	5	1.1
other	27	5.2	19	4
Burins:				
carinated	9	1.7	2	0.4
other	46	8.9	13	2.7
Nongeometrics:				
Dufour	10	1.9	4	0.8
Font Yves	9	1.7	4	0.8
other	60	11.6	15	3.2
Geometrics	3	0.6	-	
Special tools:				
sidescrapers	108	20.8	134	28.2
truncated faceted	2	0.4	7	1.5
other	1	0.2	-	-
Borers	9	1.7	23	4.8
Backed Pieces	1	0.2	1	0.2
Notch-denticulates	113	21.8	115	24.2
Truncations	7	1.4	7	1.5
Multiple tools*	13	2.5	10	2.1
Retouched pieces	80	15.4	108	22.7
Varia	-	-	3	0.6
Total	518		475	

*includes nine sidescrapers and one endscraper on a blade in Levels AA–EE; includes four sidescrapers and one endscraper on blade in Levels FF–LL.

terms—is the "sudden" appearance of the Aurignacian across much of Europe. In this case, the explanatory framework for most researchers is couched in the migration of AMHs, rather than local transitions (see Straus 1990:276 or Oliva 1993:48–49 for an alternative view).

On the other hand, the "gradual change" notion enjoys wide currency. The difficulties inherent with this perspective lie in several areas yet to be fully addressed. The first is the degree to which researchers expect to see some combination of Middle and Upper Paleolithic characteristics in the respective lithic assemblages. The strongest presumed contrasts between these two chronological periods are in technology—Levallois technology during the Middle Paleolithic and prismatic blade technology in the Upper Paleolithic—and in general tool classes—sidescrapers are widely regarded as primarily a Middle Paleolithic focus, while endscrapers and burins are seen as typical of Upper Paleolithic assemblages. Identifying Levallois technology is not always straightforward, as witnessed by numerous publications (e.g., Boëda 1988; articles in Dibble & Bar-Yosef 1995, among many others). A discussion of the implications for the "transition" in how Levallois is both defined and recognized is beyond the scope of the current paper.

Tool typologies are no less fraught with inconsistencies, especially when they are compared across a major arbitrary divide such as the Middle to Upper Paleolithic (Clark 2001:43–44). Transverse sidescrapers in the Middle Paleolithic would in some Upper Paleolithic typologies be classified as flake endscrapers. Single and double sidescrapers are present in the Upper Paleolithic, but tend to be uncommon in most assemblages and are thus relegated to tool classes such as Special Tools or are subsumed under Scrapers and counted along with endscrapers on blades—Upper Paleolithic typologies thus tend to mask the presence of sidescrapers compared to Middle Paleolithic typology. "Unique" tool types such as Emireh points, chamfered pieces, and Umm el-Tlel points are found at one or a small number of sites in the Levant. Their applicability to the "transition," other than marking a chronological placement to the interval from about 45,000–40,000 B.P., is far from clear.

Naming the shift from the MP to UP also creates problems. Some, for example, Kuhn et al. (1999:505–506), have proposed using the terminology of the Initial Upper Paleolithic (IUP) in order to avoid the semantic baggage attached to the notion of "transitional." This, however, merely begs the question, because researchers still search for elements of the "transition," but now within the IUP (or EUP, depending on their classification proclivities). Creating yet another label to proliferate in the archaeological literature is, in the words of Fox and Coinman (2004: 111), "likely to obscure, rather than clarify the issue." Of great interest, nevertheless, is the fact that so many researchers (across much of Eurasia) are apparently documenting shifts in either technology or tool typology, or both (including the Early Zagros Aurignacian presented in this paper), within a context that is semantically usually portrayed as EUP. Possible exceptions include Boker Tachtit and Umm el-Tlel (but see below).

Dichotomizing the MP and UP, and recognizing "transitions" between these two monolithic entities, essentially forces one to place "transitional" assemblages into one or the other techno-tradition. Given the presence of some aspects of EUP lithics, and the lack of Levallois (at least for some researchers), the Levantine "transitional" assemblages could all be called EUP at some level. This is certainly the argument that has been made by Marks (1983) and Ohnuma and Bergman (1990) for the "transitional" assemblage at Ksar Akil, by Olszewski (1999, 2001) and Olszewski and Dibble (1994) for the "transition" at Warwasi, and by Coinman and Fox (2000), Fox and Coinman (2004), and Fox (2000) for the "transitional" sequence at Tor Sadaf. This leaves the case of Boker Tachtit, which is seen by Marks (1983, 1990, 1993), Marks and Volkman (1983), Fox and Coinman (2004), and others, as representative of a shift from Levallois to Levallois-like in the manufacture of points, but by Bar-Yosef (1998), Kuhn et al. (1999), and others, as entirely within the EUP. Finally, there is the very odd sequence at Umm el-Tlel, in which the "transition" is stratigraphically between the EUP Ahmarian and the EUP Levantine Aurignacian. Close scrutiny of the published materials from both Boker Tachtit and Umm el-Tlel suggests that the issues here are highly related to the recognition and/or definition of Levallois used at these sites, and to inconsistency in how criteria are applied.[5] It is quite likely that both should be conceptualized as EUP sequences in a manner analogous to Ksar Akil and Tor Sadaf, but perhaps less so to the Warwasi example.

Despite the conceptual allure of a solution that sees a one-time invention of technology or typological elements in one particular region (and extrapolates this "finding" to be a record of AMH behavior), the archaeological record has actually already documented the fact that "transitions" from the MP to the EUP are far more complex and varied. They follow a number of different trajectories, which can be partially seen in the lithic differences that constitute the monolithic archaeo-cultures of entities (whatever these really mean) such as the Aurignacian, the Bohunician, the Châtelperronian, the Ahmarian, and so forth. In so far as researchers can actually describe these, they are all subject to the fact that control of time and the coarse-grained nature of the archaeological record operate at levels that obscure an understanding of individual groups of prehistoric hunter-gatherers (Clark 2001:44). The continued linkage of these monolithic lithic entities to such groups (at times, even to groups of "different" hominins), because it is assumed (but never proven) that they can be used in such a fashion, does a disservice to advancing our understanding of past behaviors across Eurasia.

Endnotes

1. Although this latter notion might be plausible to a limited degree under certain circumstances, for example, the radiocarbon date wave for the "Aurignacian movement" from east to west in Europe (Mellars 1999). Radiocarbon dates for the Aurignacian, however, are beyond the current range of calibration, and have standard deviations ranging from ca 500 to 1,800 years (Mellars 1999:345)—rendering them less suitable than it might seem for "tracking" movement of any kind across the landscape.

2. For example, if one assumed that the blade and bladelet debitage in Levels FF–LL was originally from overlying layers, then one would also expect that blade and bladelet cores would have become mixed with the Levels FF–LL lithics, but these are, for all intents and purposes, absent from these levels.

3. In contrast, in the overlying Late Zagros Aurignacian of Levels P–Z, sidescrapers drop to 4.9% of the tool assemblage (Olszewski 1993:199).

4. In Levels AA–EE, there are 59.2% single, 5.6% double, and 35.2% convergent sidescrapers. In Levels FF–LL, there are 61.5% single, 8.9% double, and 29.6% convergent sidescrapers.

5. For example, Marks (1983:65) uses the total loss of MP points and sidescrapers at Ksar 'Akil, and the appearance of endscrapers, chamfered pieces, and burins, to argue for an EUP designation. In the case of Boker Tachtit, the tool assemblage clearly does not contain MP types, except for the debated attribution of Levallois to points from Levels 1, 2, and 3, while it does contain endscrapers and burins. The "standard" used for Ksar 'Akil is not carried over to the interpretation of Boker Tachtit.

References

Allsworth-Jones, P.
1986 *The Szeletian and the Transition from Middle to Upper Palaeolithic in Central Europe.* Clarendon Press, Oxford.

Andrefsky, Jr., W.
1998 *Lithics. Macroscopic Approaches to Analysis.* Cambridge University Press, New York.

Azoury, I.
1986 *Ksar Akil, Lebanon. A Technological and Typological Analysis of the Transitional and Early upper Palaeolithic Levels of Ksar Akil and Abu Halka, Volume I: Levels XXV–XII.* BAR International Series 289(i), Oxford.

Bamforth, D. B.
1991 Technological organization and hunter-gatherer land use: a California example. *American Antiquity* 56:216–234.

Bar-Yosef, O.
1991 Stone tools and social context in Levantine prehistory. In *Perspectives on the Past: Theoretical Biases in Mediterranean Hunter-Gatherer Research* (G. A. Clark, ed), pp. 371–395. University of Pennsylvania Press, Philadelphia.
1998 On the nature of transitions: the Middle to Upper Palaeolithic and the Neolithic Revolution. *Cambridge Archaeological Journal* 8:141–163.

Bar-Yosef, O., & B. Vandermeersch
1972 The stratigraphic and cultural problems of the Passage from Middle to Upper Paleolithic in Palestinian Caves. In *The Origin of Homo Sapiens* (F. Bordes, ed.), pp. 221–226. UNESCO, Paris.

Baumler, M.
1988 Core reduction, flake production, and the Middle Paleolithic industry of Zobiste (Yugoslavia). In *Upper Pleistocene Prehistory of Western Eurasia* (H. L. Dibble & A. Montet-White, eds.), pp. 255–274. University Museum Press, Philadelphia.

Bernaldo de Quiríos, F., V Cabrera, M. Lloret, & A. Pike-Tay
2001 New kids on the block? Some comments on the Middle–Upper Paleolithic Transition in Cantabrian Spain. In *Questioning the Answers: Resolving Fundamental Problems of the Early Upper Paleolithic* (M. A. Hays & P. T. Thacker, eds.), pp. 27–38. BAR International Series 1005. Archaeopress, Oxford.

Boëda, E.
1988 Le concept Levallois et évaluation de son champ d'application. In *l'Homme de Néandertal 4. La Technique* (L. Binford & J. Ph. Rigaud, eds.), pp. 13–26. ERAUL 31, Liège.

Boëda, E., & S. Muhesen
1993 Umm El Tlel (El Kowm, Syrie): Étude préliminaire des industries lithiques du Paléolithique moyen et supérieur. *Cahiers de l'Euphrate* 7: 47–91.

Bourguignon, L.
1998 Les industries du Paléolithique intermédiaire d'Umm el Tlell: Nouveaux éléments pour le Passage entre Paléolithique moyen et supérieur. In *Préhistoire d'Anatolie. Genèse de Deux Mondes. Anatolian Prehistory at the Crossroads of Two Worlds, Vol. II* (M. Otte, ed.), pp. 709–730. ERAUL 85, Liège.

Clark, G. A.
1994 Migration as an explanatory concept in Paleolithic archaeology. *Journal of Archaeological Method and Theory* 1:305–343.
2001 Discussion: the logic of inference in transition research. In *Questioning the Answers: Re-solving Fundamental Problems of the Early Upper Paleolithic* (M. A. Hays & P. T. Thacker, eds.), pp. 39–47. BAR International Series 1005. Archaeopress, Oxford

Coinman, N. R., & J. B. Fox
2000 Tor Sadaf (WHNBS 8): The Transition to the Upper Paleolithic. In *The Archaeology of the Wadi al-Hasa, West-Central Jordan, Vol. 2: Excavations and Research at Middle, Upper and Epipaleolithic Sites* (N.R. Coinman, ed.), pp. 123–142. Anthropological Research Papers No. 52. Arizona State University, Tempe.

Copeland, L.
1975 The Middle and Upper Paleolithic of Lebanon and Syria in the light of recent research. In *Problems in Prehistory: North Africa and the Levant* (F. Wendorf & A. E. Close, eds.), pp. 317–350. SMU Press, Dallas.
1986 Introduction to Volume 1. In *Ksar Akil, Lebanon.*

A Technological and Typological Analysis of the Transitional and Early Upper Paleolithic Levels of Ksar Akil and Abu Halka. Vol. I: Levels XXV–XII (I. Azoury, ed.), pp. 1–24. BAR International Series 289, Oxford

Dibble, H. L.
1995 Middle Paleolithic Scraper Reduction: Background, Clarification, and Review of the Evidence to Date. *Journal of Archaeological Method and Theory* 2:299–368.

Dibble, H. L., & O. Bar-Yosef. (editors)
1995 *The Definition and Interpretation of Levallois Technology.* Monographs in World Archaeology No. 23. Prehistory Press, Madison, WI.

Fox, J. R.
2000 *The Rockshelter of Tor Sadaf: A Middle to Upper Paleolithic Transitional site in the Wadi al-Hasa, West-Central Jordan.* Unpublished M.A. thesis, Department of Anthropology, Iowa State University.

Fox, J. R., & N. R Coinman,
2004 The Emergence of the Levantine Upper Paleolithic: Evidence from the Wadi al-Hasa, Jordan. In *The Early Upper Paleolithic Beyond Western Europe* (J. Brantingham, S. Kuhn, & K. Kerry, eds.), pp. 97–112. University of California Press, Los Angeles.

Garrod, D. A. E.
1951 The transitional industry from the base of the Upper Paleolithic in Palestine and Syria. *Journal of the Royal Anthropological Institute* 81: 121–129.
1955 The Mugharet el-Emireh in Lower Galilee: type station of the Emiran industry. *Journal of the Royal Anthropological Institute* 85:141–162.

Hahn, J.
1988 Fiche Sagaie´ Base Fendue. In *Fiches Typologiques de l'Industries Osseuse Préhistorique. Cahier I: Sagaies* (H. Delporte, J. Hahn, L. Mons, G. Pinçon & D. de Sonneville-Bordes, eds.), pp. 1–21. Université de Provence, Aix-en-Provence.

Harrold, F. B.
1989 Mousterian, Chatelperronian and Early Aurignacian in Western Europe: Continuity or Discontinuity? In *The Human Revolution: Behavioural and Biological Perspectives in the Origins of Modern Humans* (P. Mellars & C. B. Stringer, eds.), pp. 677–713. Edinburgh University Press, Edinburgh.
2000 The Châtelperronian in historical context. *Journal of Anthropological Research* 56:59–75.

Harrold, F. B., & M. Otte
2001 Time, Space, and cultural Process in the European Middle–Upper Paleolithic Transition. In *Questioning the Answers: Re-solving Fundamental Problems of the Early Upper Paleolithic* (M. A. Hays & P. T. Thacker, eds.), pp. 3–11. BAR International Series 1005. Archaeopress, Oxford.

Iovita, R. I.
2002 *A Comparison of the Earliest Upper Palaeolithic at Ksar Akil (Lebanon) and Haua Fteah (Libya).* Unpublished M.A. thesis, Department of Archaeology, Cambridge University.

Karavanič, I.
1995 Upper Palaeolithic occupation levels and late-occurring Neandertal at Vindija Cave (Croatia) in the context of Central Europe and the Balkans. *Journal of Anthropological Research* 51:9–35.

Knecht, H.
1993 Splits and wedges: the techniques and technology of early Aurignacian antler working. In *Before Lascaux: The Complex Record of the Early Upper Paleolithic* (H. Knecht, A. Pike-Tay & R. White, eds.), pp. 137–162. CRC Press, Boca Raton.

Kuhn, S. L., M. C. Stiner, & E. Güleç
1999 Initial Upper Paleolithic in south-central Turkey and its regional context: a preliminary report. *Antiquity* 73:505–517.

Leroy-Prost, C.
1975 L'industrie osseuse aurignacienne. Essai régional de classification: Poitou, Charente, Périgord. *Gallia Préhistoire* 18:65–156.
1979 L'industrie osseuse aurignacienne. Essai régional de classification: Poitou, Charente, Périgord (suite). *Gallia Préhistoire* 22:205–370.

Mann, A.
1995 Modern human origins: evidence from the Near East. *Paléorient* 21:35–46.

Marks, A. E.
1983 The Middle to Upper Paleolithic Transition in the Levant. *Advances in World Archaeology* 2: 51–98.
1990 The Middle and Upper Paleolithic of the Near East and the Nile Valley: the problem of cultural

transformations. In *The Emergence of Modern Humans: An Archaeological Perspective* (P. Mellars, ed.), pp. 56–80. Edinburgh University Press, Edinburgh.

1993 The Early Upper Paleolithic: The view from the Levant. In *Before Lascaux: The Complex Record of the Early Upper Paleolithic* (H. Knecht, A. Pike-Tay & R. White, eds.), pp. 5–22. CRC Press, Boca Raton.

Marks, A. E., & D. Kaufman

1983 Boker Tachtit: the artifacts. In *Prehistory and Paleoenvironments in the Central Negev, Israel. Vol. III. The Avdat/Aqev Area, Part 3* (A. E. Marks, ed.), pp. 69–125. SMU Press, Dallas.

Mellars, P.

1999 The Neanderthal Problem Continued. *Current Anthropology* 40:341–350

Mussi, M.

1990 Le peuplement de l'Italie a la fin du Paléolithique moyen et au début du Paléolithique supérieur. In *Paléolithique Moyen Récent et Paléolithique Supérieur Ancien en Europe* (C. Farizy, ed.), pp. 251–262. Association pour la Promotion de la Recherche Archéologique en Île-de-France, Nemours:

Monigal, K.

2001 The Eastern Szeletian at Buran Kaya III (Crimea, Ukraine) and its place in the Middle to Upper Paleolithic Transition. In *Questioning the Answers: Re-solving Fundamental Problems of the Early Upper Paleolithic* (M. A. Hays & P. T. Thacker, eds.), pp. 51–64. BAR International Series 1005. Archaeopress, Oxford.

Ohnuma, K.

1988 *Ksar Akil, Lebanon: A Technological Study of the Earlier Upper Palaeolithic Levels at Ksar Akil: Levels XXV–XIV.* BAR International Series 426, Oxford

Ohnuma, K., & C. A. Bergman

1990 A technological analysis of the Upper Palaeolithic levels (XXV – VI) of Ksar Akil, Lebanon. In *The Emergence of Modern Humans. An Archaeological Perspective* (P. Mellars, ed.), pp. 91–138. Cornell University Press, Ithaca.

Oliva, M.

1993 The Aurignacian in Moravia. In *Before Lascaux: The Complex Record of the Early Upper Paleolithic* (H. Knecht, A. Pike-Tay & R. White, eds.), pp. 37–55. CRC Press, Boca Raton.

Olszewski, D. I.

1993 The Late Baradostian Occupation at Warwasi Rockshelter, Iran. In *The Paleolithic Prehistory of the Zagros-Taurus* (D. I. Olszewski & H. L. Dibble, eds.), pp. 187–206. University Museum Symposium Series Volume 5. University of Pennsylvania Press, Philadelphia.

1999 The Early Upper Palaeolithic in the Zagros Mountains. In *Dorothy Garrod and the Progress of the Palaeolithic. Studies in the Prehistoric Archaeology of the Near East and Europe* (W. Davies & R. Charles, eds.), pp. 167–180. Oxbow Books, Oxford.

2001 Ruminations on the Early Upper Paleolithic and a consideration of the Zagros Aurignacian. In *Questioning the Answers: Re-solving Fundamental Problems of the Early Upper Paleolithic* (M. A. Hays & P. T. Thacker, eds.), pp. 79–89. BAR International Series 1005. Archaeopress, Oxford.

Olszewski, D. I., & H. L. Dibble

1994 The Zagros Aurignacian. *Current Anthropology* 35:68–75.

Pelegrin, J.

1990 Observations technologiques sur quelques séries du Châtelperronien et du MTA B du sud-ouest de la France: Une hypothèse d'évolution. In *Paléolithique Moyen Récent et Paléolithique Supérieur Ancien en Europe* (C. Farizy, ed.), pp. 195–201. Association pour la Promotion de la Recherche Archéologique en Île-de-France, Nemours.

Rigaud, J.-Ph.

2001 Discussion: re-solving questions of typological significance and technological organization. In *Questioning the Answers: Re-solving Fundamental Problems of the Early Upper Paleolithic* (M. A. Hays & P. T. Thacker, eds.), pp. 117–119. BAR International Series 1005. Archaeopress, Oxford.

Smith, F. H., & J. C. Ahern

1994 Brief communication: additional cranial remains from Vindija Cave, Croatia. *American Journal of Physical Anthropology* 93:275–280.

Soriano, S.

1998 La production de lamelles torses dans les niveaux du Paléolithique supérieur ancien d'Umm el Tlel (Syrie). Exploration théorique et expérimentale de ses modalités. In *Préhistoire d'Anatolie. Genèse de Deux Mondes. Anatolian Prehistory at the Crossroads of Two Worlds, Vol. II* (M. Otte, ed.), pp. 731–748. ERAUL 85, Liège.

Straus, L. G.
1990 The Early Upper Paleolithic of southwest Europe: Cro-Magnon adaptations in the Iberian peripheries, 40,000–20,000 BP. In *The Emergence of Modern Humans: An Archaeological Perspective* (P. Mellars, ed.), pp. 276–302. Edinburgh University Press, Edinburgh.
1999 The Neanderthal Problem Continued. *Current Anthropology* 40:352–355.

Svoboda, J., V. Ložek, & E. Vlček
1996 *Hunters Between East and West. The Paleolithic of Moravia.* Plenum Press, New York.

Volkman, P.
1983 Boker Tachtit: core reconstructions. In *Prehistory and Paleoenvironments in the Central Negev, Israel. Vol. III. The Avdat/Aqev Area, Part 3* (A. E. Marks, ed.), pp. 127–190. SMU Press, Dallas.

Whittaker, J. C.
1994 *Flintknapping: Making and Understanding Stone Tools.* University of Texas Press, Austin.

11

Putting Transition Research in a Broader Context

Geoffrey A. Clark
Arizona State University

Recent work in eastern and south-central Europe, chronological revision of long-known sites and levels, and continent-wide syntheses in Africa have sparked new interest in industries claimed to represent a transition between the Middle and Upper Paleolithic in western Eurasia. Questions arising from these claims have underscored, among other things, the very limited explanatory potential of typological systematics to increase our understanding of the nature of these transitions, and how the biological and behavioral characteristics of the hominids involved might have contributed to our perceptions of pattern shifts across the transition interval, generally regarded as the ten millennia bracketing 40 kyr B.P. The participants in the SAA symposium that resulted in this volume were asked to address one or several of the following questions; these questions and issues formed the backbone around which the session was organized.

(1) How do we establish a sufficiently general and powerful conceptual framework that will allow us to examine the idea of 'transitional' industries?

(2) How do current questions about transitional industries differ from those asked in the past? What preconceptions do they reflect?

(3) What discoveries and methodological advances question conventional wisdom about such industries?

(4) What biological and behavioral hypotheses are transition questions designed to assess?

(5) What are the implications of these questions for hominid biological and cultural evolution over the OIS 5a–3 interval?

(6) What new methods are being developed and implemented to tackle these questions?

(7) What are some of the preliminary conclusions reached by the application of these methods, and what are their implications?

(8) What new questions are generated from the application of these new methodologies?

(9) How widely applicable are these new methods to geographically and temporally distinct 'transitional' industries?

The session itself was very successful, with over 100 people in attendance throughout. It was punctuated by lively debates and assessments of the credibility of recent discoveries and new methodologies, and the conceptual frameworks invoked to assign meaning to pattern. This level of interest is one of two reasons that motivated us to publish a set of proceedings. The other was that, despite the appearance of several Middle–Upper Paleolithic transition volumes over the past few years (e.g., Hays & Thacker 2001, Zilhão & d'Errico 2003), relatively little effort has been expended to synthesize current views from the perspective of the various national and regional intellectual traditions involved in transition research. Despite the appearance of a number of 'middle ground' scenarios, the question of our biological and behavioral origins remains somewhat polarized into 'continuity' and 'replacement' camps. One objective of these papers is to try to get away from this polarization, and to deconstruct or unpack the logic behind the views of the symposiasts. I outline below what I believe to be the salient points made by the contributors, and then expand these perspectives with a consideration of the views of other prominent workers.

Lawrence Straus
Essentialism and Typological Thinking

The essay by Lawrence Straus (Chapter 2) constitutes a thoughtful critique of the essentialist analytical units commonly used in Europe to describe transition interval archaeological assemblages (e.g., Châtelperronian, Olschewian, Szeletian, Uluzzian, Bachokirian, Bohunician, etc.). Straus focuses on the best known of them, the Châtelperronian of southwestern France and northern Spain; he underscores general problems with these units that are, after all, 'accidents of history' (Clark 1989:28, 2002a). Early prehistorians invented the paleolithic and its various subdivisions, including 'transitional' units like the 'Aurignaco-Mousterian' and the 'Châtelperronian'. As Binford pointed out long ago (e.g., 1973, Binford & Sabloff 1982), these labels were originally defined on the basis of the appearance of supposedly time-sensitive 'marker' types *(fossiles directeurs)* analogous to the index fossils used in geological and paleontological research to order strata in time. They were created—not discovered—by several generations of French prehistorians between c. 1880 and 1940 in order to solve chronological problems in the years before absolute dating techniques were available. Although all science depends upon constructs, those used in paleolithic archaeology are 'enshrined by tradition,' and are especially problematic because we do not know what they mean, or represent, in behavioral terms. Since they are seldom subjected to critical scrutiny, they tend to reduce variability and oversimplify complex situations in ways that lead to 'ethnic' or 'identity-conscious' interpretations of pattern. Although reconstructing ethnicity is widely held to be possible by many European workers, from an Americanist standpoint it is perhaps the most elusive aspect of paleolithic research (Clark 2002b,c). Ethnic identity might indeed have existed between 50–30 kya, but studies of artifact typology and technology are unlikely to shed much light on it. Straus argues that the Middle–Upper Paleolithic transition should be studied as a mosaic of biocultural change played out against the backdrop of dynamic environmental change during OIS 3 (57–24 kya).

Francis Harrold
How did Modern Humans Replace the Neandertals?

In a lucid, balanced essay (Chapter 3), Frank Harrold focuses on the fate of the neandertals in Europe, and upon what aspects of behavior might have conferred a selective advantage upon modern humans that would have allowed them to displace or replace them. The question of whether or not neandertals were a different species is essentially moot; Harrold regards them as a "distinctive, continental-scale human population with deep roots in the European Middle Pleistocene," derived from populations like those of the Sima de los Huesos hominids at Atapuerca. Focusing on the middle part of the last glacial in Europe, and confining the discussion to the Châtelperronian of France and northern Spain, Harrold addresses the fate of the neandertals as inferred from changes in the archaeological record over the Middle–Upper Paleolithic transition. The 'usual suspects' are examined (subsistence, technology, evidence for symbol systems), and no strong case can be made for dramatic pattern shifts that might shed light on the suspected adaptive advantage. Much is made of the likelihood of low neandertal population densities, and the consequences they would have had for neandertal technology and subsistence. Small coresident groups and the rarity of encounters with other such groups would not have created the conditions that would have selected for durable symbols of individual or group identity, thus accounting for their rarity or absence in neandertal contexts. Replacement or absorption in which minor behavioral advantages on the part of modern humans might have given them the edge in terms of reproductive success appears to Harrold to be the most likely scenario. Although those advantages remain elusive, they might have allowed moderns to gradually displace the neandertals from the niches they had occupied for c. 250 kyr. The apparent climatic instability of OIS 3 perhaps contributed to their demise (see papers in van Andel & Davies 2003).

Geraldine Lucas, Jean-Philippe Rigaud, Jan Simek & Marie Soressi
Assessing Contextual Integrity at Grotte XVI

Lucas and colleagues focus on the transitional interval at Grotte XVI, a site in southwestern France that includes the later phases of the Mousterian and all of the Upper Paleolithic industries (Chapter 4). These assemblages can be compared on a variety of dimensions, including technology, subsistence, activity organization, and regional spatial structure. Focusing on the Châtelperronian (Couche B), they show by example that we can assess empirically the crucial question of mixture with bracketing levels. In the case of Couche B, the conclusion is that the assemblage is homogeneous, with little contamination from above and below, and that it probably represents an early—even initial—phase of the Châtelperronian. This assessment is based upon a comparison of Couche B with Levêque's 'evolutionary sequence' from Grande Roche de la Plematie. At Grotte XVI, some transitions do occur at the Middle–Upper Paleolithic boundary, but a mosaic pattern of behavioral change is characteristic overall, with some major changes occurring during the Mousterian, others wholly within the Upper Paleolithic, and some gradual trajectories encompassing the entire sequence. While it is clearly important to assess contextual integrity, why we should expect any site to express a normative progression from an 'early' or 'archaic', through 'classic', to 'post-classic' (or 'regressive', as Levêque puts it) is, in my view, problematic. Evolution is directionless, shaped only by context and history.

Amilcare Bietti & Fabio Negrino
The Transition in Italy

In Chapter 5, Amilcare Bietti and Fabio Negrino identify three kinds of transitional industries on the Italian

Peninsula: (1) late Mousterian in northern and central Italy, (2) the Uluzzian of southern Apulia, and (3) assemblages in central and southern Italy that display a distinct regional character. All these precede the early Aurignacian and are distinct from it; the latter two overlie the late Mousterian wherever they occur in a stratigraphic sequence. Aspects of the lithic industries considered 'transitional' are already present in the late Mousterian, both in northern (e.g., San Francesco, in Liguria) and south-central (e.g., Grotta Breuil, in Latium) Italy. The authors conclude that well-stratified transitional sites are very rare, and are so far confined to Castelcivita and the sites on the Salentine peninsula, and that what differentiates the Aurignacian from the transitional industries is (1) a very specialized blade and bladelet technology, and (2) a major difference in raw material procurement. Exotic flints become more common in the Aurignacian. The late Mousterian and the transitional industries are characterized by similar patterns of raw material procurement, with exotic flints and cherts being relatively rare. Regardless of time period, sites with high quality raw material in close proximity do not exhibit many exotics. The authors' interpretation is contrasted with that of Palma di Cesnola (1993), who sees an in situ evolutionary sequence of late Mousterian, Uluzzian, Uluzzo-Aurignacian and Aurignacian proper. Bietti and Negrino regard this unilineal sequence as unlikely and unsupported empirically. Inextricably linked to the two contrasting models are presuppositions about who made these industries, although in fact there is no skeletal material associated with any of them.

C. Michael Barton & Julien Riel-Salvatore
Curation and Expediency

C. Michael Barton and Julien Riel-Salvatore develop an index of the extent to which assemblages were curated or expedient (Chapter 6). It uses three simple but robust variables to analyze site sequences on a level-by-level basis. Those variables, which are either commonly reported or which can often be reconstructed from a typical site report are (1) volume of sediment excavated, (2) total number of retouched tools, and (3) total number of lithic pieces. These variables operate at the macroscale to produce explicit sets of expectations about assemblage composition that track changing forager mobility strategies. Given a tendency to maximize lithic utility under different kinds of land-use strategies, humans would either conserve lithic resources by producing fewer flakes and extending their use lives (curation—where raw material is scarce), or produce and discard more flakes while investing little effort in maintaining them (expediency—where raw material is abundant). Thus the ratio of retouched tools to lithic totals is inversely related and negatively correlated. This model is then applied to (1) Gorham's Cave (where expectations are met, r = -0.92), (2) three rockshelters from southeastern Spain (where expectations are again met for two of the three sites—all three tend toward the 'curated' end of the spectrum), and (3) eight southern Italian sites. In this last case, the model doesn't hold up very well, but there is much local knapping of limestone, leading to the suggestion that there were two different kinds of tool kits—a small, curated one made exclusively on flint procured from afar, and a much larger expedient one made on the local siliceous limestone. It is interesting to note that in no case is there evidence for any kind of a breakpoint coinciding with the Middle–Upper Paleolithic transition interval, which suggests that these strategies are of great temporal duration and spatial extent, and have nothing whatever to do with the alleged cognitive differences of the hominids producing them.

Ivor Karavanič
A Balkan Mosaic of Transitional Adaptations

Ivor Karavanič discusses the statigraphic sequences from three Croatian sites in Chapter 7. Vindija cave in northwestern Croatia has produced neandertal fossils associated with a possible 'transitional industry', the Olschewian, which exhibits some Aurignacian characteristics. Although both the human remains and the archaeology of Vindija have been relatively well published, the Middle–Upper Paleolithic transition from the Adriatic coast of Croatia is not well known. Mujina Pečina cave, in Dalmatia, contains a well-dated Mousterian industry helpful in understanding late Middle Paleolithic adaptations in the eastern Adriatic. Significant numbers of Upper Paleolithic tools in Level B (and possibly in C) point out the dangers of normative thinking based upon typological differences. The penecontemporaneous site of Candalja II, in Istria, is the only early Upper Paleolithic site in that region. It has a fairly typical Aurignacian sequence. The Montenegran site of Crvena Stijena has both Middle and Upper Paleolithic levels, but lacks a transition between them. Karavanič concludes that late-surviving neandertals and modern humans probably coexisted in the region for c. 16,000 years (45–29 kyr B.P.). The neandertal bones at Vindija are 'late', dating to c. 29–28 kyr B.P. The archaeology thus exhibits a mosaic pattern, with a good transitional sequence at Vindija, but no transitional sites in the eastern Adriatic. The absence of a transitional industry in Istria provides further evidence of the complex pattern of adaptive change in Europe during OIS3. If these patterns withstand further scrutiny, they might be explained by environmental differences between the two regions that affected resource distributions, forager group size and composition, mobility and, eventually, the 'signatures' left in the archaeological record by distinct trajectories of human adaptation.

Brian Adams
Evolutionary Stage, Acculturation Industry, or Activity Facies?

In Chapter 8, Brian Adams analyzes the allegedly 'transitional' Szeletian industries of Hungary. The term is derived from early Upper Paleolithic assemblages discovered at Szeleta Cave in the northeastern part of the country. Retouched tools from Szeletian assemblages are dominated by 'leaf points', 'leaf scrapers', bifaces and bifacial foliates; these 'type fossils' are considered diagnostic of a Middle–Upper Paleolithic 'transitional' industry. Since bifacial technologies also occur in the regional Mousterian, two hypotheses have emerged to account for the Szeletian: (1) in situ cultural evolution, and (2) acculturation. Advocates of in situ evolution argue that the Middle Paleolithic foragers who produced the bifacial pieces evolved directly into those who made the Upper Paleolithic, while acculturation advocates contend that indigenous Mousterians interacted with allochthonous, Upper Paleolithic Aurignacians to produce assemblages with mixed Middle and Upper Paleolithic characteristics. Adams thinks these assessments rely too heavily on typological data, and fail to take other sources of information (e.g., lithic raw material, chronology) into account. Although he concedes that Allsworth-Jones' views of an 'acculturation' are the orthodox ones (e.g., 1986, 1990), he makes a fairly strong case that the Szeletian is probably a functional variant or activity faces of the central European Aurignacian. There is a tendency for Szeletian sites to be located at higher elevations in north central Europe (e.g., at 300–350 m asl in the Bükk Mountains), whereas contemporary Aurignacian sites tend, with two exceptions, to be lower (235–120 m asl). The differences in elevation are essentially non-overlapping. Although there is a lot of overlap in the composition of the industries themselves, the compositional characteristics of the two assemblage types are somewhat discrete, which would also support the idea of a functional difference. Like many workers, Adams considers the Aurignacian to be intrusive, the product of modern humans. If this equation is maintained, however, Adams' functional argument would seem to imply that the Szeletian was also made by moderns. I remain skeptical of the equation of the Aurignacian with any particular hominid biological type, although the weight of evidence so far tends to support the widespread belief that it was manufactured only by modern humans (see, esp., Churchill & Smith 2000).

Michael Bisson
Implications of Middle Paleolithic Hafting

Michael Bisson's essay (Chapter 9) focuses on the evidence for hafting of levallois points in the Middle Paleolithic, and examines collections from Skhūl (an early 'anatomically modern' site) and Tabūn (which has so far yielded only neandertal remains). Although very hard to pin down, there are apparently some differences in the incidence and mode of hafting between Skhūl B1 and the sample from Tabūn (platform modification, mainly). Whether these slight differences conferred any adaptive advantage on anatomically modern humans vis à vis neandertals is arguable, however. A broad ranging comparison with other regions in western Eurasia and Africa seems to indicate episodic or sporadic evidence for hafting, which shows up at various times and places, only to disappear, and then—sometimes—reappear again. The initial development of more effective hafted tools and weapons does not appear to have triggered any discernible changes in lifestyle or adaptation. Bisson makes the important point that these differences in hafting are likely situational, and are related to behavioral context, rather than inherent differences in cognitive evolution (indeed, anything intrinsic to a hominid taxon). Sampling error also plays a role—probably a big one.

Deborah Olszewski
Typology, Technology and the Zagros Aurignacian

Noting the plethora of contemporary lithic industries recognized as 'transitional' across western Eurasia, Olszewski (Chapter 10) examines the technological and typological parameters of the EUP levels (AA–LL) at the Warwasi rockshelter, in the Zagros Mountains of northwestern Iran. Both technology and typology here show an interesting association of classic 'Middle' and 'Upper' Paleolithic elements: prismatic blade(let) technology; small, radially prepared flake cores, carinated endscrapers and sidescrapers of various types. These industries are evaluated in terms of the criteria necessary and/or sufficient to document a 'transition', and are compared and contrasted with Aurignacian developmental sequences elsewhere. The important issue of typological comparability on either side of the transition is identified as perhaps the most important factor leading to perceptions of relatively abrupt or punctuated change. How typologies (including typologies of technology) are defined and used by workers in different research traditions lies at the heart of why most Middle and Upper Paleolithic industries are seen as so characteristically dissimilar. These perceptions of pattern shifts are inextricably linked to preconceptions and assumptions about the cognitive abilities of the hominids thought to be responsible for their manufacture, even though there is almost no fossil evidence bearing on authorship. In Olszewski's view, these preconceptions about the authorship of 'transitional' industries have distracted workers from the much more important problem of understanding changing human adaptation across space and time during the OIS 3 transition interval.

Other Views of the Archaeological Transition

As might be expected, the essays published here express a range of opinion about the nature of the behavioral changes that accompanied the Middle–Upper Paleolithic transition. There are, of course, many other views—probably as many as the sum total of researchers interested in the topic. Although by no means exhaustive (and I apologize in advance for not attempting a complete review), some other workers who have relatively well-defined, published positions on the nature of the archaeological transition include Donald Grayson and Françoise Delpech, Eudald Carbonell, Paul Mellars, Steven Kuhn, Janusz Kozlowski, Ofer Bar-Yosef, Anthony Marks, Jean-Jacques Hublin, Richard Klein, João Zilhão and Francesco D'Errico, William Davies, Luís Raposo and João Luís Cardoso, and Jean-Pierre Bocquet-Appel and Pierre-Yves Demars. Their views are summarized here, preliminary to a final assessment.

Donald Grayson & Françoise Delpech

Although not included in this collection, the SAA session upon which this volume is based also contained a paper by Donald Grayson and Françoise Delpech (since published elsewhere—Grayson & Delpech 2003). These workers used two statistical measures (evenness, dominance) to assess Paul Mellars' frequently reiterated statement that highly specialized hunting is characteristic of the Upper Paleolithic (esp. the Aurignacian) and that it distinguishes the Middle from the Upper Paleolithic. Grayson and Delpech analyzed the NISP counts from 131 Mousterian, Châtelperronian and Aurignacian assemblages from southwest France to try to answer a single question: *Is there empirical support for Mellars' assertion that most Aurignacian faunas are dominated by a single taxon?* The answer they come up with is an unequivocal 'no'. On both statistical criteria, most Aurignacian assemblages overlap extensively with most Mousterian and Châtelperronian ones. They also dissect Mellars' argument according to his statistical criteria (based on percentages at 5% intervals), and find that measures of evenness and dominance provide a much more sensitive instrument for assessing assemblage differences and similarities than do the percentages of NISP counts used by Mellars. They note that Mellars used only cave and rockshelter faunas, and that once open sites are included, there are no significant differences in the distributions of Aurignacian and Mousterian assemblages of economic faunas. This last is a good example of paradigmatic bias. Some French workers are disinclined to include open sites under the expectation that they have minimal contextual integrity when compared to caves and rockshelters (White 1985, Barton et al. 1994). In fact, given the well-documented concentration of activities in closed spaces like cave mouths and rockshelter overhangs, contextual integrity may be equally (or more) problematic there than in open sites (see, e.g., papers in Goldberg et al. 1993).

Eudald Carbonell

In a series of co-authored papers, the Catalan archaeologist Eudald Carbonell and his colleagues (e.g., Carbonell et al. 2000) focus on the Iberian Peninsula, considered of crucial importance because of the spectacular sample of unambigous 'pre-neandertals' represented by the Sima de los Huesos hominids (e.g., Bermúdez de Castro et al. 2004), the late survival of Mousterian industries (and, arguably, neandertals) at the southern Iberian sites of Zafarraya (Hublin et al. 1995) and Columbeira (Delibrias et al. 1986) and the Lagar Velho neandertal-modern 'hybrid' child, associated with a Gravettian archaeological industry dated to c. 24.5 kyr B.P. (Zilhão & Trinkaus 2002). In a nutshell, Carbonell has argued for Middle–Upper Paleolithic continuity north of the Ebro-Douro line (at c. 42° N. lat.), citing the evidence for continuity represented in levels 17–19 at El Castillo, in Cantabria (Cabrera et al. 1997, 2000). However, in Catalunya, there is a rapid, abrupt substitution of Mousterian assemblages by Aurignacian ones, argued from sites like Mollet, l'Arbreda, Reclau Viver and Abric Romaní (see also Camps 2004). In contrast with Cantabria, there is no evidence for the Châtelperronian. The Mousterian persists for a long time in southern, central and western Iberia, as exemplified by Zafarraya (with neandertal remains), Jarama VI, Columbeira and Salemas, all dated to < 30 kyr B.P. Taking the peninsula as a whole, there are very few sites with transitional industries. The only one recognized is the Châtelperronian, and it is effectively confined to the north Spanish coastal strip and the Basque country (A Valiña in La Coruña, Morín and El Pendo in Cantabria, Ekain and Labeko in Guipúzcoa). The Middle Paleolithic essentially extends up in time in the south to the Epi-gravettian and the Solutrean, at < 25 kyr B.P.

Paul Mellars

Paul Mellars is another prominent worker with well-defined views on the nature of the archaeological transition. His interpretation of pattern shifts over the Middle–Upper Paleolithic transition has not changed markedly since the early 1970s (cf. Mellars 1973, 1989) although it has become generalized for all of western Eurasia (i.e., extended beyond southwestern France, where it originated). Mellars focuses on a series of behavioral changes and their material correlates that supposedly accompany the transition interval (45–35 kya) in Europe. Taken to indicate the appearance of modern humans and (usually) Aurignacian industries, they include (1) a general shift in the pattern of stone tool technology from predominantly 'flake' technologies to more regular, standardized 'blade' technologies, achieved by means of more economic techniques of core preparation; (2) a simultaneous increase in the variety and complexity of stone tools involving more standardization of shape, and a higher degree of 'imposed form' in the various stages of production; (3) the appearance of relatively complex and extensively shaped bone, antler and ivory artifacts; (4) an increase in the tempo of technological change (apparent both in lithic and in bone technology), accompanied by increased regional diversification of tool forms produced in different geographical areas; (5) the appearance of wide range of beads, pendants, and other 'personal ornaments' made from perforated animal teeth and marine shells, and from more elaborately shaped bone, stone and ivory blanks; (6) the appearance of sophisticated and highly complex forms of representational or 'naturalistic' art, and (7) associated changes in both the economic and social organization of human groups, marked by (i) a much more specialized pattern of animal exploitation, based on systematic hunting, as opposed to opportunistic scavenging of game, (ii) a sharp increase in the overall density of human population, (iii) an apparent increase in the maximum size of local co-residential groups, and (iv) the appearance of more highly 'structured' sites, including more evidence for features (e.g., well-defined hearths, pits; huts, tents, and other habitations). There is a remarkable consistency in his assertion of these criteria, even though the empirical record of the transition interval has become richer and more complex over the past 30 years. Although there was strong empirical support for them during the 1970s, many workers now question whether or not they actually monitor 'modernness' (however defined), and the extent to which they can be generalized over large areas.

Partly because of these observations, Mellars has recently (2004, 2005) revised and expanded some of his views of the transition, deemphasizing some of his diagnostic criteria (esp. blades), adding others (symbolic notation systems, musical instruments, improved missile technology), and expanding the discussion to include more on the (admittedly equivocal) genetic evidence and on patterns in the MSA archaeology of south and east Africa. Following the landmark paper by McBrearty and Brooks (2000), which showed that archaeological indicators of modern behavior in Africa are not clustered in time or space, but are instead distributed over c. 200,000 years, he makes a case that many of the features of the Upper Paleolithic 'revolution' in Europe are present in Africa by about 70 kya, at least 30,000 years before they show up in Europe. While he acknowledges that differences in the scale and resolution of the archaeological records of Europe and Africa are likely to play a role in any comparative interpretation (see also Henshilwood & Marean 2002), he overlooks early evidence for UP diagnostics in Europe itself (e.g., Bilzingsleben, dated at c. 400 kya), LP evidence for organic technologies (e.g., Clacton, Torralba, Schoeningen, Boxgrove [dated c. 500–350 kya) and 'highly structured occupation sites' (e.g., Terra Amata, Lazaret [c. 150–130 kya), LP and MP blade-dominated assemblages (e.g., Seclin, Riencourt, Rhinedalen), and appears to dismiss the early dates for Aurignacian industries in the west (e.g., El Castillo Lev. 18, at c. 40 kya). In Mellars' view, the symbolic evidence is key, and that to describe the UP revolution in Europe as "an explosion of explicitly symbolic behavior is in no sense an exaggeration" (2005:12), an 'impossible coincidence' best explained by an influx of cognitively modern humans that differ in no important respects from ourselves (2004).

Largely as a reaction to the 'indigenist' model of Zilhão and D'Errico (e.g., 1999), which he rejects as implausible, Mellars has also focused recently on the radiocarbon chronologies of the Châtelperronian and the Aurignacian in southwestern France and northern Spain, arguing effectively that there is no question that the two overlap in these well studied areas. An acculturationist, he attributes the appearance of 'modern' behavioral features (i.e., the beads) at Grotte du Renne as the result of contact between moderns and neandertals, and equates the Aurignacian with modern humans of (ultimately) African origin, who entered western Europe from the east over an interval between c. 40 and 35 kyr B.P. To Mellars, " ... the survival of the Aurignacian and Châtelperronian industries as clearly identifiable and sharply separated technological traditions over a period of several thousand years must imply some fundamental barrier to communication and interbreeding between the two groups which prevented their total assimilation and integration over this span of time" (2000:44). The nature of that barrier remains uncertain, although Mellars suggests that it might have involved fully modern language capabilities (see Klein, below).

Richard Klein

A prominent Stanford archaeozoologist, Klein is identified with what has sometimes been called 'Out-of-Africa 2' (Stringer & Gamble 1993), a scenario heavily dependent upon a particular construal of genetic evidence for language that has yet to acquire any testable, scientific support. In publications dating back to the early 1990s, he suggested that a hypothetical mutation in an east or south African population less than 200 kya led to a cascade of neurological changes in the human brain that culminated in the appearance of modern language. Language then triggered the socioeconomic and cognitive changes we associate with behavioral modernity and Africans, armed with behavioral modernity, then spread out of that continent, out-competing, displacing, extirpating, outbreeding or—most generally—replacing the neandertals and other archaic humans throughout the middle latitudes of the Old World.

Until 2002, OA2 was founded primarily on mtDNA data extracted from the Feldhofer neandertal (Krings et al. 1997) that implied "that the last shared ancestor of neandertals and living humans lived 500,000 to 600,000 years ago" (Klein 2003:1525, 1999:507). Discovery of a mutation on the FOXP2 gene that resulted in a speech pathology (Enard et al. 2002—the normal function of FOXP2 is unknown) was recruited by Klein both for the suggested time range it implied for language origins (< 200 kya), which he thought could be squared with the archaeological evidence (see below), and for its functional implications, which implied selective fixation of the present human allele at around that time. However, Klein treats language acquisition as though it were an event, rather than a process. As Wolpoff et al. (2004:536) have pointed out, this is an oversimplistic view both of language evolution and of the genetic evidence. Language is a highly polygenic trait, and scores of selective sweeps must have occurred as it evolved over the course of the Pleistocene. No single genetic change can thus account for the appearance of modern language, and it is a mistake to assume that its evolution was primarily saltational (Hauser et al. 2002). Arguments from cognitive neuroscience, psychology, non-human primate capacity, early hominid intelligence and the successful, long-term radiation of early hominids in the absence of large-scale genetic change all suggest that language is an emergent property of human cognition deeply rooted in our evolutionary history as social primates (Clark 2003).

Because he thinks they lacked modern language, Klein also contends that neandertals lacked the cognitive ability to compete successfully with Cro-Magnons in Europe, as indicated by what he perceives to be a sharp divide between Middle and Upper Paleolithic artifact inventories there. However, pattern searches on the commonly accepted monitors of human adaptation (i.e., lithic technology, typology; raw material variation, reduction strategies, blank types and frequencies, organic technologies, ornaments, subsistence practices, settlement patterns) have consistently shown a temporal and spatial mosaic wherever the transition interval is recorded. This mosaic is inconsistent with the idea that these archaeological monitors of behavior constitute a 'modern'package' (Clark 1992, 1997, 2002; Straus 1996, 1997, 2003).

Steven Kuhn

Like Bietti (see Chapter 4), with whom he has collaborated, Kuhn views the transitional Uluzzian on the Italian peninsula as a development out of the local Mousterian. Divided into lower, middle and upper stages on the basis of an 'evolutionary' development in the tool forms (Palma di Cesnola 1993), it always overlies the Mousterian (if the latter is present in the same stratigraphic sequence) and is, in turn, overlain by a bladelet-rich Aurignacian. At no site on the peninsula is this order reversed, although only two sites, Castelcivita and Fabbrica, contain all three industries in the same stratigraphic sequence (Kuhn & Bietti 2000:66). Generally speaking, Uluzzian assemblages are quite rare (four relatively well-documented sites) and poorly dated (they appear to post-date 35 kyr B.P., thus overlap chronologically with the Aurignacian in the north and the Mousterian in the south), and are confined to the southernmost third of the peninsula. The earliest Aurignacian assemblages are found in the north, and date from c. 38–37 to 32 kyr B.P., thus antedating the Uluzzian in the south by several millennia. The Aurignacian in the south (Salento) is quite late. There are parts of the peninsula (e.g., Latium) that lack altogether an early Aurignacian or an Uluzzian, and where the local Mousterian (micromousterian or Pontinian) may date as late as 33 kyr B.P. In many ways the Uluzzian squares with the conventional notion of a transitional industry (i.e., it has lots of sidescrapers and denticulates, no prismatic blade technology, but a few Upper Paleolithic types (esp. *pièces esquillées,* backed lunates, flake endscrapers). However, as Kuhn and Bietti remark, this characterization begs the question as to what is implied in the notion of a 'transitional' industry (2000:70, 71). What does a transitional industry—however defined—mean in behavioral terms? Most of the evidence of changes in adaptation occur during the Mousterian itself (Stiner 1994, Kuhn 1995), rather than in the 10 millennia bracketing 35 kyr B.P. During that interval, there are three relatively distinct industries that exhibit a degree of spatial separation, but overlap extensively in time. Who made them? The scarcity of human fossils pertaining to OIS 3 precludes an answer. As a number of workers have remarked, the Italian situation is strikingly similar to that of Spain. If the Aurignacian is equated with *H. sapiens sapiens,* and the Mousterian and Uluzzian with *H. sapiens neandertalensis,* the sparse chronological information could be taken to imply that neandertals were successful in resisting an incursion of moderns from the north, and survived for several millennia in the southern and central parts of the country. However, the assignment of particular hominids to particular archaeological assemblages is a very dubious practice because—as is evident from the Middle East—no universal correlations exist between the two.

Janusz Kozlowski

Central and eastern Europe are often taken to be the entry point of Aurignacian-making modern humans to the subcontinent, and the problem of cultural continuity (or lack thereof) between the Middle and Upper Paleolithic has been addressed at length by Janusz Kozlowski (2000). One might expect transitional industries to occur substantially earlier than 40 kyr B.P. in the east, and Kozlowski makes a case for an autochthonous transition between 45 and 29 kyr B.P. Over that interval, Mousterian industries with levallois technologies co-occur with developed Upper Paleolithic blade industries, and with Mousterian industries showing no signs of levallois technique. The technological substrate for the transitional industries is a Mousterian variant or 'facies', the eastern Micoquian. Perhaps the best documented of the transitional industries, the Szeletian, is characterized by leaf shaped points and is a viewed as a local development out of a Micoquian base. The earliest dated Szeletian assemblage is the lower level at the eponymous site (Szeleta Cave), with a radiocarbon date of >41.7 kyr B.P. There are other transitional industries (Bohunician, Jerzmanovician, Bryndzenian) which are defined typologically, and which have localized distributions in north central Europe between 35 and 30 kyr B.P. These are all believed to be indigenous developments, little influenced by the Aurignacian. In contrast to the transitional industries, the earliest Aurignacian (or Aurignacian-like) industry is represented by the Bachokirian, dated to >43 kyr B.P. in level 11a at Bacho Kiro (Kozlowski 1982). In the Middle Danube and the southern Balkans, Aurignacian industries with split-based bone points occur, whereas the western Carpathians exhibit a mosaic of Bohunician and Szeletian leaf-point industries. To the north, in southern Poland, Jerzmanovician and Szeletian assemblages follow a brief expression of the Bohunician after c. 36 kyr B.P., and co-occur with a rather late Aurignacian after c. 32 kyr B.P. Northeast Hungary, Slovakia and the Ukraine show a persistence of eastern Micoquian-derived industries (late Mousterian, Szeletian) that co-occur, after c. 40 kyr B.P., with the Aurignacian. The Aurignacian is regarded as an allochthonous development throughout all of Europe, unrelated to the local Middle Paleolithic and its derivatives (Kozlowski 2000:99). The scarcity of human skeletal remains from central Europe precludes any assignment of authorship, although there is a general belief that modern humans were the makers of the Aurignacian, neandertals the makers of the Micoquian, Mousterian, and the Mousterian-derived transitional industries.

Ofer Bar-Yosef

Another major arena of transition research is the Middle East and the adjacent regions of southern Turkey. The Harvard archaeologist Ofer Bar-Yosef has published extensively on this area; he sees the Middle–Upper Paleolithic transition there and elsewhere as nothing short of a technological 'revolution', analogous in some respects to the appearance and spread of domestication economies (Bar-Yosef 2002:379–381). Like the Neolithic revolution, the transition originates in a 'core area' and then spreads throughout western Eurasia, either by diffusion, population movement, or a combination of both. He identifies the transition by about a dozen criteria analogous to, and partly overlapping with, those of Paul Mellars (see esp. Bar-Yosef 2002). And, like Mellars, he regards these as the material components of the archaeological record that are taken to signal important (and general) behavioral shifts identified by *en bloc* comparisons between the Middle and the Upper Paleolithic. Included are (1) systematic production of prismatic blades (and, very quickly, bladelets and other microlithic technologies—the latter signaling the appearance of multicomponent tools); flake-dominated technologies are seen as uncommon (cf. Clark & Lindly 1989); (2) a high degree of morphological standardization coupled with lots of regional variation in tool forms (cf. Mellars 1989), although it is also acknowledged that there is also formal convergence (Dibble 1995, Bisson 2000) and the possibility that perceptions of standardization and variety in the retouched tool components might be due, at least in part, to the use of different typological systems (see, e.g., Grayson & Cole 1998, Marks et al. 2001); (3) common use of bone and antler as raw materials for the production of utilitarian and ritual tools and objects; (4) systematic appearance of grinding and pounding stones probably used in the processing of plant foods; (5) systematic use of bodily adornment (beads, pendants) made from organic materials (shell, bone, teeth, ostrich eggshells), taken to indicate individual self-awareness; (6) long-distance exchange networks in lithics, raw materials and marine shell; (7) significant technological developments that improved the efficiency of hunting (dart throwers, bows and arrows, boomerangs), (8) *art mobilier et parietal*—human and animal figurines, decorated and carved bone, antler, ivory and stone objects, and representational and abstract images carved or painted on fixed surfaces (cave and rockshelter walls, exposed rocky surfaces)—become common after c. 36 kya; (9) widespread modalities in lithic technology identified by a relatively limited number of *chaînes opératoires,* or operational sequences, taken to represent shared transmission of ideas amongst individuals and perhaps indicative of the extent of mating networks; (10) storage facilities (esp. in northern latitudes), apparently unknown in the Middle Paleolithic (however, cf. Jelinek 1992); (11) hearths constructed of rocks and designed to conserve warmth, (12) functional segregation of space within sites (cf. Henry et al. 2004), and potential differences in subsistence activities (cf. Chase 1989). In aggregate, Bar-Yosef takes pattern shifts in these criteria to indicate rapid technological change, the emergence of self-awareness and group identity, increased social differentiation, and the formation of long-distance alliances—all coinciding with the ten millennia bracketing 40,000 years ago. Whether these criteria change in concert with one another has, however, been vigorously contested by Straus (e.g., 1996, 1997), Clark (e.g., 1997, 1999, 2002) and many others. Interestingly, and unlike White (1989, 1993, 1997) or Davidson (1997, Davidson & Noble 1989), Bar-Yosef does not regard the European evidence for symbolism as especially relevant or of critical importance to explaining the transition. He sees it in much more basic, technological terms. The origin point of the Upper Paleolithic remains in dispute. It could be East Africa, the Nile Valley, the Levant, central Asia or Anatolia, with evidence for an extremely early initial Upper Paleolithic perhaps best documented in Anatolia (Bar-Yosef 2000:142).

Anthony Marks

A second worker active in Levantine and Crimean research is Anthony Marks, who has written extensively on the archaeological transition (e.g., 1983a, 1990, 1993) and/or on aspects thereof (e.g., Marks et al. 2001). Based on surveys and excavations in the central Negev highlands (Marks 1976, 1977, 1983b) and on the excavations at the important transitional site of Boker Tachtit (see papers in Marks 1983b), he emphasizes pattern in the record of lithic technology and human settlement, and concludes that the transition interval in the Levant exhibits a robust pattern of its own (an emphasis on points made on lamellar blanks), and that traditional European markers of the Upper Paleolithic either do not occur there, or show up scattered over the last 100,000 years of the Pleistocene. The transition is definable mainly through shifts in the way lithic materials were reduced from cores into blanks. These technological changes are taken to monitor human adaptation to a series of changing Upper Pleistocene environments and a kind of progressive improvement in efficiency: early Mousterians produced a wide range of morphological blanks using levallois, discoidal and (rarely) pyramidal blade cores (Crew 1976, Munday 1976); the EUP Ahmarian is recognizable by a single reduction strategy capable of producing the entire range of morphological blanks desired. The work at the stratified and dated M/UP transitional site of Boker Tachtit, in the central Negev highlands showed, through the extensive refitting of cores, that typologically identical levallois points were made early on from classic 'MP' levallois point cores (level 1) and, toward the end of the sequence, from classic 'UP' single platform blade cores (level 4), the one production stream gradually replacing the other (numerous publications, but esp. Marks 1983b). The major implication of this work is that typology can vary independently of technology (something few European workers would acknowledge). This very significant conclusion is practically always obscured in studies that emphasize the typology of re-touched stone tools. Driven by long term climatic change, the major behavioral shift in human land use implied by the dichotomy between the radiating (logistical) settlement patterns of the Middle Paleolithic and the circulating patterns found during the Upper Paleolithic is now seen to taken place *within* the Middle Paleolithic, after about 70 kya, and not during the transition interval at c. 45 kya (Marks & Freidel 1977, Marks 1993:17). Because modern humans were either in the Levant, or in and out of it, for approximately 60,000 years before the archaeological record there can be used to infer 'fully cultural behavior', Marks concludes that the archaeological monitors of modern behavior in Europe (where the biological and cultural transitions supposedly coincide) are unrelated to one another in the Levant, and occurred at quite different times (Marks 1993). Although arguably present by the EUP Ahmarian, the full *repertoire* of material indicators of 'moderness' as defined in Europe only occur in the Natufian, after c. 12 kya (Marks 1994). The work in the Crimea (e.g., Marks 1998, Marks & Monigal 1998, Monigal 2001), elsewhere in eastern Europe, and in central and eastern Asia (see papers in Brantingham et al. 2004), documents an astonishing variety of late Middle and early Upper Paleolithic industries that simply cannot be accommodated by existing conceptual frameworks developed in the west.

Jean-Jacques Hublin

Paleoanthropologist Jean-Jacques Hublin focuses on the circum-Mediterranean fossil evidence, which he divides into three geographical domains—Europe, the Near East and North Africa. He believes that, for the Middle and most of the Upper Pleistocene, the Mediterranean Sea constituted a formidable barrier to hominid migration or range extension, and that hominids evolved in Europe in relative geographical isolation from the rest of western Eurasia, and from Africa (this view has been contested by Tobias [2000] who argues that hominids could easily have crossed the Straits of Gibraltar throughout the Pleistocene during pleniglacial maxima). Pre-neandertals, best represented by the Sima de los Huesos hominids, were thinly distributed on the landscape (even in southern Europe) and underwent periodic bottlenecks. Colonization across the Straits of Gibraltar is regarded as unlikely, at least on a large enough scale so that hy-brid populations would have resulted. So far as the Near East is concerned, where early anatomically modern humans and neandertals co-occur at Skhūl and Tabūn on Mount Carmel, the chronology does not support con-temporaneity over OIS 5 and 4, neandertals having moved south during cold intervals (OIS 6, 4), displacing moderns, only to retreat northward during episodes of climatic amelioration (OIS 5e, 5c, 5a). The Auri-gnacian, of unknown geographical origin, is taken to be an archaeological 'signal' of modern humans, and is regarded as intrusive in Europe. Although neandertals may or may not be a valid species (Hublin thinks they are), the Aurignacian scenario appears to include the Levantine variant, which doesn't resemble the European Aurignacian either technologically or typologically (Coinman 1998). A better Levantine analogue to the European Aurignacian would be the Ahmarian, a blade(let) dominated assemblage which occurs at Boker Tachtit after c. 45 kyr B.P. (Marks 1993), after c. 40 kyr B.P. at K'sar Akil Level XVII in Lebanon (Ohnuma & Bergman 1990), after c. 34 kyr B.P. at Boker A, Ein Aqev East and other open sites in the Sinai Desert and in the central Negev highlands (Phillips 1988, Marks 1993), and after 32 kyr in

Jordan (Coinman 1998). There are no human fossils associated with the early Upper Paleolithic in the Levant.

João Zilhão & Francesco D'Errico

Sparked by the discovery of a neandertal-modern 'hybrid' child in a Gravettian level at Lagar Velho cave in Portugal, Francesco d'Errico and João Zilhão developed what has sometimes been called 'the indigenist model' in 1998. Construals of pattern in Iberian archaeological and paleontological data are central to the formulation of the model, although it has been generalized over the past five years to encompass much of western Europe, wherever Châtelperronian industries are found (e.g., d'Errico et al. 1998, Zilhão & d'Errico 1999; Zilhão 2000, 2001). These provocative essays have generated a lot of discussion, not only in the scholarly community (see, e.g., comments on d'Errico et al. 1998; Mellars 1999, Straus 1999, Otte 1999), but also in the media (e.g., the BBC series 'The Dawn of Man', 2000), and in popular and semi-popular magazines devoted to 'big issues' in science and paleoanthropology (prehistoric archaeology, human paleontology) (e.g., Norris 1999, Zilhão 2000, Wong 2000). Although specific to Europe, the indigenist model is inextricably tangled up in the complex package of issues and data that make up the ongoing and now-global debate about our origins.

The 'indigenist model' was the name given the interpretations of Zilhão and d'Errico by Francis Harrold in an essay co-authored with Marcel Otte (Harrold & Otte 2001). Zilhão and d'Errico adopt without question the conventional, normative, typological analytical units defined by several generations of European prehistorians, which they believe detect pattern at the level of undefined 'technocomplexes'. They also accept the credibility of 'neandertals' and 'moderns' as valid taxonomic units. The basic tenets of the indigenist position are (1) that neandertals making Châtelperronian industries underwent a separate, independent, and earlier cultural transition from the Middle to the Upper Paleolithic (at c. 38 kyr B.P.), (2) that contemporaneity between the Aurignacian (dated to < 36.5 kya B.P.) and the Châtelperronian (c. 38 kyr B.P.), cannot be established; (3) that a careful review of the radiocarbon evidence discounts the possibility of Aurignacian-Châtelperronian contemporaneity (thus, interaction), (4) that there are no convincing cases of interstratification between the two, and that wherever they are found together in a single site in unambiguous sequences, the former overlies the latter; (5) that the authorship of these industries is known with certainty, can be generalized, and is 'hominid specific' (i.e., neandertals uniquely manufactured Châtelperronian and other EUP 'transitional' industries, moderns were uniquely responsible for the Aurignacian), (6) that neandertals persisted for a long time in southern European refugia, as indicated by St. Césaire (c. 33 kyr B.P.) and Vindija (c. 28 kyr B.P.); (7) that they persisted very late in Iberia south of the Ebro-Douro 'frontier' (e.g., Zafarraya at c. 28 kyr B.P.)—for unknown reasons and despite their 'precocity,' moderns were apparently unable to cross these particular drainages; (8) that the morphology of the Lagar Velho child, dated at c. 24.5 kyr B.P., demonstrates long-term hybridization between neandertals and moderns (the latter originating in Africa c. 130 kyr ago), (9) that neandertals and moderns differed from one another only at the level of subspecies (if they differed at the species level, hybridization would have been impossible, regardless of the duration of contact), (10) that neandertals were eventually replaced by moderns, probably through a relatively rapid process of genetic 'swamping,' and (11) that neandertals were fully modern (or at least 'fully Upper Paleolithic') cognitively and behaviorally, engaging in the same behaviors and producing the same kinds and ranges of artifacts as those associated with EUP moderns. The many contradictory cases are essentially 'explained away' by arguing that the relevant data were collected from archaeological palimpsests that confused cultural attribution, that the analytical units themselves (esp. the Aurignacian, and EUP equivalents of the Châtelperronian—Jerzmanovician, Bohunician, Szeletian, Uluzzian, etc.) were ignored or defined inappropriately, and that there was contamination of (conflicting) radiocarbon determinations (or that the dates were derived from confused stratigraphic contexts).

William Davies

A student of Mellars, William Davies, has recently proposed a two-phase dispersal model for the origins and spread of the Aurignacian in Europe (1999, 2001). His approach focuses on an assessment of patterning across the whole of the spatial and temporal spread of the Aurignacian (c. 1.5 million km^2, 45–27 kya) and what it might tell us about human behavior, particularly in regard to mobile, dispersing forager populations. Like his mentor, Davies equates the Aurignacian with a suite of behaviors connected to a modern human dispersal from the East. Noting the relative scarcity of Aurignacian diagnostics in the early phases of the industry, he divides the Aurignacian into 'Pioneer' (45–37 kya) and 'Developed' (37–27 kya) phases, each characterized by a set of expectations predicated on the reduced material inventory early colonists might have been expected to carry with them. The Pioneer Phase is marked by (1) relatively simple, small assemblages (< 700 pieces) and (2) little typological diversity; (3) it underlies more complex Aurignacian assemblages in any given region or sequence, and is often assigned to the Aurignacian 0 or I facies as defined in France; (4) sparse, scattered geographical distribution along coasts or rivers, (5) low population densities, with (presumably) minimal impact on neandertal groups, (6) extensive use of local raw material sources and a predominance of 'expedient' types, (7) little or no evidence of raw material or artifact caching; and (8) little or no evidence for symbolic activity.

The Developed Phase is defined by (1) more complex, larger assemblages, with (2) more typological diversity; (3) stratified above Pioneer Phase assemblages (when present), often equated with Aurignacian I–IV facies; (4) denser site distributions, more frequently found along ecotones, (5) higher population densities with decreased mobility, impacting local neandertal adaptations; (6) continued dependence on local raw materials, but with slightly greater use of exotics; (7) more evidence for caching behavior (a consequent of greater residential stability), and (8) frequent evidence of symbolic activity. Davies' construal of pattern in the Aurignacian is predicated almost entirely upon a variety-minimizing, typological, essentialist conceptual framework, and the assertion that it was manufactured only by modern humans, despite the fact that virtually no skeletal remains are associated with its earliest phases.

Luís Raposo & João Luís Cardoso

Two other scenarios that also involve population dispersals should be mentioned here. Like those of Mellars and Davies, these scenarios have in common the notion that the Aurignacian is an allochthonous industry that originated or developed outside of Europe and subsequently was carried there by modern humans. Heavily dependent upon perceptions of abrupt change supported to some extent by radiocarbon dates, Raposo and Cardoso (1998, Raposo 2000) divide the transition into three phases. In the earliest phase (c. 42–40 kya), Aurignacian industries originate in assemblages like that of Bacho Kiro lev. 11 (in Bulgaria), occupy a narrow 'wave front' along the Danube Valley (Istállöskö lev. 9, Willendorf II, Geißenklösterle IIIa) and the southern Alps (Fumane), and extend as far west as Catalunya (l'Arbreda) and Cantabria (Castillo lev. 18c). According to these authors, there were only foliate point industries outside these areas, derived from the local Mousterian substrate prior to the arrival of the Aurignacian. The second phase (35–30 kya) is characterized by a multitude of transitional industries (Lincombian, Altmühlian, Jerzmanovician, Szeletian, Châtelperronian, Uluzzian) on the margins of the Aurignacian 'front,' supposedly representing acculturation of the indigenous Neandertals and producing 'Moustero-Aurignacian' industries with mixed Mousterian and Aurignacian characteristics. The third phase (c. 30 kya) sees the transformation of these transitional industries into the Gravettian complex (including the late Aurignacian) throughout most of western Europe, except for the persistence of a late Mousterian in Iberia south of the Ebro-Douro line. The presence of an early Aurignacian in the Danube Valley has been dubbed the 'Danube Corridor Hypothesis' by Conard (2002, Conard & Bolus 2003), who envisioned a rapid replacement of indigenous Neandertals, citing the AMH remains (Stetten 1–5) thought to be associated with an art-rich, early Aurignacian industry from Vogelherd Cave as the best evidence in support of this hypothesis. Stetten 1–4 have recently been dated directly and are now known to be late Neolithic, with dates ranging from 3.6–5.2 kya (Conard et al. 2004).

Jean-Pierre Bocquet-Appel & Pierre-Yves Demars

The second dispersal scenario is also based on pattern searches using a statistical analysis of 468 radiocarbon dates and their geographical co-ordinates (Bocquet-Appel & Demars 2000). The intent is to nail down the latest Mousterian industries (= neandertals) and the earliest Aurignacian ones (= modern humans) found on the European subcontinent, and to identify times and areas of coexistence. The Aurignacian is thought to have evolved in the Middle East, although the authors acknowledge that the flake-dominated Levantine Aurignacian is considerably later than the highly laminar Ahmarian there (pg. 544; see Coinman [2003] for a current overview of the Levantine Upper Paleolithic). Dated archaeological sites and levels are used as a proxy variable for the two hominin taxa, and the credibility of doing this is not questioned. Six phases are identified.

Before 40 kya, only neandertals are present within the limits of the study area (c. 10° W lat.– c. 26° E lat.). Between c. 40 and 37.5 kya, a process of neandertal range contraction begins, with neandertals abandoning Britain, an East European band c. 5–600 km in width, northern Italy and northwestern coastal Spain, while AMHs are present only south of the Carpathian Mountains and at one point north of the Danube, after entering Europe via the North European plain. From 37.5–35 kya, neandertal contraction accelerates over large areas of western Europe. In Iberia, they become confined to the Atlantic coasts of the Peninsula while AMH occupies much of east Europe (from 13–24° E, 51–41° N, 3–5° W). By c. 32.5 kya, neandertals are found only in an isolated strip corresponding to the southern half of France from the Atlantic to the western Pyrenees, and in the western third of the Iberian Peninsula, along the Atlantic façade. These two refugia themselves contract over the next five millennia, and disappear altogether by c. 27.5 kya. Geographical areas of coexistence include (1) two very small regions corresponding to the northernmost and southernmost AMH 'entry points' at 37.5–35 kya, (2) southern France between 35–32.5 kya, and (3) southwest Iberia between 32.5–30 kya. The latter two regions persist for 7500 and 2500 years respectively, identifying neandertal refugia from which they were slowly and inexorably displaced, probably by a process of competitive exclusion.

Problem Areas in Paleolithic Research

Over the past 15 years, the literature on the Middle–Upper Paleolithic archaeological transition has grown to immense proportions (let alone that of the now-global topic of our biological and cultural origins). It is evident from even a cursory examination of that literature that there is an enormous diversity of informed opinion about the nature of the archaeological transition, riddled with preconceptions and assumptions about (1) who made what, how they made it, when they made it and where; (2) how different perceptions of chronology influence the interpretation of pattern at the level of analytical units; (3) how modernity might be defined (morphologically, cognitively, behaviorally); (4) what processes were involved in transition mechanisms like acculturation, replacement, displacement, hybridization and genetic swamping, and (5) what the impetus for the modern human exodus from Africa might have been (assuming it occurred at all). All these partly contrastive, partly consilient views are 'fuzzy sets'; they turn on vague notions implicit in the conceptual frameworks adopted by those involved in MHO research. Despite nominal acknowledgment of the power and generality of Darwinian evolution, no one can claim to control all of its aspects. So we tend to become consumers of one another's research conclusions, inevitably affected by assumptions about which particular construals of pattern in, for example, human paleontology appear most credible to us as archaeologists, which particular interpretations of pattern in the genetic evidence appear best supported to us as biological anthropologists, etc. While it is clear that any generally accepted explanation of our origins must reconcile pattern in the archaeology, human paleontology and molecular biology, no one has so far been able to do this very successfully and there is no easy solution to the problem. There is a commonly expressed hope for the appearance of pattern so robust that it will unequivocally support a particular hypothesis, but it is a vain hope in the absence of any consensus about criteria for the definition of modernity—this despite the powerful, overarching neo-Darwinian (NDT) paradigm. Although we clearly need 'more data', acquiring data alone will not resolve MHO questions like the nature of the archaeological transition because data are not 'neutral' or intuitively obvious in terms of the meanings we assign to them. They only acquire meaning in the context of a particular conceptual framework (be it archaeological, paleontological or genetic), and many alternative meanings are possible under the 'big tent' of Darwinian evolution (Clark 2002).

Four problem areas appear to afflict transition archaeology (and paleolithic archaeology in general): (1) essentialism, (2) reductionism (more accurately, inappropriate reductionism), (3) post-hoc accommodation, and (4) the absence of an evolutionary ecology perspective of sufficient scope and generality to deal with process questions related to adaptation (Clark 1992).

Essentialism

In archaeology, essentialism is most often linked to typology because typological systematics play such an important role in the definition of Stone Age analytical units (so important, in fact, that typology can sometimes 'trump' patterns defined on the basis of more objective criteria like radiocarbon dates). Essentialism is a philosophical standpoint that originated in classical antiquity based on the concept of essence, and subscribing to the idea that metaphysical essences really exist (or subsist) in nature and are intuitively accessible, resident in the mind. Essentialism is often juxtaposed with realism, the philosophical doctrine that universals exist outside the human mind.[1] This juxtaposition goes back to Plato (d. 347 BC) and the idea that ultimate reality consists of transcendent, eternal universals which are the true objects of knowledge, that knowledge consists of the 'reminiscence' of these universals under the stimulus of sense perception, and that objects of sense perception are not completely real, but participate to varying degrees in the reality of ideas. His student, Aristotle (d. 322 BC), modified Platonic idealism by arguing that *all reality is of particular things,* each of which is the union of observable, tangible matter and an essentialist, intangible form that is characteristic of its kind (the archetype). Essentialist thinking permeates typological systematics and the archaeological 'index fossil' tool types *(fossiles directeurs)* derived from it through the Aristotelian notion that every natural group of objects has an archetype—an invariant, generalized or idealized form shared by all members of the group. Whether artifacts are 'natural groups' in the same sense that organisms are is, of course, arguable.

The history of classification in paleolithic archaeology can account for the importance of essentialism in its research protocols. Paleolithic archaeology originated in, and developed in parallel with, the kind of natural history (esp. biology, geology, paleontology) that dominated much of 19th century European intellectual life. Until Darwin (or, more accurately, until the rediscovery of Mendel's laws in 1900), classification in the life sciences consisted of the systematic arrangement of organisms into groups or categories according to established criteria. Linnaean species were held to be the immutable products of divine creation, and the process of classification simply involved the assignment of the proper species identification to each individual organism. A type specimen (= archetype) was used to define the species and served as the unique standard of comparison for identifying and categorizing other specimens. With the realization, in the first half of the 20th century, that populations, rather than individuals, are the units of classification, the concept of variation somehow had to be accommodated in biological systematics. After the1930s, classification became a descriptive preliminary to life scientists, who began to look for explanations in genetics, ecology and development, using principles derived ultimately from the work of Darwin, Wallace and Mendel. Unfortunately, for reasons discussed elsewhere (Clark 1993), *European paleolithic archaeology never made this crucial conceptual transition.* Archaeological sequences in 'key' caves and rockshelters were, and in many cases continue to be, seen through a typological filter or lens as analogous to geological type sections with time-sensitive index fossils and sequences transferred more or less directly from the earth and life sciences to the study of human culture history (Clark 1993: 220, 221). In sum, the (1) absence of an overarching conceptual framework below the level of the metaphysic that is specific to 'paleoarchaeology' (Clark 2003b:51, 52), (2) the European tendency to view prehistory as 'history-like,' replete with processes and analytical units analogous to those of the tribes, nations and peoples of history, and (3) the near-total absence in European university curricula of what might be called an explicit concern with the logic of inference—the foundation for archaeological theory building—are the major contributing factors. These issues are more extensively discussed elsewhere (Clark 1993, 1994, 1997, 1999a, 2002).

Reductionism

Linked to essentialism is a rather naïve kind of reductionism that also serves to normalize or minimize variation in the perception of pattern in paleolithic archaeology. Reductionism is a complex subject, treated at length by the British cognitive biologist Steven Rose, from whose book, *Lifelines—Biology Beyond Determinism* (1998), much of what follows is abstracted (see also Clark 1999b). In a reaction to the brand of genetic determinism most often associated with workers like Richard Dawkins, Edward Wilson and Daniel Dennett, Rose argues (1) that we can only understand complexity at different scales (genes, cells, organisms, populations) by looking at their current locations in an historical context, (2) that, with our emphasis on populations, we need to restore the organism to its central place in the conceptual framework of evolution, and assign to it an active role in its own fate; and (3) that a new paradigm is required that allows for the capacity of organisms to organize and direct their own life histories (1998:1–20).

Like essentialism, reductionism has a long history. It is based on a coherent philosophical standpoint that sees modern science as materialist, and the inheritor of 19th century mechanical materialism (itself tightly linked ideologically to a particular phase in the development of industrial capitalism). Criticisms of reductionism come from many quarters. New Age philosophers argue, for example, that the human experience is uniquely multivalent and richly textured, and that reductionism drains the life out of that richness and texture. Feminist philosophers of science contend that reductionism typifies the masculine, cognitive, objectifying approach to the world taken by modern science, and that it fails to respect the validity of personal, subjective experience. Some ecologists criticize reductionism because it appears to deny the interconnectedness of the living world. Reductionism is not a unified concept, however, and there are many construals of it, depending upon context and standpoint. Almost invariably, these different notions are predicated on perceptions of where a particular scientific discipline falls in a hierarchy or pyramid of the sciences.

In the pyramidal model, physics is regarded as the most fundamental of the sciences because it is thought to deal with the most general principles by which nature is organized. It provides explanations of natural phenomena and predicts outcomes. Moreover, the laws of physics apply to biology, and to everything else, whereas the converse is not true (e.g., biological laws and principles do not apply to non-living systems). Physics deals with linear causal relationships,[2] is experimental in terms of its research protocols, and is concerned with a few, simple relationships amongst a very few basic analytical units. Physics is thus a 'hard' science, whose principles can be expressed mathematically. By contrast, the social sciences are 'soft' because they are concerned with complex multivariate relationships at many different levels, because these relationships are historically contingent and often vectored (or directional) in time, because they are least capable of precise mathematical expression, and because they cannot predict outcomes (Clark 1982, 1987). Orthodox philosophy is still mainly a philosophy of physics, however, premised on the view that the task of science is ultimately to collapse biology into chemistry and chemistry into physics. As James Watson (of double helix fame) put it, "There is only one science, physics; everything else is social work" (1994). But most social, and practically all life scientists will not concede such reductionism (see Mayr [1976:221–478; 1991, 1997] for a history and philosophy of biology). A hybrid natural and social science, archaeology is among them.

Kinds of Reductionism

Rose divides reductionism into methodological, theoretical, and philosophical categories (1998:73–97). *Methodological reductionism* is fundamental to all science (one might even say it is a practical cognitive necessity for all sentient organisms in order to cope with the bombardment of perception).[3] The experiential world is characterized by complexity, flux and by a multitude of interacting processes. We find it easier to understand the phenomena we wish to study if we can isolate them from the rest of the world and alter potential variables one at a time. Reductionist methodology simplifies, and enables us to generate seemingly linear chains of cause and effect. Partly because of the way western science has developed, to 'capture' a phenomenon mathematically is to control it by logic and symbol, so it is no surprise that reductionist methodology has been so powerful and attractive in hard science research over the past 300 years. It works. It has given us unrivalled insights into the mechanics of the universe, at least for relatively simple systems. But living systems are not simple—they involve many interacting variables. Parameters are not fixed; properties are non-linear. And the living world is highly non-uniform and historically contingent, fundamentally different ontologically from the world of the physicist. As I pointed out some years ago, experimental disciplines can afford to rely on reductionist methodologies essentially because they are nearly perfectly congruent with their research protocols, but observational or hybrid fields like biology and archaeology lack consensus about research protocols, and have to be very careful about how they reduce complexity in order to understand its components (Clark 1982).

Theory reduction is the second of Rose's categories. The goal of theory reduction (sometimes called parsimony) is to arrive at a maximally satisfying description of the world, or some part thereof, with a minimum number of laws, axioms and variables. Theory reduction can only take place *within* metaphysical paradigms, since the logic of inference between or among them can be very different (Clark 1993). Put another way, parsimonious theory comprises the body of concepts that goes along with the practice of strategic simplification (Cowgill, pers. comm.). The quintessential example is the quest of physics to come up with a theory that will encompass all the forces in the universe—strong and weak interactions among subatomic particles, electromagnetic radiation, and gravity—the so-called Grand Unification Theory, or GUT. It is widely assumed that the GUT exists, that there is some (still hypothetical) basic force or field of interaction of which the known forces are but aspects. While perhaps attainable in highly experimental disciplines like physics that rely heavily upon grounded theory (i.e., theory already well tested and confirmed), theory reduction is problematic in biology and archeology, essentially because of observational research protocols, time's arrow (i.e., historical contingency), the underdetermination of theory, the theory-laden nature of facts, and the complexity of the processes we seek to understand (Clark 1982, 1987).

Although it lacks the universal, time-reversible, laws of physics, modern evolutionary biology is founded on a relatively small number of concepts and principles that constitute a powerful metaphysical paradigm with vast explanatory power (e.g., Mayr 1997). In respect of theory reduction in the life sciences, Mayr (1982:67–71) makes a distinction between proximate cause, concerned with the function of an organism and its parts, and its development, and ultimate cause, which attempts to explain why an organism is the way it is as a product of its evolutionary history. He contends that no biological phenomenon is ever fully explained until both proximate and ultimate causation are understood.

Whereas methodological reductionism is a fundamental part of all science, and theory reduction desirable and attainable to some extent in the life and social sciences, *philosophical reductionism* is deeply problematic in all science. In the full-blown philosophical reductionist view of the hierarchy of the sciences, science is unitary and one approaches the hierarchy 'from the bottom up'. Because physics is the most fundamental of the sciences, philosophical reductionism would reduce chemistry to a special case of physics, biochemistry to chemistry, sociology to physiology, and so on, with the social sciences the most epiphenomenal of all.[4] However, the notion of 'level' in the hierarchy is very ambiguous, and some would say that the various sciences identified therein are an accident of history, a consequence of how universities are organized in the West over the millennium or so of their existence. Nor is it clear whether the levels in the hierarchy are ontological (concerned with the nature of reality, each level corresponding to a distinct organization of matter) or epistemological (concerned with how we go about understanding reality).[5]

Carried to an extreme, philosophical reductionism would argue that the levels in the hierarchy are only an accident of the history of how we go about studying the world—that is, they are entirely epistemological. Most of us, however, would stop short of that, insisting that there is an ontological aspect to the levels, that they really do 'cleave nature at its joints', and that they represent successively more complex relationships that cannot be discerned simply by taking the system apart. As Rose puts it, the key feature distinguishing a lower level in the hierarchy from those above it is that at each level new interactions and relationships appear among the component parts (pg. 93). Philosophical reductionism also implies that, whatever higher-order properties emerge and however they do so, they are always somehow explicable in terms of, and secondary to, lower order ones. Yet, whatever the case may be for physical or chemical systems, the nature of evolutionary and developmental processes in the life and social sciences means that there is no such necessary primacy. Wholes, emerging, may in themselves constrain or demand the appearance of parts (see also Koestler & Smythies 1969).

Reductionism in Paleolithic Archaeology

How is reductionism manifest in transition archaeology? Although conceptually quite distinct, reductionism and essentialism are inextricably entwined in typological (and sometimes technological) systematics because typology is privileged in much paleolithic research, especially on the Continent. The distinctions Rose makes are not evident in paleolithic archaeology, let alone transition research. When there is no overarching conceptual framework, when theory building is largely implicit, and when pattern in the remote past is treated as analogous to, and explicable by, pattern in recent history, it doesn't make much sense to talk about philosophical or theory reduction. Appropriately, most reductionism takes place at the methodological level, as different workers relentlessly 'pattern search', emphasizing different suites of variables differentially. This is evident in the use of trait lists for behavioral (e.g., Mellars 2004) and morphological modernity (e.g., Stringer 1984), but is problematic in paleoanthropology because there is no consensus about just what 'modernity' is (either in the past or the present). A compelling archaeological example, albeit one in which essentialism, reductionism and post-hoc accommodation (Binford 1981) are all hopelessly entangled with one another, is Zilhão and d'Errico's 'indigenist model'.

In a long paper published in the *Journal of World Prehistory* (1999), Zilhão and d'Errico dismiss anything (1) that might suggest that the Aurignacian and the Châtelperronian are not discrete, well-defined entities tightly bounded in space and time, (2) that they are not 'the same thing' everywhere they are found, (3) that they are not exclusively associated with particular kinds of hominids (moderns and neandertals, respectively); (4) they dismiss any dates that do not support their hypothesis, and take utterly for granted an influx of moderns—the bearers of the Aurignacian—'from the East' after c. 37 kyr B.P. (cf. Clark & Lindly 1991). As a consequence of these mental gymnastics, Zilhão and d'Errico end up defining the 'real' Aurignacian strictly on the basis of the presence of Dufour bladelets (!) (pg. 56), dismissing all the other conventionally accepted indicators of 'Aurignacianness' in order to accommodate the Levantine Aurignacian, a flake industry that bears little resemblance to the classic French Aurignacian (Marks 1983, 1994).

Selective reduction of complexity is particularly evident in Zilhão and d'Errico's treatment of the north Spanish evidence. They dismiss the Castillo Lev. 18 dates at c. 40 kyr B.P., and similar dates from l'Arbreda and Abric Romaní, as derived from mixed deposits—post-hoc accommodations that stand in sharp contrast to the well-argued interpretations of the excavators (e.g., Cabrera & Bischoff 1989, Bischoff et al. 1994, Cabrera et al. 1997). Zilhão and d'Errico also accuse Mellars (1999) of circular reasoning based on uncritical acceptance of radiometric dates and claim he doesn't pay sufficient attention to site formation processes (particularly sedimentation rates which, they suggest, are more reliable than radiocarbon chronologies!)—all this to dismiss the possibility that an idiosyncratically defined Aurignacian could be earlier than a conventionally defined Châtelperronian. They take Straus and Otte to task for the recent excavations at Trou Magrite (Otte & Straus 1995), claiming that the formation processes of Stratum 3 (AMS dated to 41 kyr B.P.) are not yet well understood and that, as a result, the problem of the cultural affiliations of the mobile art from the old excavations remains unsolved, going on to characterize the attribution of that level to the Aurignacian as "unwise" (pg. 41). There are many other examples. In a *tour de force* of post-hoc accommodation, the primary 'modern era' excavators (Karavanič, Svoboda, Kozlowski, Ginter, Straus, Otte, Cabrera, etc.) are criticized for failing to take site formation process questions into account, for an uncritical acceptance of radiometric evidence, for confounding mixed with pristine deposits, and for failing to detect palimpsests, thus allowing Zilhão and d'Errico to conclude that "nowhere in Europe does the true Aurignacian...date to before c. 36,500 radiocarbon years ago" (pg. 43). Throughout, it never occurs to them to question the credibility, utility or reality of the basic analytical units themselves, nor the logic of inference underlying typological systematics, nor the notion that the analytical units represent, in some fundamental sense, the tangible remains of identity-conscious social units, nor the association—across vast reaches of space and time—between those units and the neandertals and moderns who supposedly created them, nor indeed any of their own preconceptions, biases and assumptions.

Evolutionary Ecology

In my opinion, the absence of a conceptual framework below the level of the metaphysic (NDT), yet specific to 'paleoarchaeology,' has impeded progress in arriving at a satisfactory solution to the question of our origins (in general) and the nature of the Middle–Upper Paleolithic transition (in particular). For such a conceptual framework to be viable, it must at least be consistent with the core tenets of neo-Darwinism, yet flexible enough to allow for investigation of the wide range of problems associated with transition research. There are a number of these conceptual frameworks (e.g., evolutionary ecology, behavioral ecology, evolutionary psychology, dual inheritance theory, evolutionary genetics, community ecology, animal ethology, decision theory, etc.) recently outlined, compared and contrasted with one another by Winterhalder and Smith (2000). Together they constitute a 'fuzzy set', overlapping, as they do, in terms of concepts, methods and problem domains. However, all are firmly grounded in NDT, share a focus on adaptation, are explicitly reductionist, and are anchored in basic principles of evolution by natural and sexual selection. In aggregate, they address research domains that, one might think, would be central to the concerns of a genuinely cross-disciplinary, integrative, paleoarchaeology (e.g., primate life history, demography, maturation, mating strategies, reproductive ecology, resource transfers, division of labor, etc.). What is so striking about the literature of this research tradition (in addition to its 'newness'—most of it post-dates 1985) is how extraordinarily fruitful it has been in terms of testability, predictive adequacy, internal coherence, external consistency, simplicity and unifying power—epistemic qualities that should be highly prized by any archaeology that aspires to be scientific. However, the extent to which most paleolithic archaeologists are even aware of these literatures is arguable. It's pretty clear that, given the possibility of species differences in the hominids involved, we can no longer afford to consider transition archaeology a kind of culture history. Over the past decade, there has been recognition of the largely unrealized potential of NDT to serve as a conceptual framework for all kinds of archaeology, from that of early hominids, to that of the very recent time frames with which New World workers are accustomed to deal. These efforts, embodied now in at least a dozen books published since 1995, seek to demonstrate at the levels of 'high' and 'middle range' theory, and at the level of case studies and applications, the shared conviction that NDT and its modern derivatives constitute the only generally acknowledged framework within which to understand human biological and cultural evolution, 'writ large' or 'small' (e.g., Teltser 1995, Maschner 1996, O'Brien 1996, Steele & Shennan 1996, Barton & Clark 1997). More precisely, evolutionary archeologists argue that natural and sexual selection (and other NDT mechanisms and processes) act on human behavior and more or less directly influence the material products of that behavior. By focusing on the requirements of NDT at the theoretical and methodological levels, the approach goes some considerable distance toward creating a novel, coherent framework for explaining all kinds of variation in the archeological record.

The Theory of Natural Selection

The cornerstone of NDT is the theory of natural selection, defined simply as changing adaptation to strictly local environments. Darwin's concept of natural selection (which he got from the Scottish economist Adam Smith) maintained that the order apparent in the natural world (the good design of organisms, the harmony of ecosystems) arose only as an accidental consequence of a universal striving for individual reproductive success (i.e., there was no principle or intelligence explicitly producing it). In the mid-19th century, natural selection was a philosophically radical idea—so radical, in fact, that Darwin himself resisted exploring its implications, although he was clearly aware of them (Gould 1994). It is a materialist theory that runs counter to the notion of dualism—the idea from classical antiquity (Plato, Aristotle) remodelled and rearticulated in the context of a philosophy of scientific inquiry by the Enlightenment philosopher and mathematician René Descartes (1596–1650), that the universe was divided into the material stuff of everyday existence (matter) and intangible (mental or spiritual) stuff (of which the spirit, since it was more closely related to God, was the higher of the two forms). A philosophically radical idea when it was first proposed, natural selection remains a philosophically radical idea today, since it implies that human existence has no intrinsic purpose or meaning (except insofar as individual humans create meaning in their own lives) and that evolution in general, and human evolution in particular, is directionless (relative to orthogenetic or teleological notions of progress implicit in some forms of sociocultural evolution) (Clark 1997, 2003).

The Extended Phenotype

Basic to the case for adopting NDT as the metaphysical paradigm for paleoarchaeology is the notion of the generalized or extended phenotype—the idea that the material remains of human behavior are as much aspects of the human phenotype as are the observable features of human biology (Dawkins 1990, O'Brien & Holland 1995). Quantitatively, the archeological record is immensely richer than that of fossil hominids, and in its later manifestations, effectively holds biological macroevolution constant, so it could be argued that it constitutes a better basis for building hypotheses about human evolution than does the exceptionally 'coarse-grained' time-space grid of the human fossil record (even though the latter is relatively uncomplicated by 'culture'). The notion of the extended phenotype is a decidedly unradical idea when it is noted that most evolutionary biologists routinely include things like spiderwebs and birds' nests in their conceptualizations of the phenotype (e.g., Dawkins 1990). Most of those involved in transition research will acknowledge that, regardless of the hominid involved, there is a large component of learned behavior that acts to generate phenotypic variety, manifest in the material remains neandertals and modern humans once manufactured, modified, lost and discarded. Evolutionary archaeologists would maintain that selection operates on the behaviors that produced this mountain of clutter, and culture—construed here as learned behavior—simply constitutes part of the human phenotype, just as it is part of the phenotype of chimpanzees and bonobos (e.g., Vogel 1999, Whiten et al. 2005).

An Adaptationist Perspective

Consistent with NDT, an evolutionary ecology perspective also entails adoption of an adaptationist view of human social behavior, conceptualized, after Binford (1962), as systemic in nature. There are many definitions of adaptation (Mithen 1993). I define it as evolutionary biology does: any structure, physiological process, or behavioral pattern that makes an organism more fit to survive and reproduce (Wilson 1975:577). Behavior can be viewed as "the dynamics of adaptation"—a strategy for survival and reproduction (Binford 1972:133). An important goal for archaeologists involved in MHO research is to develop a perspective for the study of the Paleolithic that emphasizes changing adaptive systems. While scarcely a novel idea from an Americanist point of view (see, e.g., Binford 1965, Flannery 1973), most Old World archaeological research traditions either (1) overemphasize the characteristics of retouched stone tools, as if these were somehow meaningful in their own right, and/or (2) treat subsistence, paleoenvironmental and site distributional information as if these were data categories independent of the lithics (Clark 2005). Integrated systems perspectives are not particularly common outside the Anglophone research traditions, and the tendency to compartmentalize aspects of the research gets in the way of a more unified approach.

For all hunter-gatherers, adaptation is a regional or local problem. It can be defined biologically (in terms of inclusive fitness) or, in the present context, culturally, by identifying particular behavioral solutions from a range of possible solutions that would allow human groups to persist over time. Put another way, human adaptation is "the possession of a valid set of solutions to a variety of problems" (Jochim 1981:19). These solutions are developed in different contexts to meet various objectives, only one of which need be reproductive success. Studies cast in a broadly ecological, systems framework seek to understand the adaptive significance of different kinds of human behavior without making the assumption that all such behavior is necessarily adaptive (i.e., some behaviors might be adaptively neutral, maladaptive, etc. over the long run). More important, adaptation has specific empirical referents that can be monitored using archaeological data and that can potentially inform us about the nature of change or process (i.e., whether change is directional, continuous, or not; whether change is occurring at similar or different rates; whether patterns of change are correlated with one another across different variables). Analyses guided by an adaptationist perspective should be able to identify both correlated sets of variables that are changing in tandem with one another, and isolate those that static or are exhibiting a different pattern of change. Paleoclimatic fluctuations are controlled by the palynological, sedimentological, and geomorphological studies that are so fundamental to the European natural science research traditions. Time, however, is a reference variable against which to measure change attributed to other causes. Whenever possible, time is controlled by absolute dating methods and, in default of samples suited to such techniques, by dated paleoclimatic information—*never* by the supposedly time-sensitive characteristics of the retouched tool components themselves.

The adaptationist program, then, demands a regional perspective and a multivariate approach to the assessment of systemic change. What this means in terms of transition research is that we must examine (1) lithic typology and technology on both sides of the transition, (2) raw material variation and how it affects lithic reduction strategies under a variety of mobility models, (3) any evidence for organic technologies, (4) taphonomic aspects of the archaeofaunal record, (4) subsistence, (5) site numbers, locations and use, and (6) settlement patterns in relation to paleotopography and resource distributions. As much of this pattern searching as possible should be quantified, in order to avoid the essentialism inherent in typological approaches.

Discussion

Recognizing that no one has so far undertaken a full-blown evolutionary ecology analysis of European archaeological transition data, it is nevertheless instructive to see what kinds of overall patterns appear to hold for the European subcontinent over the transition interval. According to many of the scenarios described above, evidence for biological and behavioral modernity should coincide over the transition interval (50–30 kya) as the appearance of modern human migrants of African origin is often invoked to account for the appearance of Aurignacian archaeological industries. The reader is reminded that the distinction between the Middle and the Upper Paleolithic is itself an 'accident of history', just as the nomina for the various Upper Paleolithic and transitional analytical units are (see Sackett [1981, 1991] for synopses of the historical development of paleolithic systematics in France).

Lithic Technology and Typology

The eminent paleoanthropologist F. Clark Howell (1999) recently summarized the ever-increasing number of spatially and temporally distinguishable variants (facies) of the Eurasian Middle Paleolithic. Recognized primarily on technological and typological grounds, the 20 Mousterian facies represent a quantum increase in qualitative and quantitative variation over the half-dozen or so variants recognized as recently as the late 1980s. Ignoring the inevitable problem of sampling error, the facies appear to vary amongst themselves according to diverse aspects of raw material (availability, package size, quality), modal production sequences *(chaînes opératoires),* the extent of reduction and utilization of particular artifact categories (esp. sidescrapers), functional constraints related to forager behavior (esp. mobility); and the nature, size, duration, integrity and intensity of use or occupation of the site context. Taken together, they document a *complex mosaic of adaptations* that, in aggregate, persists for c. 250,000 years (c. 280 – <30 kya), overlapping extensively with both the Lower and Upper Paleolithic over the entire geographical extent of western Eurasia. Even without the 12–15 'transitional' industries now recognized in the same area, it appears likely that Mousterian formal variation, site characteristics, and faunal inventories rival (perhaps even exceed) those of the early Upper Paleolithic. Below we discuss eight aspects of human forager adaptations (technology, typology, raw material transfers, subsistence, settlement patterns, site structure and organization, organic technologies and art) where alleged Upper Paleolithic 'diagnostics' or patterns show up in sites generally regarded as Middle, and even Lower, Paleolithic (see also Clark & Riel-Salvatore 2006).

Technology

Until the early 1990s, the European Mousterian was thought to be confined to the Upper Pleistocene (i.e., < 130 ky old), but is now known to extend as far back in time as c. 300 kya (as, e.g., in TD 11, 10 at the Gran Dolina site at Atapuerca (Bermúdez de Castro et al. 1999:203). The Mousterian is sometimes identified historically with levallois reduction strategies of various kinds (classic, lineal, recurrent with subdivisions, etc.) but the existence of many alternative technologies was recognized from the very beginning of Middle Paleolithic research, employed either alongside levallois methods, or to their exclusion. Perhaps the best known example of the latter is the so-called 'Quina subfacies' of the Charentian Mousterian, which appears to lack levallois technique altogether. Blade technologies have also long been known in the eastern Mediterranean (e.g., in the Lower Paleolithic 'Pre-Aurignacian' at the Haua Fteah [Libya], which contains blades, burins, endscrapers, and backed knives resembling those found in Upper Paleolithic industries like the Aurignacian). Other early paleolithic examples are the Yabrudian (Syria), the Amudian (= the Pre-Aurignacian), the Acheuleo-Yabrudian or Mugharan (all in the central and southern Levant), and the Tabūn D-type Mousterian (Israel, Jordan). Regionally, these 'bladey' industries occur interstratified with those rich in flake sidescrapers, and with those containing small handaxes (by virtue of which they are considered Acheulean—there is no MTA in the Levant). More recently, blade-dominated early paleolithic assemblages have been reported in at least nine sites in France, Germany and Belgium (Seclin, Riencourt lès-Bapaume, Rheindahlen, Tönchesberg, St. Germain-des-Vaux, Vallée de la Vanne, St. Valery-sur-Somme, Coquelles) and at Crayford, in southeastern England (Amerloot-van der Heijden 1993). Both levallois and non-levallois methods are recognized by Boëda (e.g., 1988), with some examples from Seclin (Pas de Calais), TL dated to c. 90 kya, constituting prismatic blade cores with detachments extending around most or all of the platform, and standardized blades indistinguishable from those of the Upper Paleolithic (Révillion 1989). Blade technologies with a distinctively 'Upper Paleolithic' cast, and associated with the crested blades often found in UP contexts, are also documented at St. Valery (Heinzelin & Haesaerts 1983) and at Riencourt (Tuffreau 1992).

Middle–Upper Paleolithic transitional industries are generally assumed to have been manufactured by neandertals but, despite the alleged cognitive deficiencies of their makers, transitional sites appear to exhibit much the same kinds of pattern variation evident in the MP and the EUP. The best known transitional industry, the Châtelperronian, is probably contemporaneous with the Aurignacian at c. 35–32 kya (Mellars 1998, 2001; *pace* Zilhão & d'Errico 1999) and, at Grotte du Renne (Arcy-sur-Cure, France), is blade-based, contains UP tools in some frequency, and is associated with neandertal skeletal remains and good evidence for ornaments and organic technologies. St. Césaire (France), best known for its 'classic' neandertal skeleton, also contains a good Châtelperronian industry with UP 'diagnostics', many of which are made on lamellar blanks (Leveque et al. 1993). Another transitional industry, the Bohunician (Moravia), contains UP endscrapers and burins made on blade blanks produced by both levallois (MP) and crested-blade (UP) techniques (Svoboda et al. 1996). Level 11 at Bacho Kiro cave (Bulgaria) is often considered an 'Aurignacian-like' industry (the 'Bachokirian'), as it has many prismatic blades and endscrapers, along with a few perforated teeth, yet is apparently contemporaneous with the Châtelperronian to the west (Hedges et al. 1994). No hominid remains were recovered from Level 11, nor from the only other Bachokirian site known, Temnata cave (also in Bulgaria), underscoring the paucity of human remains associated with these transitional assemblages. A third transitional industry, the Szeletian (central Europe), also lacks human fossils but exhibits strong technological similarities with the preceding Micoquian, including blattspitzen-derived bifacial foliates (leaf points), flake and blade blanks made by non-levallois techniques, and bone and ivory points and rods. In a book-length treatment of the Szeletian, Allsworth-Jones (1986) invokes this evidence for technological continuity to argue that the industry was probably made by neandertals.

Finally, there is the recent work at the Buran-kaya III rockshelter in the Crimea (Chabai et al. 2004). Level C, dated to 32–38 kya, yielded bifacial leaf-shaped points; preforms and blanks for these and other tools, endscrapers on flakes and blades, *pièces ésquillées;* sidescrapers, denticulates and notches, along with small trapezoids resembling Mesolithic pieces (and gun flints, and trillium stones...), and evidence for organic technologies, including five sawn tubes made from hare long bones and an enigmatic piece made from a horse metatarsal (also worked by sawing). There is apparently no evidence of mixing. Attributed to the 'Early Streletskayan' by the excavators, Layer C cannot be accommodated by any Middle or Upper Paleolithic analytical unit defined in the West (Chabai et al. 2004). In fact, and despite 33 radiocarbon, ESR and U-series dates, the Crimean paleolithic as a whole defies any neat characterization according to western schemata.

In a concluding chapter, Chabai, Marks and Monigal (2004: 419–460) synthesize what is known about the east-central

European Middle and early Upper Paleolithic in space and time, and place the Crimean Paleolithic in that context. It makes fascinating reading. They divide the Upper Pleistocene according to three chronological periods (125–60, 55–38 and 38–c. 26 kya). Over this interval, three Middle Paleolithic entities are represented (a Micoquian, a Levallois Mousterian, and a blade-dominated Mousterian [Chabai 2003]), along with no less than five transitional and/or early Upper Paleolithic units (the Streletskayan, Spitsynskayan, Gorodtsovskayan, Aurignacian and early Gravettian). Of the former, only the Micoquian is found throughout east-central Europe; of the latter, a rather late (< 30 kya) Aurignacian is the most widespread. In the Crimea, the Middle Paleolithic entities appear, disappear and reappear again over most of the Upper Pleistocene, with the oldest Micoquian levels at Kabazi II and Zaskalnaya V perhaps as old as 125 kya, and the youngest ones at Zaskalnaya VI, Prolom I and Burankaya III dated to c. 28 kya. The Aurignacian is represented at Siuren I, and is fully contemporary with the latest Micoquian. Human skeletal material shows up in the later (post-50 kya) Micoquian; parts of 10 different individuals were recovered, all classified as neandertals. Isolated teeth, along with bits and pieces of adults, juveniles and infants, were recovered from levels assigned to the Spitsynskayan (Kostenki 17), Gorodtsovskayan (Kostenki 12, 14, 15), Aurignacian (Siuren I) and Gravettian (Kostenki 8); all are considered modern human. What emerges from the synthesis is an extremely valuable 'first approximation' of a very wide range of (admittedly patchy) paleoenvironmental, chronological, faunal, floral and archaeological data not previously available, and a plethora of 'colonization' models, most of which cannot be evaluated yet because of a dearth of skeletal material linking the archaeology to the hominids involved.

In this brief review, it is impossible to do justice to the wide diversity and variety of primary reduction strategies exhibited by industries generally regarded as Middle Paleolithic or 'transitional'. About a half-dozen different kinds of levallois technique recur over the >250 ky allotted the European Middle Paleolithic; a similar number of non-levallois techniques that are, perhaps, more common in the Upper Paleolithic have also been extensively documented, along with the blade technologies just mentioned (see also Bar-Yosef & Kuhn 1999). The sheer complexity of some of the strategies described by Boëda (1988, 1993) and others for the Middle and even Lower Paleolithic (e.g., Maastricht-Belvédère in Holland, Grotte Vaufrey in France, Bilzingsleben in Germany) rival or exceed those inferred from reconstructions of Upper Paleolithic technologies. The recurrent levallois technology at the 250,000 year old site of Biache-Saint-Vaast (France), for example, involved at least six sequential operations to produce three kinds of primary blanks that were, in turn, used selectively to make equally specific (and evidently hafted) retouched tools (Beyries 1988). Because it implies mental processes that differ little, if at all, from our own, technological complexity like that at Biache raises serious questions about the allegedly different cognitive abilities of modern and pre-modern European hominids—at least so far as lithic technology is concerned.

Typology

A great deal has been written about Middle Paleolithic tool morphology, function and typology, much of it having to do with the implications of Dibble's sidescraper reduction model, and the extent to which it undermines the notion of shaped or standardized artifact form (e.g., Dibble & Rolland 1992, Dibble 1995). The major distinctions in Bordes' Lower and Middle Paleolithic typology (1961) are based on the position of the retouched edges in relation to the axis of detachment of the blank, and on the shape of the retouched portion itself, regardless of the overall shape of the piece. These criteria differ sharply from the shape-dependent, time-sensitive, stylistic marker types (e.g., Dufour bladelets, Solutrean points, keeled scrapers) that supposedly identify the various European Upper Paleolithic subdivisions (de Sonneville-Bordes & Perrot 1954, 1955, 1956). The retouched tool types commonly found in European Middle Paleolithic sites (sidescrapers, points, notched and denticulated pieces, backed knives, bifacial foliates [leaf points] and handaxes) vary considerably among themselves in the extent to which one could argue that they were intentionally shaped, with a pretty good consensus that notches and denticulates, at least, were expediently produced on a wide range of blanks as needed to shape wood (more generally, plant material [Anderson-Gerfaud 1990]). While acknowledging the 'overfine' shape distinctions evident in Bordes' 17 sidescraper types, and conceding the role played by formal convergence in overall morphology, Mellars (1996:95–140) amasses considerable evidence that the more common sidescraper morphs and, especially, bifacial pieces almost certainly were conceptually distinct tool forms that existed as mental templates in the minds of their makers. Paradoxically, though, he also maintains that the 'imposed form' (large scale reduction of blanks affecting not only working edges but overall shape) clearly present in the extensively shaped bifacial points and handaxes is, nevertheless, mostly absent in the Middle Paleolithic (albeit evident in the Châtelperronian and the Uluzzian, thought by many to have been made by neandertals).

In addition to the common MP retouched types, a very considerable variety of 'typical' Upper Paleolithic tools show up in most of the MP facies, including perforators, *becs, rabots* (planes), truncated pieces, endscrapers and burins morphologically indistinguishable from their UP counterparts (these often end up in Type 62—various). Although Mellars claims such pieces are rare (< 2% of the retouched tool totals), Aurignacian lev. 18 at El Castillo (Cantabria) yielded—in addition to 'classic diagnostics' like keeled and nosed endscrapers, perforators, burins, blades with scalar retouch (Aurignacian, strangled blades), split-based antler points, and other UP types (some of them present in the under-lying Mousterian)—large numbers of typically MP side-scrapers, notches and denticulates. Nor did raw material (quartzite, limestone, ophite) change much in kind or relative frequency over the transition (Cabrera et al. 2000, 2002; Straus 2005:57). The early last-glacial French site of Riencourt has produced a spectacular array of single and multiple dihedral burins made on blade blanks. Whatever the case for Europe, UP tool types are quite common in some Levantine Mousterian sites (e.g., dihedral burins account for 10–20% of the retouched pieces at Rosh Ein Mor in Israel [Crew 1976]). Like Bordes (1963), Mellars (1996:122–124) tends to dismiss these anomalous sites and levels as the accidental products of taphonomic processes (e.g., trampling, cryoturbation) and/or as unfinished pieces broken or discarded in manufacture.

Raw Material Transfers

The procurement and distribution of raw materials in paleolithic sites has been used over the past 20 years to monitor patterns of movement of human groups, possible relations amongst them, as a proxy measure for technological differences, and an indication of how lithic technologies were organized. Again, much of this work has been undertaken in France, and French scholars (esp. Jean-Michel Geneste [e.g., 1988, 1989a,b] and Alain Turq [e.g., 1988, 1989, 1992]) have played a leading role. Lithic provenience studies have been admirably summarized by Mellars (1996:141–168) who points out that, although most MP raw materials are derived from local sources (<5 km from a site), there is a very considerable presence of material derived from more distant sources (20–30 km away), and the occasional appearance in almost all MP sites of small quantities of high quality material derived from much greater distances (80–100 km). When quantities of raw materials are plotted against the distances over which they were transported, the patterns usually correspond to a roughly exponential distance decay curve, recalling the fall-off curves exhibited by later sites, up to and including those of the Neolithic (e.g., Renfrew 1969). The patterns of raw material procurement in France during the Middle Paleolithic appear to be broadly similar to those of the Upper Paleolithic, raw materials are traveling along the same, E/W trending river valleys (and up and down their N/S trending tributaries) in both periods, come from similar sized catchment areas, and exhibit the same kind of strong correlation between high quality stone and complex tools (Mellars 1996:165–168). The major differences have to do with the quantities of high-quality raw material transported over long distances (c. 1–2% in most MP sites, as high as 20–25% in a few UP sites) and in the form in which the more distant materials were transported (finished pieces are more common in the MP, prepared cores in the UP). Unambiguous quarry, extraction, or 'workshop' sites *(ateliers de taille)* were long thought to be confined to the Upper Paleolithic, although recent research has demonstrated their presence in the Middle Paleolithic of Italy, Egypt and the Levant (Barkai et al. 2002, del Lucchese et al. 2000–2001,Vermeersch et al. 1997).

Subsistence

Although the aspects of lithic technology and typology just described are, in some sense, 'primary', because many MP sites do not preserve organic remains, much the same pattern of underacknowledged variability within the Middle Paleolithic also appears to be documented for subsistence, site placement within the landscape and intra-site spatial organization (i.e., 'well-organized campsites'). Although MP foragers probably scavenged when the opportunity to do so presented itself (Stiner 1994), they could also be (1) effective hunters of prime-age adult animals (Chase 1986, 1988, 1989), (2) capable of highly selective predation on large bovids (as, e.g., at Mauran, La Borde, Coudoulous, Le Roc [see summary in Mellars 1996:217–219, 231–236]); (3) there is good evidence for game drives, with all of the complex cognitive and organizational abilities they entail (as, e.g., at La Cotte de St. Brelade [Scott 1986, 1989], at La Quina [Chase 1989]), (4) for the systematic transport of carcasses or parts thereof (as, e.g., at Mauran, Champlost [Farizy & David 1992, David & Farizy 1994]), (5) for complex butchery practices, including filleting, marrow extraction and other 'modern' kinds of animal processing (Jaubert & Brugal 1990, Farizy et al. 1994), and (6) for the intensive, seasonal exploitation of large, gregarious bovids (e.g., aurochs at La Borde [Jaubert & Brugal 1990, cf. Slott-Möller 1990], bison and aurochs at Mauran [Farizy & David 1994]), implying cooperative intercept hunting linked tightly to migration routes. Work in Latium (west-central Italy) at Grotta dei Moscerini (Stiner 1993, 1994), and at Vanguard and Gorham's caves on Gibraltar (Barton 2000) also show neandertal exploitation of limited numbers of shellfish (limpets, mussels, clams) and, at Moscerini, tortoises—the first well-documented cases of Middle Paleolithic shellfish exploitation in mid-latitude Europe.

Settlement Patterns

A similar pattern of underacknowledged variation appears to apply to MP site placement in the landscape, although the tendency to ignore or to minimize the importance of the numerous but usually untested and undated open sites introduces a source of bias, as does the tendency for European scholars to deemphasize survey research. In what is perhaps the best-known European region, southwestern France, there is an apparent (and possibly universal) dichotomy between cave and rockshelter sites, on the one hand, and open sites, on the other, in respect of site placement with regard to basic environmental and topographical features. Most of the former are located in cliff faces in valley walls offering extensive and wide-ranging views of the local landscape, almost invariably in close proximity to abundant and high quality raw materials (Mellars 1996:251–252). As Turq has pointed out (1989:196), they tend to overlook, or to be located on or adjacent to ecotones, with a diversity of habitats that could be easily and efficiently exploited from those locations. Although this view has been contested by Stiner (1991), it has led to the notion that many MP caves and rockshelters were, in some sense, 'central places' from which diverse economic, social and technological activities were carried out (see also Duchadeau-Kervazo 1984).

A striking feature of Middle Paleolithic open sites is their sheer abundance, greatly exceeding their Upper Paleolithic counterparts (e.g., Geneste 1985; Duchadeau-Kervazo 1982, 1986). Whereas the locations of cave and rockshelter sites are tightly constrained by the calcareous bedrock formations in which they have formed, which in France extend in a NW/SE trending belt through the Périgord, open sites tend to occur on exposed interfluvial *plateaux*, at higher average elevations than the caves, albeit with the densest concentrations also associated with limestones containing high-quality flints (Turq (1989:182–196). All the 'classic' Bordesian facies are supposedly represented, although MTA sites are most commonly recorded, probably because of the relatively high visibility of the handaxes and their obvious appeal to generations of local collectors. Based on the relative importance of lithic extraction and production activities vs more generalized patterns in tool use and discard, Turq (1988, 1989) has developed a four-part functional classification of MP open sites in the Périgord that replicates almost exactly that used to classify their UP counterparts. His types are (1) extraction and exploitation (= quarry) sites, (2) extraction and production (= workshop, *atelier*) sites, (3) 'mixed strategy' sites (mostly MTA sites with rich and varied lithic assemblages (= domestic or residential sites), and (4) episodic or ephemeral sites (= small, limited activity stations with sparse and restricted artifact assemblages).

Site Structure and Organization

There is compelling evidence for 'well-organized' Middle (and even Lower) Paleolithic campsites that exhibit all of the features supposedly uniquely associated with the structured use of space noted in ethnographic contexts. These characteristics include (1) open, constructed, paved and excavated hearths (e.g., Grotte Vaufrey, Combe Grenal, Pech de l'Azé II, Grotte du Bison (Arcy), Terra Amata, Hauteroche; Kebara, Ain Difla, Shanidar—dozens of examples), (2) stone pavements (*pavages*—e.g., Baume-Bonne, Terra Amata, Aldène, Biache-Saint-Vaast, La Ferrassie), (3) stone walls (e.g., Lazaret, Cueva Morín, Baume des Peyrards, Saint-Vaast-la-Hougue, Terra Amata), (4) intentionally excavated pits (e.g., Combe Grenal, Le Moustier, La Quina, Morín and other sites in southwest France and northern Spain; Ain Difla, Kebara in the Levant), and (5) huts, lean-tos, shelters, postholes and other evidence for intentional, highly patterned behavior indistinguishable from that generated by modern foragers, and extending from western Europe to the southern Levant (e.g., Bilzingsleben, Lazaret, Terra Amata, Grotte du Renne, Les Canalettes, Tor Faraj). Although claims for 'modern-like' early paleolithic spatial organization should not be accepted uncritically (Villa 1982, 1983), the weight of evidence seems to indicate that differences between the Middle and Upper Paleolithic are essentially differences in quantity, rather than kind.

Burials

Sparked by Robert Gargett's claim that *all* evidence of Middle Paleolithic burials could be explained in terms of natural processes (1989, 1999), a debate ensued in the literature of the 1990s as to whether or not there are significant pattern differences between neandertals and modern humans in respect of their treatment of the dead. While Gargett's views engendered considerable skepticism (e.g., Hayden 1993, Belfer-Cohen & Hovers 1992), many paleoanthropologists had come to accept the existence of at least some credible MP burials, not different in any substantial way from their UP counterparts. Gargett's contention that intentional burial first appeared in the Upper Paleolithic squared well with the 'symbolic explosion' postulated by White (e.g., 1989) and Mellars (e.g., 1989), however—a proliferation of evidence for modern human cognition that first appeared in Europe after c. 40 kya. While *en bloc* comparisons of Middle and Upper Paleolithic burial data showed clear differences between the two periods (e.g., Harrold 1980), it could be argued that the numerical preponderance of late Upper Paleolithic burials (those dated after 20 kya) was amplifying differences that would otherwise appear clinal were the Upper Paleolithic to be subdivided into 'early' (40–20 kya) and 'late' (20–10 kya) units. Although sampling error was (and always will be) a factor, a pattern search comparing mortuary data from the Middle and Early Upper Paleolithic of western Eurasia showed that, contrary to previous assessments, there is much evidence for continuity in pattern over the two periods (Riel-Salvatore & Clark 2001). Evidence for continuity extended to the practice itself (i.e., some neandertals were intentionally buried (and some were not), some EUP hominids buried their dead (and some did not), some LUP hominids buried their dead, etc.), to the kinds and quantities of grave goods, and to structures and features associated with grave sites. This, in our view, reinforced conclusions derived from lithic and faunal analyses and site locations that the Upper Paleolithic as a reified category is replete with implicit assumptions and preconceptions about the cognitive abilities of the hominids who created it, masks much formal variation in the archaeological record, and is therefore too 'coarse grained' to be of much analytical utility. Substantively, the work showed that there were no correlations in mortuary behavior between neandertals and the Middle Paleolithic, on the one hand, and moderns and the Upper Paleolithic, on the other.

Organic Technologies and 'Art'

The picture is less clear in respect of 'art', beads, and organic technologies, although, again, there are a steadily growing number of early paleolithic examples (e.g., the Tata nummulite (Hungary), the Berekhet Ram figure (Palestine), the La Ferrassie cupules (France), the perforated and engraved teeth from Prolom 2 (Crimea), the incised elephant vertebra from Stránská skála (Czech Republic), etc.—see Duff et al. [1992] for a summary of views on the origins of symbolic thought). Evidence for pre-Upper Paleolithic organic technologies and art-like phenomena (what he calls 'concept-mediated marking') available through the middle-1990s has been summarized by Robert Bednarik (1995:605–616). The main points of Bednarik's essay are the following: (1) evidence for concept-mediated marking not qualitatively different from that of the European Upper Paleolithic is found in many parts of the world, and (2) some of it, at least, is demonstrably much older than the Middle–Upper Paleolithic transition in Europe, conventionally dated at ca. 40 kyr B.P. The evidence occurs in well-dated contexts that extend back to the Lower and Middle Paleolithic, and the appearance of significant quantities of 'art' in Europe after the transition interval is attributed to a kind of time-factored, universal taphonomic threshold that destroyed most earlier examples.

Two spectacular Lower Paleolithic examples are Bilzingsleben and Schoeningen, both in Germany. If correctly interpreted by Mania (1990, 1991; cf. Gamble 1999:153–173), the c. 350,000 year old Bilzingsleben open site has yielded evidence of several huts or shelters, an 'organized campsite', stone and bone anvils, wooden and bone artifacts, abundant worked stone, hearths, a stone pavement, symbolically engraved bones, specialized predation on rhinos, even human fossil remains, all in a low-energy lakeshore environment, and with only traces of bone-modifying carnivores. Schoeningen refers to several localities at an operational lignite mine in eastern Germany monitored for years by Hartmut Thieme. The recovery of five wooden spears, including three complete examples c. 2 m long, made from hard, dense spruce with fire-hardened tips, associated with the partly articulated remains of horses, bifacial handaxes and other lithics, and hearths, documents organic technologies, fire and effective predation on big game by *H. erectus (= H. heidelbergensis?)* by around 400 kya (Thieme 1997, Thieme & Maier 1995). Other examples of early organic artifacts include the Clacton yew spear tip (England), dated to c. 300 kya, and associated with the eponymous flake-based, unifacial lithic industry; the Lehringen (Germany) spear, tentatively dated to early OIS 7 (c. 240 kya), also of fire-hardened yew, and bits and pieces of worked wood and bone from the Middle Pleistocene Acheulean sites of Torralba and Ambrona (Spain), and from another lakeshore camp discovered in a lignite mine, Neumark-nord (Germany) (Mania et al. 1990, 1991). Although Bednarik's taphonomic threshold has been questioned as an explanation for the paucity of worked organic remains prior to c. 40 kya, there is little doubt that evidence for pre-Upper Paleolithic organic technologies is accumulating. It has implications for the archaeological transition since it implies a gradual evolution of cognitive faculties that long predates the appearance of modern humans (however defined).

Conclusions

What do we end up with? *A temporal and spatial mosaic of different human adaptations that long precedes and long postdates the Middle–Upper Paleolithic transition.* This, I suggest, is exactly what one would expect, given that adaptation is—always and everywhere—historically contingent and context specific. Leaving aside the issue of just what they might mean, the commonly invoked criteria for modern behavior show *no correlation whatsoever* with the appearance of morphological moderns anywhere, including their alleged homeland, Africa (McBrearty & Brooks 2000). Some of them originated long before the appearance of neandertals, became elaborated in neandertal contexts, and were either lost or became still more marked features of the human condition during and after the Upper Paleolithic. Preconceptions about the authorship of the transitional industries, and typological myopia, have caused some to overlook the ecosystemic context in which Upper Pleistocene hominids, as social animals, evolved. Modern humans are not the 'end product' of that evolution, and are only 'unique' in the sense that any species is unique—by virtue of possessing a unique evolutionary history. The point is that we can no longer afford to approach the problem of the transition in the purely inductive, piecemeal, atheoretical fashion that has been the practice of paleolithic archaeology for more than a century.

Evolutionary Ecology An Example from Latium

Earlier I proposed that evolutionary ecology is a suitably broad and powerful conceptual framework for the analysis of the archaeological transition. In what is probably the most successful study of Upper Pleistocene human adaptation to date, Mary Stiner (1994) and Steven Kuhn (1995) use one of its variants, community ecology, to analyze and interpret changing neandertal and modern human forager adaptations along the coast of Latium during the Upper Pleistocene (c. 120–30 kya). What did they do? Why do I think it is so successful? I beg the indulgence of the reader for an essay already overlong to try to answer these questions (Clark 1995).

Mary Stiner—Archaeotaphonomy

Stiner is an archaeotaphonomist. She uses ecological niche theory to analyze and interpret neandertal hunting, scavenging and foraging behavior, which is then compared and contrasted with that of other large social predators living in the region at the time (esp. wolves, spotted hyenas), and with latest Pleistocene human foragers. The work is organized around three broad, interlocking research questions: (1) Did neandertal subsistence adaptations, teased from the complex patterns discernible in a series of cave faunas, resemble those of other, contemporary members of the carnivore guild, or those of late Upper Paleolithic foragers? Is there something unique about hominid subsistence adaptations in Latium that allows them to be distinguished from those of other canid and felid bone accumulators? (2) What was the nature and extent of variability in neandertal foraging patterns in west-central Italy over time, given local fluctuations in food supply? Is vectored change apparent? Are there discontinuities in adaptation that might correspond to the hypothetical appearance of 'moderns'? And finally, (3) What were the relationships between hominid foraging patterns and technology in the study area?

After discussing the basic tenets of the community ecology approach and the premises of niche theory, Stiner addresses taphonomic issues that affect all cave faunas (e.g., bone damage patterns documented in modern contexts, sampling problems encountered in making inferences about taxonomic representation). The intent here is to develop relatively unambiguous 'signatures' for various bone modifiers, which are then used to determine authorship of bone assemblages, and the prey species consumed by hominids, hyenas and wolves. Of particular interest here is her work on small species, frequently overlooked or dismissed as economically inconsequential in paleolithic research. On the basis of damage patterns and actualistic studies, Stiner shows that most of the terrestrial small species were prey elements in the diets of a variety of small carnivores and raptors, whereas most of the marine molluscs and reptiles (tortoises) were collected and eaten by humans. Most prior work assumed that neandertals were not actively exploiting shellfish.

Ungulate body part representation is also analyzed to compare prey procurement and transport strategies of hominids and other large predators. The comparisons show that body part representation is not diagnostic of predator species, but rather that it distinguishes broad classes of predator strategies that cross-cut genus and species. These analyses also suggest that the distribution of carcasses in the landscape and the nutritional quality of scavenged body parts set further limits on the kinds of bone debris that eventually ends up in the caves.

Stiner also looks at ungulate and carnivore mortality patterns as inferred from tooth eruption and wear data. Mortality patterns are used to determine prey age selection habits of predators, and season of death in carnivores and ungulates. Contrary to some previous claims, it would seem that most mortality patterns cannot distinguish the causes of death in fossil assemblages. However, mortality patterns do sometimes identify broad classes of predatory strategies and niches, provided that many assemblages are compared, and that assemblage comparisons can allow for relatively unambiguous predator attribution (Stiner suggests that the best way to do this is by making maximal use of convergent lines of taphonomic evidence tailored to the particular circumstances of the assemblages being analyzed).

Integrating the results of these pattern searches with the archaeology, Stiner is able to examine the relationships among hominid foraging patterns, Mousterian technology, mobility strategies and raw material procurement. These comparisons put the hominids back into the picture, so to speak, and allow for the identification of Mousterian resource and land-use strategies. In aggregate, the diverse and creative analytical approaches (which Stiner essentially 'invented') suggest that Mousterian foraging patterns were organized according to a somewhat different set of rules than those of modern humans, and that significant shifts occurred in the human resource niche *during the Mousterian itself* (around 55 kya), and not coincident with the Middle–Upper Paleolithic transition at c. 40 kya, as is often assumed. There is little support for the notion of a purely scavenging niche. Both archaic and modern humans were clearly mobile foragers, but neandertal foraging strategies appear to have been more variable over time than those of modern humans. This variability underscores differences in the scale of territory use relative to (late) Upper Paleolithic peoples.

Steven Kuhn—Paleolithic Archaeology

A paleolithic archaeologist, Steven Kuhn addresses the ecological, economic and strategic factors that shaped the behavior of Mousterian tool makers. The work is guided by three emergent themes in paleolithic research that post-date the 1970s: (1) analyses of *chaînes opératoires*—entire sequences of technological acts involving human use of chipped stone, with special emphasis on core reduction; (2) studies of stone tool 'life histories', which show that what had traditionally been regarded as functional and/or stylistic artifact types actually represent points along a continuum of edge reduction and modification, with final artifact form determined by the relative intensity of use, and by the shape and size of the original blank (i.e., Dibble's 'reduction argument'), and (3) the investigation of patterns of raw material exploitation, with an emphasis on the differential procurement, transport and conservation of workable stone as these are affected by region-specific contextual variables like forager mobility, resource distributions and the shape and scale of human economic territories. The research focuses on the material correlates of modal variation in stone artifact production, maintenance and discard, and shows how neandertals and moderns selected among alternative technological strategies in order to keep themselves supplied with the artifacts they used on a daily basis.

Kuhn first makes explicit the factors he thinks govern pattern formation in ancient archaeological palimpsests: the sites were created over millennia by the complex interactions of a variety of human, non-human and natural agents—in no case is human agency solely responsible, nor do the sites represent 'moments frozen in time'. In Latium, stone suitable for knapping is available only in the form of small pebbles and cobbles, resulting in a 'micromousterian' (the Pontinian) that replicates many classic Mousterian core and tool forms on a scale at which the largest artifacts are generally < 3–4 cm long. Mindful of the frequency shift in scavenging and hunting documented by Stiner, and some of its implications for mobility, Kuhn identifies two alternative responses to the problem of keeping foragers supplied with tools and raw materials. One is to supply individuals with a flexible, portable toolkit, useful under conditions of uncertainty and in a variety of contexts. The other is to provision places in the landscape where technological activities are likely to be performed. Although foragers practice both provisioning strategies over the long term, which strategy is selected in a particular situation will depend largely upon perceptions of resource distributions in the local and regional context, and upon patterns of mobility and the duration of site occupation. The basic argument is that if foragers are highly mobile, and if the location of target resources cannot be predicted (as, for example, would be the case if locating scavenged meat was an objective), they must depend largely on the limited array of things they can carry with them. This in turn puts constraints on the sizes and kinds of artifacts that are likely to end up in what, in functional terms, would have been small, ephemeral campsites. If foragers were more stable residentially—if they moved less over the course of an annual round and if occupations tended to be of longer duration (as, for example, might be the case if intercept hunting were emphasized)—there would be more predictable reuse of specific locations in the landscape, and tool makers could stockpile artifacts and raw materials at these places in anticipation of needs likely to arise. Again, mobility and duration of site occupation, linked ultimately to the distribution of food in the environment, should allow for prediction of the composition of artifact assemblages left behind at the two kinds of sites.

Core reduction technology and lithic raw material economy as mediated by differences in mobility constitute the basis for Kuhn's approach. Using simple attribute and metric measures largely of his own devising, Kuhn shows that at least four distinct approaches to flake production are represented in the Pontinian Mousterian collections analyzed. Three resulted in recognizable core types (centripetal or radial, pseudo-prismatic, and prepared platform); the fourth method, bipolar (or hammer-on-anvil) percussion rarely left cores, but was indicated by diagnostic traces on the ventral faces of flakes and on their platforms. Analyses of core and flake blank metrics suggested that these four methods constituted distinct reduction sequences, rather than a series of techniques applied sequentially to the reduction of a single core, as a prominent Italian archaeologist, Amilcare Bietti, had argued.

Although the various core and flake forms are found in all assemblages, there appear to be temporally vectored frequency differences for some of them. Radial and bipolar cores and their products are found throughout the sequence, but were clearly the dominant methods of flake production before c. 55 kya. Only the most recent Mousterian assemblages (dating to < 40 kyr B.P.) contain significant numbers of pseudo-prismatic cores. Clear evidence of parallel flaking, prismatic cores would have yielded the largest number of usable flakes and blades per unit volume, and the flakes produced would have tended to be long and relatively narrow (i.e., to approach 'blade' dimensions). Centripetal or radial reduction would have resulted in fewer flakes per core than parallel reduction, but those flakes would have tended to be larger and wider, with a greater potential for resharpening and renewal.

When Kuhn integrates the lithic data with Stiner's study of the Latium faunas, a bimodal pattern emerges in which (1) heavily reduced assemblages dominated by core technologies yielding large, durable blanks, suggestive of a strategy for provisioning mobile individuals, are associated with the largely scavenged faunas found in the earlier sites. (2) Artifact assemblages found with faunas obtained by ambush hunting are distinguished by less modification and reduction of tools and blanks, fewer transported artifacts, and a shift to pseudo-prismatic core technologies and longer, narrower blanks less suited to subsequent modification. These assemblages, which tend to be found in the later sites, are most compatible with a strategy of provisioning places.

Linked changes in Mousterian foraging patterns and technology are long-term changes, operative at an evolutionary time-scale (i.e., we cannot know with certainty that they were the products of rational decision-making). In coastal Latium, they are ultimately driven by changes in sea level, and resultant changes in the landscapes and resource distributions in the immediate vicinity of the cave and open sites studied. Both Kuhn and Stiner argue that sea level changes altered the foraging opportunities available in the vicinity of the sites, which affected mobility. Changes in mobility had, in turn, a direct (albeit only quantitatively discernible) impact on the material correlates of mobility—the kinds of gear required to extract energy from the environment. Change is directional, they suggest, only because the key underlying environmental variable—sea level—changed in a quasi-directional fashion between 120,000 and 30,000 years ago.

In sum, the work in Latium gives us our first realistic portrait of neandertals as long-standing, successful, predators—highly intelligent, flexible and cunning, with more than a little dependence upon technology, competing more or less directly with other predators early on, and eventually coming to occupy a niche which differs only in detail from that of human foragers today. The research is guided by the general conceptual framework of community ecology, and more specifically by ecological niche theory, and combines elements of an innovative and powerful, yet fundamentally simple, model of human economic behavior. Well-grounded empirically in the particulars of the biophysical environment of Latium, it makes use of the kinds of archaeopaleontological data that are most commonly found in the museums of Europe. Although both Kuhn and Stiner reiterate that changes in foraging behavior and land use cannot easily be extended to regions outside west-central Italy, it is also clear that the variables monitored in both studies—conceptually simple, straightforward, and with relatively clear-cut empirical referents—are applicable to many archaeological problem domains, including those involving modern humans. They show that (1) strong covariation exists between aspects of technological systems and basic kinds of hominid foraging strategies, (2) that Mousterian foragers, while in the scavenging mode, nevertheless exhibited behaviors that were quite possibly unique to hominids (e.g., transporting fat-rich head parts, probably in seasons of fat/carbohydrate depletion), (3) that Mousterian lithic typology, much beloved of European workers, tells us absolutely nothing about even the most fundamental aspects of subsistence (e.g., scavenging vs hunting), and (4) that covariant patterns in lithic technology and animal exploitation can most readily be explained by relative levels of mobility and by changing land use practices. While clearly members of the carnivore guild, and thus subject to many of the constraints applicable to large social carnivores, the makers of the Mousterian artifacts behaved over the long term in economically rational ways, adjusting patterns of resource procurement to the exigencies of a changing local environment, and manipulating their technologies to fit the problems inherent in differing patterns of land use.

* * *

Writing about our origins, Cambridge paleoanthropologist Rob Foley remarked recently that "a lack of theory is a far greater problem than a lack of consensus about what that theory might be" (2001:13). While all those involved in modern human origins research subscribe to (or at least nominally acknowledge) the metaphysical paradigm of post-synthesis Darwinism, there is a 'muddle in the middle' (Isaac 1975) about what kinds of middle range theory can best address the many sub-issues involved in this research. This is especially true of the archaeology of the Middle–Upper Paleolithic transition, where two competing perspectives—those of culture history and evolutionary ecology—contend with one another in practically every forum of the debate. The culture history paradigm, which emphasizes patterns defined by typological systematics, is a legacy of the 'golden age' of Paleolithic research, when our analytical units were first defined, usually with the intention of erecting a chronotypological framework based on time-sensitive stylistic marker types long before any absolute dating methods had become available. These units, and the perspective that underlies them, are with us yet, even though they have never been explicitly defended, warranted or justified as useful to the resolution of any particular problem, large or small. Like beauty, their behavioral meaning lies in the eye of the beholder. The general absence of an explicit concern with epistemology (how we know what we think we know about the remote human past), particularly marked in some of the European research traditions, means that this fundamental ambiguity is never confronted, addressed, nor resolved. This is not a particularly good way to 'do science'. The absence of critical self-consciousness in the research protocols of the culture history 'school', and its failure to make explicit the logic of inference underlying its knowledge claims, results in an interminable debate about what variables are considered relevant or important in the definition of pattern, and the meaning assigned to pattern.

In the evolutionary ecology perspective exemplified by the work of Stiner and Kuhn, there is an explicit conceptual framework (that of community ecology) that identifies variables considered important to measure, and what pattern defined on the basis of those variables is supposed to mean in behavioral terms. The focus is on adaptation, and (1) ethnographic models favored by earlier generations of prehistorians are rejected , (2) paleolithic foragers are studied as members of guilds comprised of social carnivores, and (3) ecological niche theory guides the analysis and interpretation of prehistoric hunting, scavenging and foraging behaviors. The community ecology approach essentially reframes questions pertaining to subsistence over the transition, and requires new kinds of data to answer them. Some examples drawn from Mary Stiner's *Honor Among Thieves* (1994):

> Did neandertal subsistence adaptations, teased from the complex patterns discernible in cave faunas, resemble those of other, contemporary members of the carnivore guild, or those of LUP foragers?
>
> Is vectored change apparent over the LMP/EUP transition? Over the EUP/LUP transition? Over the LUP/Mesolithic boundary?
>
> Are there discontinuities in adaptation that might correspond to the hypothetical appearance of 'moderns'?
>
> Is there something unique about hominid subsistence adaptations at particular places (regions) and moments (intervals) in time that allows them to be distinguished from those of felid, ursid, hystricid and canid bone accumulators?
>
> What was the nature and extent of variability in paleolithic foraging patterns in any region over time, given local fluctuations in the time–space distribution of the available food supply?
>
> What were the relationships between hominid foraging patterns, technologies, mobility strategies and raw material procurement in a given region, over a given interval of time?

How much more satisfying than the endless noodling of the Aurignacian by the culture history enthusiasts so prominent in MHO research! The point, of course, is that qualitatively different questions require different *kinds* of answers. Although still in its infancy, these approaches have many implications for modern human origins research in general, and the nature of the Middle–Upper Paleolithic transition in particular. They show unequivocally that Upper Pleistocene hominids could adjust or adapt to changing environmental circumstances without necessarily turning into modern humans, and that the Middle Paleolithic could show patterned or vectored change without necessarily becoming the Upper Paleolithic.

Endnotes

1. Epistemological realism underpins modern science. It is defined as the philosophical conception that objects of sense perception (sometimes of cognition in general) are real in their own right and exist independently of our perceptions of them.

2. Cowgill (pers. comm.) points out that 'linear' has at least two meanings in common use. One, favored by many social scientists, is the sense of *unidirectional causality*. A causes B, B causes C, and so on, and that is all there is to it; feedback and other cybernetic processes, and systemic networking, are not taken into account. Such concepts or models are undoubtedly too simplistic. The other meaning, the one taken for granted by mathematicians and those in the more mathematical sciences, is that relationships can be sufficiently well 'captured' or described by equations of the form $y = a+bx_1+cx_2+ ...$ *etc.* That is, there are no non-linear terms (e.g., x_1^2, x_1x_2, $log(x_1)$, *etc.*).

3. A more accurate synonym for methodological reductionism is 'strategic simplification,' as opposed to 'un-strategic messiness' and 'unstrategic oversimplification'. Getting the 'strategic' amount of simplification is a very tricky business, and involves a combination of luck and intuition, with exceptional 'good intuition' sometimes amounting to 'genius'. Both unstrategic messiness and unstrategic oversimplification are rife in the social sciences. Some of the more extreme forms of Darwinism also unstrategically oversimplify it, both methodologically and conceptually (Cowgill pers. comm.).

4. I am again indebted to my colleague, George Cowgill, for pointing out that perhaps a better way to look at philosophical reductionism is to *reverse* the direction implied by the reduction—chemistry is generated by physical and chemical processes, biology is generated by physical and chemical processes, etc. In other words, biology can happen without anything sociocultural happening, chemistry can happen without any biology happening, etc. But nothing sociocultural can happen if biology doesn't happen. Looking at it this way also facilitates consideration of emergent properties. At each level there are properties that are not properties at lower levels, even if these properties can be deduced by a sufficiently powerful and well-informed intelligence from the properties of lower levels. Typically, such emergent properties require new concepts in order to think well about them. The point is that concepts that apply very well at one level do not necessarily 'travel well' to other, bracketing levels.

5. It is my opinion that the sciences really do 'cleave nature at its joints' (i.e., pattern variation in nature is discrete; the sciences are not purely arbitrary constructs). They appear to constitute a 'fuzzy set' that corresponds to an ontological reality that exists independent of our perception of it, divided up the way it is for historical and contextual reasons. Modern science is, of course, increasingly interdisciplinary, suggesting that the conventional divisions are breaking down, a process accelerated by the democratization of science, and a proliferation in the number of its practitioners, after World War 2. However, like all large, complex bureaucracies, the universities tend to be very conservative; most of them continue to cling to the traditional organizational structure, derived in large part from German universities during the second half of the 19th century.

References

Allsworth-Jones, P.
1986 *The Szeletian and the Transition from the Middle to Upper Palaeolithic in Central Europe.* Oxford University Press, Oxford.
1990 The Szeletian and the stratigraphic succession in central Europe and adjacent areas. In *The Emergence of Modern Humans* (P. Mellars, ed.), pp. 160–242. Edinburgh University Press, Edinburgh.

Amerloot-van der Heijden, N.
1993 l'Industrie laminaire du niveau CA. In *Riencourt-les-Bapaume (Pas-de-Calais): un Gisement du Paléolithique Moyen* (A. Truffeau, ed.), pp. 26–52. Maison des Sciences de l'Homme, Paris.

Anderson-Gerfaud, P.
1990 Aspects of behaviour in the Middle Palaeolithic: functional analysis of stone tools from southwest France. In *The Emergence of Modern Humans: an Archaeological Perspective* (P. Mellars, ed.), pp. 389–418. Edinburgh University Press, Edinburgh.

Barkai, R., A. Gopher, & C. La Porta
2002 Palaeolithic landscape of extraction: flint surface quarries and workshops at Mt. Pua, Israel. *Antiquity* 76:672–680.

Barton, C. M., & G. A. Clark (editors).
1997 *Rediscovering Darwin: Evolutionary Theory and Archaeological Explanation.* Archeological Papers of the American Anthropological Association No. 7, Washington.

Barton, C. M., G. Clark, & A. Cohen
1994 Art as information: explaining Upper Paleolithic art in western Europe. *World Archaeology* 26: 185–207.

Barton, C. M., N. Coinman, & D. Olszewski
1996 Beyond the graver: reconsidering burin function. *Journal of Field Archaeology* 23:111–125.

Barton, R. N. E.
2000 Mousterian hearths and shellfish; late Neanderthal activities on Gibraltar. In *Neanderthals on the Edge* (C. Stringer, R. N. E. Barton & C. Finlayson, eds.), pp. 85–93. Oxbow Books, Oxford.

Bar-Yosef, O.
2000 The Middle and early Upper Paleolithic in southwest Asia and neighboring regions. In *The Geography of Neandertals and Modern Humans in Europe and the Greater Mediterranean* (O. Bar-Yosef & D. Pilbeam, eds.), pp. 107–156. Harvard University, Peabody Museum Bulletin No. 8, Cambridge, MA.
2002 The Upper Paleolithic revolution. *Annual Review of Anthropology* 31:363–393.

Bar-Yosef, O., & S. L. Kuhn
1999 The big deal about blades: laminar technologies and human evolution. *American Anthropologist* 101:322–338.

Bednarik, R.
1995 Concept mediated marking in the Lower Palaeolithic. *Current Anthropology* 36:605–634.

Belfer-Cohen, A., & E. Hovers
1992 In the eye of the beholder: Mousterian and Natufian burials in the Levant. *Current Anthropology* 33:463–471.

Beyries, S.
1988 Functional variability of lithic sets in the Middle Paleolithic. In *The Upper Pleistocene Prehistory of Western Eurasia* (H. L. Dibble & A. Montet-White, eds.), pp. 215–230. University of Pennsylvania Museum Press, Philadelphia.

Bermúdez de Castro, J. M. et al.
1999 *Atapuerca—Nuestros Antecesores.* Junta de Castilla y León, Salamanca.

Bermúdez de Castro, J. M. et al.
2004 The Atapuerca sites and their contribution to the knowledge of human evolution in Europe. *Evolutionary Anthropology* 13:25–41.

Binford, L. R.
1965 Archaeological systematics and the study of culture process. *American Antiquity* 29:425–441.
1973 The Mousterian and the functional argument. In *The Explanation of Culture Change* (C. Renfrew, ed.), pp. 227–254. Gerald Duckworth, London.
1981 *Bones: Ancient Men and Modern Myths.* Academic Press, New York.

Binford, L. R., & J. Sabloff
1982 Paradigms, systematics and archaeology. *Journal of Anthropological Research* 38:137–153.

Bischoff, J. et al.
1994 Dating the basal Aurignacian 'sandwich' at Abric Romaní (Catalunya, Spain) by radiocarbon and uranium-series. *Journal of Archaeological Science* 21:541–551.

Bisson, M.
2000 Nineteenth century tools for twenty-first century archaeology? Why the Middle Paleolithic typology of François Bordes must be replaced. *Journal of Archaeological Method and Theory* 7: 1–48.

Bocquet-Appel, J.-P., & P.-Y. Demars
2000 Neanderthal contraction and modern human colonization of Europe. *Antiquity* 74:544–552.

Boëda, E.
1988 Le concept Levallois et évaluation de son champ d'application. In *l'Homme de Néandertal 4: La Technique* (L. Binford & J.-Ph. Rigaud, eds.), pp. 13–26. ERAUL No. 31, Liège.
1993 *Le Concept Levallois: Variabilité des Méthodes*. CNRS, Paris.

Bordes, F.
1961a *Typologie du Paléolithique Ancien et Moyen.* Delmas, Bordeaux.
1961b Mousterian cultures in France. *Science* 134: 803–810.

Brantingham, P. J., S. Kuhn, & K. Kerry (editors)
2004 *The Early Upper Paleolithic Beyond Western Europe.* University of California Press, Berkeley.

Cabrera Valdés, V., & J. Bischoff
1989 Accelerator ^{14}C dates for the early Upper Paleolithic (Basal Aurignacian) at El Castillo Cave, Spain. *Journal of Archaeological Science* 16: 577–584.

Cabrera Valdés, V., M. Hoyos, & F. Bernaldo de Quirós
1997 The transition to the Upper Paleolithic in the cave of El Castillo (Cantabria, Spain). In *Conceptual Issues in Modern Human Origins Research* (G. Clark & C. Willermet, eds.), pp. 177–190. Aldine de Gruyter, New York.

Cabrera Valdés, V., A. Pike-Tay, M. Lloret, & F. Bernaldo de Quirós
2000 Continuity patterns in the Middle–Upper Paleolithic transition in Cantabrian Spain. In *Neanderthals on the Edge* (C. Stringer, R. N. E. Barton & C. Finlayson, eds.), pp. 85–93. Oxbow Books, Oxford.

Camps i Calbet, M.
2004 *Revisiting the 40,000* B.P. *Crisis in Iberia: A Study of Selected Transitional Industries and their Significance.* D.Phil. dissertation, Institute of Archaeology, University of Oxford.

Chabai, V.
2003 The chronological and industrial variability of the Middle to Upper Paleolithic transition in eastern Europe. In *The Chronology of the Aurignacian and of the Transitional Technocomplexes* (J. Zilhão & F. d'Errico, eds.), pp. 71–88. Instituto Português de Arqueologia, *Trabalhos de Arqueologia* 33, Lisbon.

Chabai, V., A. Marks, & K. Monigal
2004 Crimea in the context of the eastern European Middle Paleolithic and early Upper Paleolithic. In *The Middle Paleolithic and Early Upper Paleolithic of Eastern Crimea, Vol. III* (Chabai et al., eds.), pp. 419–460. ERAUL No. 104, Liège.

Chabai, V. P., & K. Monigal (editors)
1999 *The Middle Paleolithic of Western Crimea Vol. II.* ERAUL No. 87, Liège.

Chabai, V.P., K. Monigal, & A. E. Marks (editors)
2004 *The Middle Paleolithic and Early Upper Paleolithic of Eastern Crimea, Vol. III.* ERAUL No. 104, Liège.

Chase, P. G.
1989 How different was Middle Paleolithic subsistence? In *The Human Revolution* (P. Mellars & C. Stringer, eds.), pp. 321–337. Edinburgh University Press, Edinburgh.

Churchill, S., & F. Smith
2000 Makers of the early Aurignacian of Europe. *Yearbook of Physical Anthropology* 43:61–115.

Clark, G. A.
1982 Quantifying archaeological research. In *Advances in Archaeological Method and Theory* (M. Schiffer, ed.), pp. 217–273. Academic Press, New York.
1987 Paradigms and paradoxes in contemporary archaeology. In *Quantitative Research in Archaeology: Progress and Prospects* (M. Aldenderfer, ed.), pp. 30–60. Sage Publications, Beverly Hills.
1989 Romancing the stones: biases, style and lithics at La Riera. In *Alternative Approaches to Lithic Analysis* (D. Henry & G. Odell, eds.), pp. 27–50. Archeological Papers of the American Anthropological Association No. 1, Washington.
1992 Continuity or replacement? Putting modern

human origins in an evolutionary context. In *The Middle Paleolithic: Adaptation, Behavior and Variability* (H. Dibble & P. Mellars, eds.), pp. 183–205. University of Pennsylvania Museum, Philadelphia.

1994 Migration as an explanatory concept in paleolithic archaeology. *Journal of Archaeological Method and Theory* 1:305–343.

1997 The Middle–Upper Paleolithic transition in Europe: an American perspective. *Norwegian Archaeological Review* 30:25–53.

1999a Modern human origins—highly visible, curiously intangible. *Science* 283:2029–2032; 284:917.

1999b Reductionism in Science, Biology, and Archaeology. Paper presented at the annual meetings of the Society for American Archaeology (24–28 March, 1999). Chicago, Illinois.

2002a Neandertal Archaeology—Implications for Our Origins. Paper presented at the annual meetings of the Society for American Archaeology (20–24 March, 2002). Denver, Colorado.

2002b Neandertal archaeology—implications for our origins. *American Anthropologist* 104:50–67.

2002c Observations on paradigmatic bias in French and American paleolithic archaeology. In *The Role of American Archeologists in the Study of the European Upper Paleolithic* (L. Straus, ed.), pp. 19–26. BAR International Series No. 1048, Oxford.

2003 American archeology's uncertain future. In *Archeology is Anthropology* (S. Gillespie & D. Nichols, eds.), pp. 51–68. Archeological Papers of the American Anthropological Association No. 13, Arlington.

2005 Modern approaches to Paleolithic archaeology in Europe: a sampler of research traditions. *American Antiquity* 70:376–384.

Clark, G. A., & J. M. Lindly

1989 The case for continuity: observations on the biocultural transition in Europe and western Asia. In *The Human Revolution: Behavioural and Biological Perspectives on the Origins of Modern Humans* (P. Mellars & C. Stringer, eds.), pp. 626–676. Edinburgh University Press, Edinburgh.

1991 Paradigmatic biases and paleolithic research traditions. *Current Anthropology* 32:577–587.

Clark, G. A., & J. Riel-Salvatore

2003 What's in a Name? An Empirical Review of the Compositional Integrity of the West Eurasian Aurignacian. Paper presented at the annual meetings of the Paleoanthropology Society (22–23 April, 2003). Tempe, Arizona.

2006 Observations on systematics in paleolithic archaeology In *Transitions before the Transition: Evolution and Stability in the Middle Paleolithic and the Middle Stone Age* (E. Hovers & S. Kuhn, eds.), pp. 29–56. Springer , New York.

Coinman, N. R.

1998 The Upper Paleolithic of Jordan. In *The Prehistoric Archaeology of Jordan* (D. Henry, ed.), pp. 39–63. BAR International Series No. 705, Oxford.

2003 The Upper Paleolithic of Jordan: new data from the Wadi al'Hasa. In *More Than Meets the Eye* (A. N. Goring-Morris & A. Belfer-Cohen, eds.), pp. 151–170. Oxbow Books, Oxford.

Conard, N.

2002 The Early Upper Paleolithic in Swabia. Paper presented at the annual meeting of the Paleoanthropology Society (30–31 March, 2002). Denver, Colorado.

Conard, N., & M. Bolus

2002 Radiocarbon dating the appearance of modern humans and the timing of cultural innovations in Europe: new results and new challenges. *Journal of Human Evolution* 44:331–371.

Conard, N., P. Grootes, & F. Smith

2004 Unexpectedly recent dates for human remains from Vogelherd. *Nature* 430:198–201.

Crew, H.

1976 The Mousterian site of Rosh Ein Mor. In *Prehistory and Paleoenvironments in the Central Negev, Israel, Vol. I* (A. Marks, ed.), pp. 75–112. SMU Press, Dallas.

David, F., & C. Farizy

1994 Les vestiges osseux: étude archéozoologique. In *Hommes et Bisons du Paléolithique Moyen à Mauran (Haute-Garonne)* (C. Farizy, F. David & J. Jaubert, eds.), pp. 177–234. CNRS, Paris.

Davidson, I., & W. Noble

1989 The archaeology of perception: traces of depiction and language. *Current Anthropology* 30: 125–137.

Davidson, I.

1997 The power of pictures. In *Beyond Art: Pleistocene Image and Symbol* (M. Conkey et al., eds.), pp. 125–159. California Academy of Sciences, San Francisco.

Davies, W.

1999 *The Aurignacian as a Reflection of Modern Human Population Dispersal in Europe.* D. Phil. dissertation, Department of Archaeology, University of Cambridge.

2001 A very model of a modern human industry: new perspectives on the origins and spread of the Aurignacian in Europe. *Proceedings of the Prehistoric Society* 67:195–217.

Dawkins, R.
1990 *The Extended Phenotype—the Long Reach of the Gene.* Oxford University Press, Oxford.

D'Errico, F. et al.
1998 Neandertal acculturation in western Europe? A critical review of the evidence and its interpretation. *Current Anthropology* 39:S1–S44.

Delibrias, G., M. Guillier, & J. La Beyrie
1986 GIF natural radiocarbon measurements X. *Radiocarbon* 28:9–68.

Dibble, H.
1995 Middle Paleolithic scraper reduction: background, clarification and review of the evidence to date. *Journal of Archaeological Method and Theory* 2:299–368.

Dibble, H., & N. Rolland
1992 On assemblage variability in the Middle Paleolithic of western Europe: history, perspectives, and a new synthesis. In *The Middle Paleolithic: Adaptation, Behavior and Variability* (H. Dibble and P. Mellars, eds.), pp. 1–28. University of Pennsylvania Museum, Philadelphia.

Duchadeau-Kervazo, C.
1982 *Recherches sur l'Occupation Paléolithique du Bassin de la Dronne.* Doctoral dissertation, University of Bordeaux I.
1986 Les sites paléolithiques du bassin de la Dronne: observations sur les modes et emplacements. *Bulletin de la Société Préhistorique Français* 83:56–64.

Duff, A., G. A. Clark, & T. Chadderdon
1992 Symbolism in the early palaeolithic: a conceptual odyssey. *Cambridge Archaeological Journal* 2:211–229.

Enard, W. et al.
2002 Molecular evolution of FOXP2, a gene involved in speech and language. *Nature* 418:869–872.

Farizy, C., & F. David
1992 Subsistence and behavioral patterns of some Middle Palaeolithic local groups. In *The Middle Paleolithic: Adaptation, Behavior and Variability* (H. Dibble and P. Mellars, eds.), pp. 87–96. University of Pennsylvania Museum, Philadelphia.

Farizy, C., F. David, & J. Jaubert, Eds.
1994 *Hommes et Bisons du Paléolithique Moyen à Mauran (Haute-Garonne).* CNRS, Paris.

Flannery, K.
1973 Archaeology with a capital 'S'. In *Research and Theory in Current Archaeology* (C. L. Redman, ed.), pp. 47–53. Wiley, New York.

Foley, R.
2001 In the shadow of the modern synthesis: alternative perspectives on the last 50 years of paloanthropology. *Evolutionary Anthropology* 10:5–15.

Gamble, C.
1999 *The Palaeolithic Societies of Europe.* Cambridge University Press, Cambridge.

Gargett, R.
1989 Grave shortcomings: the evidence for neandertal burial. *Current Anthropology* 30:157–177.
1999 Middle Paleolithic burial is not a dead issue. *Journal of Human Evolution* 37:27–90.

Geneste, J.-M.
1988 Les industries de la Grotte Vaufrey: technologie du débitage, économie et circulation de la matière première lithique. In *La Grotte Vaufrey: Paléoenvironnement, Chronologie, Activités Humaines* (J.-Ph. Rigaud, ed.), pp. 441–517. Mémoires de la Société Préhistorique Française No. 19, Paris.
1989a Économie des ressources lithiques dans le Moustérien du sud-ouest de la France. In *l'Homme de Néandertal 6: La Subsistence* (M. Patou & L. G. Freeman, eds.), pp. 75–97. ERAUL No. 33, Liège.
1989b Systèmes d'approvisionnement en matières premières au paleolithique moyen et au paléolithique supérieur en Aquitaine. In *l'Homme de Néandertal 8: La Mutation* (J. Kozlowski, ed.), pp. 61–70. ERAUL No. 35, Liège.

Goldberg, P., D. Nash, & M. Petraglia (editors).
1993 *Formation Processes in Archaeological Context.* Prehistory Press, Madison.

Gould, S. J.
1994 *Darwin's Revolution in Thought.* Into the Classroom Video, Inc., Northampton, MA.

Grayson, D., & S. Cole
1998 Stone tool assemblage richness during the Middle and early Upper Paleolithic in France. *Journal of Archaeological Science* 25:927–938.

Grayson, D., & F. Delpech
1998 Changing diet breadth in the early Upper Paleo-

lithic of southwestern France. *Journal of Archaeological* Science 25:1119–1125.

2002 Specialized early Upper Paleolithic hunters in southwestern France? *Journal of Archaeological Science* 29:1439–1449.

2003 Ungulates and the Middle-to-Upper Paleolithic transition at Grotte XVI. *Journal of Archaeological Science* 30:1633–1648.

Harrold, F.
1980 A comparative analysis of Eurasian Paleolithic burials. *World Archaeology* 12:195–211.

Harrold, F., & M. Otte
2001 Time, space and cultural process in the European Middle–Upper Paleolithic transition. In *Questioning the Answers: Re-solving Fundamental Problems of the Early Upper Paleolithic* (M. Hays & P. Thacker, eds.), pp. 3–12. BAR International Series No. 1005, Oxford.

Hauser, M., N. Chomsky, & W. Fitch
2002 The faculty of language: what is it, who has it, and how did it evolve? *Science* 298:1569–1579.

Hayden, B.
1993 The cultural capacities of neandertals: a review and re-evaluation. *Journal of Human Evolution* 24:113–146.

Hays, M., & P. Thacker (editors)
2001 *Questioning the Answers: Re-solving Fundamental Problems of the Early Upper Paleolithic.* BAR International Series No. 1005, Oxford.

Hedges, R. E. M. et al.
1994 Radiocarbon dates from the Oxford AMS system: *Archaeometry* datelist 18. *Archaeometry* 36:337–374.

Heinzelin, J. de, & P. Haesaerts
1983 Un cas de débitage laminaire au paléolithique ancien: Croix-l'Abbé à Saint-Valery-sur-Somme. *Gallia Préhistoire* 26:189–201.

Henry, D. et al.
2004 Human behavioral organization in the Middle Paleolithic: were Neanderthals different? *American Anthropologist* 106:17–31.

Henshilwood, C. S., & C. W. Marean.
2003. The origin of modern human behavior. *Current Anthropology* 44:627–652.

Howell, F. C.
1999 Paleo-demes, species clades, and extinctions in the Pleistocene hominin record. *Journal of Anthropological Research* 55:191–243.

Hublin, J.-J. et al.
1995 The Mousterian site of Zafarraya (Andalucía, Spain): dating and implications on the palaeolithic peopling processes of western Europe. *Comptes-Rendus de l'Académie des Sciences de Paris* 321:931–937.

Isaac, G.
1975 Sorting out the muddle in the middle: an anthropologist's post-conference appraisal. In *After the Australopithecines: Stratigraphy, Ecology and Culture Change in the Middle Pleistocene* (K. Butzer & G. Isaac, eds.), pp. 875–877. Mouton, The Hague.

Jaubert, J., & J.-P. Brugal
1990 Contribution à l'étude du mode de vie au Paléolithique moyen: les chasseurs d'aurochs de la Borde. In *Les Chasseurs d'Aurochs de La Borde: Un Site du Paléolithique Moyen (Livernon, Lot)* (J. Jaubert et al., eds.), pp. 128–145. Maison des Sciences de l'Homme, Paris.

Jelinek, A. J.
1988 Technology, typology and culture in the Middle Paleolithic. In *Upper Pleistocene Prehistory of Western Eurasia* (H. Dibble & A. Montet-White, eds.), pp. 199–212. University of Pennsylvania Museum, Philadelphia.

Jelinek, A., A. Debénath, & H. Dibble
1989 A Preliminary Report on Evidence Related to the Interpretation of Economic and Social Activi-ties of Neanderthals at the Site of La Quina (Charente), France. In *l'Homme de Néandertal 6: La Subsistance* (M. Patou & L. G. Freeman, eds.), pp. 99–106. ERAUL No. 33, Liège.

Jochim, M.
1981 *Strategies for Survival.* Academic Press, New York.

Klein, R. G.
1999 *The Human Career* (2nd Edition, revised and expanded). University of Chicago Press, Chicago.

2003 Whither the neanderthals? *Science* 299:1525–1527.

Koestler, A., & A. Smythies
1969 *Beyond Reductionism: New Perspectives in the Life Sciences.* Hutchinson, London.

Krings, M. et al.
1997 Neandertal DNA sequences and the origin of modern humans. *Cell* 90:1–20.

Kozlowski, J.
1982 *Excavation in the Bacho Kiro Cave.* Panstwowe Wydawnictwo Naukowe, Warsaw.
2000 The problem of cultural continuity between the Middle and Upper Paleolithic in central and eastern Europe. In *The Geography of Neandertals and Modern Humans in Europe and the Greater Mediterranean* (O. Bar-Yosef & D. Pilbeam, eds.), pp. 77–107. Harvard University, Peabody Museum Bulletin No. 8, Cambridge, MA.

Kuhn, S.
1995 *Mousterian Lithic Technology: an Ecological Perspective.* Princeton University Press, Princeton.

Kuhn, S., & A. Bietti
2000 The late Middle and early Upper Paleolithic in Italy. In *The Geography of Neandertals and Modern Humans in Europe and the Greater Mediterranean* (O. Bar-Yosef & D. Pilbeam, eds.), pp. 49–76. Harvard University, Peabody Museum Bulletin No. 8, Cambridge, MA.

Lévêque, F., A. Backer, & M. Guilbaud (editors)
1993 *Context of a Late Neandertal.* Prehistory Press Monographs in World Archaeology No. 16, Madison, WI.

Mania, D.
1990 *Auf den Spuren des Urmenschen: Die Funde von Bilzingsleben.*Theiss, Berlin.
1991 The zonal division of the Lower Palaeolithic open-air site Bilzingsleben. *Anthropologie* 29: 17–24.

Marks, A. E.
1983a The Middle to Upper Paleolithic transition in the Levant. *Advances in World Archaeology* 2: 51–98.
1990 The Middle and Upper Paleolithic of the Near East and the Nile Valley: the problem of cultural transformations. In *The Emergence of Modern Humans* (P. Mellars, ed.), pp. 56–90. Edinburgh University Press, Edinburgh.
1993 The early Upper Paleolithic: the view from the Levant. In *Before Lascaux* (H. Knecht, A. Pike-Tay & R. White, eds.), pp. 5–21. CRC Press, Boca Raton, FL.
1994 A Levantine perspective. *Cambridge Archaeological Journal* 4:104–106.
1998 A new Middle to Upper Paleolithic 'transitional' assemblage from Buran-Kaya III, Level C, eastern Crimea: a preliminary report. In *The Prehistory of Anatolia and Adjacent Regions* (M. Otte, ed.), pp. 353–366. ERAUL No. 85, Liège.

Marks, A. E. (editor).
1976 *Prehistory and Paleoenvironments in the Central Negev, Israel, Vol. I.* SMU Press, Dallas.
1977 *Prehistory and Paleoenvironments in the Central Negev, Israel, Vol. II.* SMU Press, Dallas.
1983b *Prehistory and Paleoenvironments in the Central Negev, Israel, Vol. III.* SMU Press, Dallas.

Marks, A., & D. Freidel
1977 Prehistoric settlement patterns in the Avdat/Aqev area. In *Prehistory and Paleoenvironments in the Central Negev, Israel, Vol. II* (A. Marks, ed.), pp. 131–158. SMU Press, Dallas.

Marks, A., & K. Monigal
1998 Starosele 1993–1995: the lithic artifacts. In *The Middle Paleolithic of Western Crimea, Vol. 1* (A. Marks & V. Chabai, eds.), pp. 117–166. ERAUL No. 84, Liège.

Marks, A., H. Hietala, & J. Williams
2001 Tool standardization in the Middle and Upper Palaeolithic: a closer look. *Cambridge Archaeological Journal* 11:17–44.

Maschner, H. (editor)
1996 *Darwinian Archaeologies.* Plenum, New York.

Mayr, E.
1976 *Evolution and the Diversity of Life.* Belknap Press, Cambridge, MA.
1991 *One Long Argument: Charles Darwin and the Genesis of Modern Evolutionary Thought.* Belknap Press, Cambridge, MA.
1997 *This is Biology: the Science of the Living World.* Belknap Press, Cambridge, MA.

McBrearty, S., & A. S. Brooks.
2000 The revolution that wasn't: a new interpretation of the origin of modern human behavior. *Journal of Human Evolution* 39:453–563.

Mellars, P.
1973 The character of the Middle–Upper Palaeolithic transition in south-west France. In *The Explanation of Culture Change* (C. Renfrew, ed.), pp. 255–276. Gerald Duckworth, London.
1989a Technological changes across the Middle–Upper Paleolithic transition: technological, social and cognitive perspectives. In *The Human Revolution* (P. Mellars, ed.), pp. 338–365. Edinburgh University Press, Edinburgh.

1989b Major issues in the emergence of modern humans. *Current Anthropology* 30:349–385.
1996a The emergence of biologically modern populations in Europe: a social and cognitive 'revolution'? *Proceedings of the British Academy* 88: 179–201.
1996b *The Neanderthal Legacy: an Archaeological Perspective from Western Europe.* Princeton University Press, Princeton.
2000 The archaeological records of the Neandertal-modern human transition in France. In *The Geography of Neanderthals and Modern Humans in Europe and the Greater Mediterranean* (O. Bar-Yosef & D. Pilbeam, eds.), pp. 35–48. Peabody Museum of Archaeology and Ethnology Bulletin No. 8, Cambridge, MA.
2004 Neanderthals and the modern human colonization of Europe. *Nature* 432:461–465.
2005 The impossible coincidence: a single-species model for the origins of modern human behavior in Europe. *Evolutionary Anthropology* 14:12–27.

Mithen, S.
1993 Individuals, groups, and the palaeolithic record: a reply to Clark. *Proceedings of the Prehistoric Society* 59:393–398.

Monigal, K.
2001 The Eastern Szeletian at Buran Kaya III (Crimea, Ukraine) and its place in the Middle to Upper Paleolithic transition. In *Questioning the Ansers: Re-solving Fundamental Problems of the Early Upper Paleolithic* (M. Hays & P. Thacker, eds.), pp. 51–64. BAR International Series No. 1005, Oxford.
2002 *The Levantine Leptolithic: Blade Production from the Lower Paleolithic to the Dawn of the Upper Paleolithic.* Ph.D dissertation, Department of Anthropology, Southern Methodist University.

Munday, F.
1976 Intersite variability in the Mousterian occupation of the Avdat/Aqev area. In *Prehistory and Paleoenvironments in the Central Negev, Israel, Vol. I* (A. Marks, ed.), pp. 113–140. SMU Press, Dallas.

O'Brien, M., Ed.
1996 *Evolutionary Archaeology: Theory and Practice.* University of Utah Press, Salt Lake City.

O'Brien, M., & T. Holland
1995 The nature and premise of a selection-based archaeologyl. In *Evolutionary Archaeology* (P. Teltser, ed.), pp. 175–200. University of Arizona Press, Tucson.

O'Brien, M., & R. L. Lyman
2000 *Applying Evolutionary Archaeology: a Systematic Approach.* Kluwer Academic/Plenum, New York.

Ohnuma, K., & C. Bergman
1990 A technological analysis of the Upper Paleolithic levels (XXV–VI) of K'sar Akil, Lebanon. In *The Emergence of Modern Humans* (P. Mellars, ed.), pp. 91–138. Edinburgh University Press, Edinburgh.

Otte, M., & L. G. Straus
1995 *Le Trou Magrite.* ERAUL No. 69, Liège.

Otte, M., & A. Derevianko
2001 The Aurignacian in Altai. *Antiquity* 75:44–49.

Palma de Cesnola, A.
1993 *Il Paleolitico Superiore in Italia.* Garlatti e Razzai, Florence.

Phillips, J.
1988 The Upper Paleolithic of the Wadi Firan, southern Sinai. *Paléorient* 14:183–201.

Raposo, L.
2000 The Middle–Upper Paleolithic transition in Portugal. In *Neandertals on the Edge* (C. Stringer, R. N. E. Barton & C. Finlayson, eds.), pp. 95–109. Oxbow Books, Oxford.

Raposo, L., & J. L. Cardoso
1998 Las industrias líticas de la Gruta Nova de Columbeira (Bombarral, Portugal) en el contexto del Musteriense final de la Península Ibérica. *Trabajos de Prehistoria* 55:39–62.

Renfrew, C.
1969 Trade and culture process in European prehistory. *Current Anthropology* 10:151–169.

Révillon, S.
1989 Le débitage du gisement paléolithique moyen de Seclin (Nord). In *Paléolithique et Mésolithique du Nord de la France: Nouvelles Recherches* (A. Tuffreau, ed.), pp. 79–90. Centre d'Études et de Recherches Préhistorique, Lille.

Riel-Salvatore, J., & G. A. Clark
2001 Grave markers: Middle and early Upper Paleolithic burials and the use of chronotypology in contemporary Paleolithic research. *Current Anthropology* 42:449–480.

Riel-Salvatore, J., & C. M. Barton
2004 Late Pleistocene technology, economic behavior and land-use dynamics in southern Italy. *American Antiquity* 69:257–274.

Rose, D.
1998 *Lifelines—Biology Beyond Determinism.* Oxford, University Press Oxford.

Sackett, J. R.
1981 From de Mortillet to Bordes: a century of French paleolithic research. In *Towards a History of Archaeology* (G. Daniel, ed.), pp. 85–99. Thames & Hudson, London.
1988 The Mousterian and its aftermath: a view from the Upper Paleolithic. In *The Upper Pleistocene Prehistory of Western Eurasia* (H. L. Dibble & A. Montet-White, Eds), pp. 413–426. University of Pennsylvania Museum, Philadelphia.
1991 Straight archaeology French style: the phylogenetic paradigm in historic perspective. In *Perspectives on the Past: Theoretical Biases in Mediterranean Hunter-Gatherer Research* (G. A. Clark, ed.), pp. 109–140. University of Pennsylvania Press, Philadelphia.
1997 Neanderthal behaviour: the archaeological evidence. *Cambridge Archaeological Journal* 7: 148–149.

Scott, K.
1986 The bone assemblages of layers 3 and 6. In *La Cotte de St. Brelade 1961–1978: Excavations by C. B. M. Burney* (P. Callow & J. M. Cornford, eds.), pp. 159–183. Geo Books, Norwich.
1989 Mammoth bones modified by humans: evidence from la Cotte de St Brelade, Jersey, Channel Isalnds. In *Bone Modification* (R. Bonnichsen & M. H. Sorg, eds.), pp. 335–346. Center for the Study of the First Americans, Orono.

Slott-Möller, R.
1990 La faune. In *Les Chasseurs d'Aurochs de La Borde: un Site du Paléolithique Moyen (Livernon, Lot)* (J. Jaubert, M. Lorblanchet, R. Slott-Möller, A Turq, & J.-P. Brugal, eds.), pp. 33–68. Maison des Sciences de l'Homme, Paris.

Smith, F., I. Jankovic, & I. Karavanic
2005 The assimilation model, modern human origins in Europe, and the extinction of Neandertals. *Quaternary International* 137:7–19.

de Sonneville-Bordes, D., & J. Perrot
1953 Essai d'adaptation des méthodes statistiques au Paléolithique supérieur. Premiers résultats. *Bulletin de la Société Préhistorique Française* 51:323–333.

de Sonneville-Bordes, D., & J. Perrot
1954 Lexique typologique du Paléolithique supérieur, outillage lithique. 1. Grattoirs. 2. Outils solutréens. *Bulletin de la Société Préhistorique Française* 51:327–335.
1955 Lexique typologique du Paléolithique supérieur, outillage lithique. 3. Outils composites, perçoirs. *Bulletin de la Société Préhistorique Française* 52:76–79.
1956. Lexique typologique du Paléolithique supérieur, outillage lithique, 4. Burins. *Bulletin de la Société Préhistorique Française* 53:408–412, 547–559.

Steele, J., & S. Shennan (editors)
1996 *The Archaeology of Human Ancestry.* Routledge, London.

Stiner, M.
1993 Small animal exploitation and its relation to hunting, scavenging, and gathering in the Italian Mousterian. In *Hunting and Animal Exploitation in the Later Palaeolithic and Mesolithic of Eurasia* (G. L. Peterkin, H. Bricker & P. Mellars, eds.), pp. 107–126. Archeological Papers of the American Anthropological Association No. 4, Washington.
1994 *Honor Among Thieves: a Zooarchaeological Study of Neandertal Ecology.* Princeton University Press, Princeton.

Straus, L. G.
1996 Continuity or rupture; convergence or invasion; adaptation or catastrophe; mosaic or monolith: views on the Middle to Upper Paleolithic transition in Iberia. In *The Last Neandertals, The First Anatomically Modern Humans* (E. Carbonell & M. Vaquero, eds.), pp. 203–218. Universitat Rovira I Virgili, Barcelona.
1997 The Iberian situation between 40,000 and 30,000 BP in light of European models of migration and convergence. In *Conceptual Issues in Modern Human Origins Research* (G. Clark & C. Willermet, eds.), pp. 235–252. Aldine de Gruyter, New York.
2003 *'The* Aurignacian'? Some thoughts. In *The Chronology of the Aurignacian and of the Transitional Technocomplexes: Dating, Stratigraphies, Cultural Implications* (J. Zilhão & F. D'Errico, eds.), pp. 11–17. Instituto Português de Arqueologia, *Trabalhos de Arqueologia* 33, Lisbon.
2005 A mosaic of change: the Middle–Upper Paleolithic transition as viewed from New Mexico and Iberia. *Quaternary International* 137:47–68.

Stringer, C., & C. Gamble
1994 *In Search of the Neanderthals.* Thames & Hudson, London.

Stringer, C., J.-J. Hublin, & P. Vandermeersch
1984 The origin of anatomically modern humans in western Europe. In *The Origins of Modern Humans* (F. Smith & F. Spencer, eds.), pp. 51–136. Alan Liss, New York.

Stringer, C., R. N. E. Barton, & C. Finlayson, Eds.
2000 *Neanderthals on the Edge.* Oxbow Books, Oxford.

Svoboda, J., V. Lozek, & E. Vlcek
1996 *Hunters Between East and West.* Plenum, New York.

Teltser, P. (editor)
1995 *Evolutionary Archaeology: Methodological Issues.* University of Arizona Press, Tucson.

Thieme, H.
1997 Lower Palaeolithic hunting spears from Germany. *Nature* 385:807–810.

Thieme, H., & R. Maier
1995 *Archäologische Ausgrabungen im Brauaohletagebau Schöningen, Landkreis helmstadt.* Verlag Habüsche Buchhandlung, Hannover.

Tobias, P.
2000 The role of water in the extra-African dispersal of humanity, with special reference to the peopling of the Iberian Peninsula. In *The Iberian Peninsula and Human Evolution* (M. Walker et al., eds.). Oxbow Books, Oxford (in press).

Tuffreau, A.
1992 Middle Paleolithic settlement in Northern France. In *The Middle Paleolithic: Adaptation, Behavior, and Variability* (H. L. Dibble & P. Mellars, eds.), pp. 59–74. University of Pennsylvania Museum, Philadelphia.

Turq, A.
1988 Le Moustérien de type Quina du Roc de Marsal à Campagne (Dordogne): contexte stratigraphique, analyse lithographique et technologique. *Dossiers d'Archéologie Périgourdienne* 3:5–30.
1989 Exploitation des matières premières lithiques et occupation du sol: l'exemple du Moustérien entre Dordogne et Lot. In *Variations des Paléomilieux et Peuplement Préhistorique* (H. Laville, ed.), pp. 179–204. Éditions CNRS, Paris.
1992 Raw material and technological studies of the Quina Mousterian in Périgord. In *The Middle Paleolithic: Adaptation, Behavior, and Variability* (H. L. Dibble & P. Mellars, Eds), pp. 75–85. University of Pennsylvania Museum, Philadelphia.

Van Andel, T., & W. Davies (editors)
2003 *Neandertals and Modern Humans in the European Landscape during the Last Glaciation.* McDonald Institute for Archaeological Research, University of Cambridge, Cambridge.

Vermeersch, P., E. Paulissen, & P. van Peer
1997 Extensive Middle Paleolithic chert extraction in the Qena area (Egypt). *In Man and Flint: Proceedings of the VIIIth International Flint Symposium* (R. Schild & Z. Sulgostowska, eds.), pp. 133–142. Institute of Archaeology and Ethnology, Polish Academy of Science, Warsaw.

Villa, P.
1982 Conjoinable pieces and site formation processes. *American Antiquity* 47:276–290.
1983 *Terra Amata and the Middle Pleistocene Archaeological Record of Southern France.* University of California Press, Berkeley.

Vogel, G.
1999 Chimps in the wild show stirrings of culture. *Science* 284:2070–2073.

White, R.
1982 Rethinking the Middle–Upper Paleolithic transition. *Current Anthropology* 23:169–192.
1985 *Upper Paleolithic Land Use in the Périgord.* BAR International Series No. 253, Oxford.
1989 Production complexity and standardization in early Aurignacian bead and pendant manufacture: evolutionary implications. In *The Human Revolution* (P. Mellars & C. Stringer, eds.), pp. 366–390. Edinburgh University Press, Edinburgh.
1993 Technological and social dimensions of 'Aurignacian-age' body ornaments across Europe. In *Before Lascaux* (H. Knecht, A. Pike-Tay & R. White, eds.), pp. 277–300. CRC Press, Boca Raton, FL.
1997 Substantial acts: from materials to meaning in Upper Paleolithic representation. In *Beyond Art: Pleistocene Image and Symbol* (M. Conkey et al., eds.), pp. 93–121. California Academy of Sciences, Memoir No. 23, San Francisco.

Whiten, A.
2005 The second inheritance system in chimpanzees and humans. *Nature* doi:10.1038/Nature04023, accessed 17 September, 2005.

Whiten, A. et al.
1999 Cultures in chimpanzees. *Nature* 399:682–685.

Whiten, A., V. Horner, & F. de Waal
2005 Conformity to cultural norms of tool use in chimpanzees. *Nature* doi:10.1038/Nature04047, accessed 21 September, 2005.

Wilson, E. O.
1975 *Sociobiology*. Harvard University Press, Cambridge, MA.

Winterhalder, B., & E. A. Smith
2000 Analyzing adaptive strategies: human behavioral ecology at twenty-five. *Evolutionary Anthropology* 9:51–72.

Wolpoff, M. et al.
2004 Why *not* the Neandertals? *World Archaeology* 36:527–546.

Zilhão, J.
2000 The Ebro frontier: a model for the late extinction of Iberian neandertals. In *Neanderthals on the Edge* (C. Stringer, R. N. E. Barton & C. Finlayson, eds.), pp. 111–121. Oxbow Books, Oxford.
2001 *Anatomically Archaic, Behaviorally Modern: the Last Neanderthals and their Destiny.* 23rd Kroon Lecture, Nederlands Museum voor Anthropologie en Praehistorie. Universiteit van Amsterdam, Amsterdam.

Zilhão, J., & E. Trinkaus, Eds.
2002 *Portrait of the Artist as a Child. Trabalhos de Arqueologia* 22. Instituto Português de Arqueologia, Lisbon.

Zilhão, J., & F. d'Errico
1999 The chronology and taphonomy of the earliest Aurignacian and its implications for the understanding of neandertal extinction. *Journal of World Prehistory* 13:1–68.
2003 *The Chronology of the Aurignacian and of the Transitional Technocomplexes: Dating, Stratigraphies, and Cultural Implications. Trabalhos de Arqueologia* 33. Instituto Português de Arqueologia, Lisbon.

List of Contributors

Brian Adams
Department of Anthropology
University of Illinois at Urbana-Champaign
109 Davenport Hall
607 S Mathews Ave.
Urbana, IL 61801
U.S.A.
e–mail: *badams4@uiuc.edu*

C. Michael Barton
School of Human Evolution & Social Change
Arizona State University
Tempe, AZ 85287-2402
U.S.A.
e–mail: *michael.barton@asu.edu*

†Amilcare Bietti
Dipartimento di Biologia Animale e dell'Uomo
Università di Roma 'La Sapienza'
Piazzale Aldo Moro 5
00185, Rome
ITALY

Michael S. Bisson
Department of Anthropology
855 Sherbrooke St. West
McGill University
Montréal, QC, H3A 2T7
CANADA
e–mail: *michael.bisson@mcgill.ca*

Geoffrey A. Clark
Department of Anthropology
Arizona State University
Tempe, AZ 85287-2402
U.S.A.
e–mail: *gaclark@asu.edu*

Francis B. Harrold
Dean, College of Natural and Social Sciences
University of Nebraska at Kearney
Kearney, NE 68849
U.S.A.
e–mail: *harroldfb@unk.edu*

Ivor Karavanič
Odsjek za arheologiju
Filozofski fakultet
Sveučilište u Zagrebu
Ivana Lučiča 3
HR-10000 Zagreb
CROATIA
e–mail: *ikaravan@mudrac.ffzg.hr*

† *deceased*

Géraldine Lucas
Institut de Préhistoire et Géologie du Quaternaire
Université de Bordeaux I
Bâtiment de Géologie B18
Avenue des Facultés
33 405 TALENCE CEDEX
FRANCE
e–mail: *geraldine.c.lucas@wanadoo.fr*

Fabio Negrino
Via Lorenzo Costa 2/6a
16136 - Genova
ITALY
e–mail: *archeoge@tin.ti*

Deborah I. Olszewski
Department of Anthropology
University of Pennsylvania Museum
33rd and Spruce Streets
Philadelphia, PA 19104
U.S.A.
e–mail: *deboraho@sas.upenn.edu*

Jean-Philippe Rigaud
Université de Bordeaux I
Institut de Préhistoire et de Géologie du Quaternaire
Avenue des Facultés
33405 - Talence
FRANCE
e–mail: *rigaud@iquat.u-bordeaux.fr*

Julien Riel-Salvatore
Department of Anthropology
Arizona State University
Tempe, AZ 85287-2402
U.S.A.
e–mail: *julienrs@asu.edu*

Jan F. Simek
Department of Anthropology
University of Tennessee
Knoxville, TN 37996-0720
U.S.A.
e–mail: *jsimek@utk.edu*

Marie Soressi
Max-Planck-Institut für evolutionäre Anthropologie Leipzig
Inselstraße 22-26
04103 - Leipzig
GERMANY
e–mail: *soressi@eva.mpg.de*

Lawrence G. Straus
Department of Anthropology
University of New Mexico
Albuquerque, NM 87131
U.S.A.
e–mail: *lstraus@unm.edu*

www.ingramcontent.com/pod-product-compliance
Lightning Source LLC
LaVergne TN
LVHW071211210926
848482LV00015B/778